COMPUTATIONAL METHODS IN THE CHEMICAL SCIENCES

ELLIS HORWOOD SERIES IN PHYSICAL CHEMISTRY

Series Editor: Professor T. J. KEMP, Department of Chemistry and Molecular Science, University of Warwick

Atherton, N.	**Electron Spin Resonance Spectroscopy**
Ball, M.C. & Strachan, A.N.	**Chemistry and Reactivity of Solids**
Buxton, G. V. & Salmon, G.A.	**Pulse Radiolysis and its Applications in Chemistry**
Coyle, J.D. & Horspool, W.	**Organic Photochemistry**
Cullis, C.F. & Hirschler, M.	**Combustion and Air Pollution: Chemistry and Toxicology**
Davies, P.B. & Russell, D.K.	**Laser Magnetic Resonance**
Devonshire, R.	**Physical Photochemistry**
Fadini, A. & Schnepel, F.-M.	**Vibrational Spectroscopy**
Harriman, A.	**Inorganic Photochemistry**
Horvath, A.L.	**Handbook of Aqueous Electrolyte Solutions**
Jankowska, H., Kwiatkowski, A. & Choma, J.	**Active Carbon**
Jaycock, M.J. & Parfitt, G.D.	**Chemistry of Interfaces**
Jaycock, M.J. & Parfitt, G.D.	**Chemistry of Colloids**
Ladd, M.F.C.	**Symmetry in Molecules and Crystals**
Mason, T.J. & Lorimer, P.	**Sonochemistry: Theory, Applications and Uses of Ultrasound in Chemistry**
Milinchuk, V.K. & Tupikov, V.I.	**Organic Radiation Chemistry Handbook**
Mills, A. & Darwent, J.R. & Douglas, P.	**Photochemical and Photoelectrical Conversion of Solar Energy**
Navratil, O., Hala, J., Kopune, R. Leseticky, L., Macasek, F. & Mikulai, V.	**Nuclear Chemistry**
Paryjczak, T.	**Gas Chromatography in Adsorption and Catalysis**
Rest, A. & Dunkin, I.R.	**Cryogenic Photochemistry**
Sadlej, J.	**Semi-empirical Methods in Quantum Chemistry**
Snatzke, G., Ryback, G. & Slopes, P. M.	**Optical Rotary Dispersion and Circular Dichroism**
Southampton Electrochemistry Group	**Instrumental Methods in Electrochemistry**
Wan, J.K.S. & Depew, M.C.	**Polarization and Magnetic Effects in Chemistry**

COMPUTATIONAL METHODS IN THE CHEMICAL SCIENCES

A. F. CARLEY, M.A., Ph.D.
School of Chemistry and Applied Chemistry
University of Wales, College of Cardiff

P. H. MORGAN, B.Sc.(Tech.), Ph.D.
Cardiff Business School
University of Wales, College of Cardiff
(formerly of School of Chemistry and Applied Chemistry
University of Wales)

ELLIS HORWOOD LIMITED
Publishers · Chichester

Halsted Press: a division of
JOHN WILEY & SONS
New York · Chichester · Brisbane · Toronto

First published in 1989 by
ELLIS HORWOOD LIMITED
Market Cross House, Cooper Street,
Chichester, West Sussex, PO19 1EB, England
The publisher's colophon is reproduced from James Gillison's drawing of the ancient Market Cross, Chichester.

Distributors:
Australia and New Zealand:
JACARANDA WILEY LIMITED
GPO Box 859, Brisbane, Queensland 4001, Australia
Canada:
JOHN WILEY & SONS CANADA LIMITED
22 Worcester Road, Rexdale, Ontario, Canada
Europe and Africa:
JOHN WILEY & SONS LIMITED
Baffins Lane, Chichester, West Sussex, England
North and South America and the rest of the world:
Halsted Press: a division of
JOHN WILEY & SONS
605 Third Avenue, New York, NY 10158, USA
South-East Asia
JOHN WILEY & SONS (SEA) PTE LIMITED
37 Jalan Pemimpin # 05–04
Block B, Union Industrial Building, Singapore 2057
Indian Subcontinent
WILEY EASTERN LIMITED
4835/24 Ansari Road
Daryaganj, New Delhi 110002, India

British Library Cataloguing in Publication Data
Carley, A.F.
Computational methods in the chemical sciences.
1. Chemistry. Applications of computer systems.
I. Title II. Morgan, P.H.
542′.8

Library of Congress data available

ISBN 0–85312–746–8 (Ellis Horwood Limited)
ISBN 0–470–21490–2 (Halsted Press)

Typeset in Times by Ellis Horwood Limited
Printed in Great Britain by The Camelot Press, Southampton

Table of contents

Preface 7

1 Some fundamental concepts and basic methods 11
1.1 Introduction 11
1.2 Approximation of functions 11
1.3 Errors 20
1.4 Method of successive approximations 24
1.5 Solution of non-linear equations 30
1.6 Monte Carlo methods 38
1.7 Arrays 49
1.8 Matrices, vectors and systems of equations 51
Programs 71

2 Interpolation 85
2.1 Filling in the gaps 85
2.2 Divided differences and Newton's forward formula 86
2.3 Aitken's Method 90
2.4 Other forms of interpolation polynomial 94
2.5 Equispaced ordinates 100
2.6 Error in interpolation 102
2.7 Other forms of interpolation function 105
2.8 Spline interpolation 108
2.9 Interpolation with more than one variable 114
Programs 118

3 Numerical integration and differentiation 125
3.1 Integration 125
3.2 Numerical approaches to derivatives and partial derivatives 142
3.3 Special cases and summary 147
Programs 149

4 Solving differential equations . . . 160
4.1 Introduction . . . 160
4.2 Tangent field diagrams . . . 160
4.3 Calculating a solution . . . 162
4.4 Euler's method . . . 165
4.5 Predictor–corrector approach . . . 168
4.6 Runge–Kutta approach . . . 170
4.7 Extrapolation approach . . . 175
4.8 Systems of equations . . . 177
4.9 Some illustrative examples . . . 179
4.10 Partial differential equations . . . 184
Programs . . . 196

5 Fitting straight lines and polynomials to experimental data . . . 216
5.1 Introduction . . . 216
5.2 Central Limit Theorem . . . 218
5.3 Goodness of fit . . . 220
5.4 Best straight line . . . 222
5.5 Polynomial curve-fitting . . . 234
5.6 Fitting non-polynomial models . . . 241
Programs . . . 246

6 Optimization methods and non-linear least squares minimization . . . 261
6.1 Introduction . . . 261
6.2 The minimization problem . . . 261
6.3 Direct search methods . . . 263
6.4 Gradient methods . . . 274
6.5 Least squares minimization . . . 281
6.6 Error estimation . . . 286
6.7 Constrained minimization . . . 291
6.8 Least squares and systems of equations . . . 293
Programs . . . 293

References . . . 323

List of programs . . . 327

Index . . . 329

Program index . . . 336

Preface

Our motivation for writing this book lies in a strongly held belief that every science graduate ought to be versed in rudimentary numerical and computational skills. Sadly, our experience is that this is often not the case, a failing which seems to be especially prevalent in the chemical sciences. Any topic with a mathematical flavour is too frequently approached with trepidation and an alarmingly defeatist attitude. In the modern scientific world instrumentation is frequently controlled by dedicated microcomputers, one side effect of which is the ease with which large quantities of high quality data may be generated. In order to cope with all of these data and to extract the maximum amount of information, the intelligent application of numerical processing techniques (a field of chemistry now known as 'chemometrics') is essential. We stress the adjective intelligent since it is only too easy to use a method either inappropriately or with inadequate data, and to be deluded by apparently good, yet in reality meaningless, results. This book, therefore, is aimed at both the novice (who should benefit from a careful reading of Chapter 1, and especially from experimenting with the programs therein) and also the more experienced worker or student who would like to develop his or her basic skills further.

In most chemical sciences laboratories one finds a host of equipment ranging from the unmodified (and perhaps unmodifiable) commercial units to more *ad hoc* 'home-brew' systems. Indeed, there is a healthy flow of innovation from the latter to the former. A similar situation exists in the software arena, where the programs range from the inviolate word-processor, database or spreadsheet packages to in-house developed products. An interesting recent development has been the application of commercial spreadsheet software, normally used for cash-flow forecasts and the like, to the acquisition and processing of experimental data. The initial impetus came from the inventive experimentalist, but there has been a gratifyingly quick response from software 'manufacturers', leading for example to the release of an acquisition module for Lotus 1-2-3. To use even the commercial version of this approach to the processing of experimental data one must give up the 'black box' idea of software, and begin to think logically about what one is doing. The benefits are well worth the effort — the process of defining step by step exactly what one wants to do with the data can provide an instructive insight into one's understanding of the problem.

It is often said that one learns best by 'doing', and this is especially true where the subject matter of this book is concerned. The reader is urged most strongly to begin experimenting at the earliest stage with the programs, modifying them when he/she feels confident about so doing, and seeing the effect of varying any parameters involved. A very instructive example of how a lack of awareness of software (and to some extent hardware) limitations can lead to the acceptance of very erroneous results is provided by the harmonic series 1, 1/2, 1/3, 1/4, ... The sum of this infinite series $S=1+1/2+1/3+1/4+\ldots$ is known mathematically to be infinite — the successive terms get smaller and smaller, but not quickly enough to stop the sum approaching infinity (slowly) as we incorporate more terms. After reading Chapter 1 the reader should be able to write a program to perform this summation, stopping and printing the result when, as is usual in this sort of solution, successive partial sums change by less than a pre-chosen small amount and convergence is deemed to have occurred. The smallest value we can take for this accuracy parameter (other than to make it identically zero) is the smallest number which can be represented by the hardware/language combination — in Applesoft BASIC this is approximately 3×10^{-39}. The program must thus always terminate (albeit after a very large number of terms) with a finite value for S, a dramatic demonstration of the limitations of digital arithmetic. Although this is a pathological case, one must always be on the look-out for less extreme but just as dangerous examples, such as in the computation of EPR spectra discussed in Chapter 3.

In order to keep this book to a manageable length, and so as not to overwhelm the less-experienced reader, several important topics have been deferred to a 'sister' volume to be published in the near future. These include advanced matrix and vector methods, including eigenvalues/eigenvectors and molecular graphics, Fourier transform techniques, convolution and correlation, and cluster analysis.

And now, a few words about the programs themselves which appear at the end of each chapter, and are written in Applesoft. This is a rather primitive dialect of the BASIC language, and as a result the programs should transport to other microcomputers with little modification. Although the rudimentary nature of Applesoft is good for portability, the lack of constructions such as IF . . . THEN . . . ELSE, REPEAT . . . UNTIL and WHILE . . . WEND results in some rather tortuous segments and simulated constructions, which can be avoided in more modern dialects of BASIC and in other languages such as FORTRAN and PASCAL. There is thus great scope for making the code shorter and more elegant, but this should not be attempted until the programs are thoroughly understood.

The only machine specific routines inevitably relate to screen handling and disk input/output. These sections of code are identified by explanatory REMARK statements where necessary. The resolution of the graphics mode used (invoked by the command HGR) is 160 points (vertical) by 280 points (horizontal), with the origin (0,0) in the top left-hand corner of the screen. The command HPLOT X,Y plots a pixel (dot) of the colour with code C (defined by a HCOLOR=C statement) at the screen coordinate (X,Y). All the graphics routines assume a monochrome screen and use HCOLOR=3 to plot white dots. The text screen consists of 24 lines of 40 characters; the sequence HTAB H:VTAB V places the text cursor at character position H on line V. Line and character numbering is in the range 1–24 and 1–40 respectively. Four lines of text (21–24) are visible below the graphics screen. As an aid to program portability, the following table compares three primitive graphics commands in Applesoft with the equivalent commands in TurboBASIC and GWBASIC, two popular BASIC dialects.

Command function	Applesoft	TurboBASIC/GWBASIC
Plot a point at (X,Y)	HPLOT X,Y	PSET (X,Y)
Draw a line from (X1,Y1) to (X2,Y2)	HPLOT X1,Y1 TO X2,Y2	LINE (X1,Y1)–(X2,Y2)
Select and clear graphics screen	HGR	SCREEN N

(In TurboBASIC the SCREEN parameter N determines the screen resolution and number of colours available; allowed values depend on the hardware configuration.)

Inevitably, the interested reader is going to reach the stage where he or she will want to speed up the programs. Two relatively straightforward ways of achieving this are the use of integer variables, and the use of a compiler. Integer variables are identified by the per cent sign, e.g. X%, and can only take integer values. Thus the statement X%=2.5 results in the variable X% containing the number 2. Many BASIC dialects (and all compiled (see later) BASICs known to the authors) perform arithmetic faster with integers than with real numbers. Applesoft does all of its arithmetic with real numbers, converting the real result to integer if necessary, and no speed advantage is obtained. However, integer numbers take up less space than reals (2 bytes compared with 5 bytes in Applesoft), a consideration which may be particularly important if large arrays need to be stored. The need to append the % symbol to all occurrences of the name of an integer variable can make programs difficult to read, less compact and harder to type in. Some implementations such as MBASIC and TurboBASIC allow an integer definition such as DEFINT I–N, which identifies all variables with names beginning with I, J, K, L, M and N as integer variables.

BASIC is usually an interpreted language, a feature which makes for convenient program development, but which limits the speed of program execution. With an interpreted language, the computer 'reads' each program line as it is encountered during program execution, and translates the commands into machine code instructions which the microprocessor can understand and carry out. For example, if the lines are

```
100 X=2.5
110 X=X+Y
```

the interpreter (e.g. BASIC) first sets aside a unit of memory, labels it X and 'writes' the number 2.5 into it. It proceeds to line 110, reads the contents of the memory labelled X and places it in a special memory location in the microprocessor called a register. Memory location Y is then located and the contents read and added to the contents of the register. The result is stored in X, overwriting the previous contents.

If the interpreter comes across a program line more than once, such as line 110 in the loop

```
80 S=0
90 FOR I=1 TO 10
100 S=S+2.5
110 NEXT I
```

then this line must be translated afresh each time round the loop, clearly an inefficient procedure. A compiler is a complex piece of software which does the translation once and for all, storing the resulting sequence of machine instructions in what is called an object file for subsequent (fast) execution. The combination of an interpreted version of the language for ease of program development, together with compilation for fast execution, ensures that BASIC is still the most popular micro language for the non-professional programmer. The emergence of BASIC implementations which embody many of the advantages of structured languages such as PASCAL, but without the more inflexible of their features, has served to reinforce this position.

ACKNOWLEDGEMENTS

We would like to thank our colleagues in the School of Chemistry for numerous helpful discussions, Jeff Roberts for drawing most of the diagrams, and our wives Barbara and Diana without whose uncomplaining support this book would never have been possible.

To Sarah, Robert and Bethan
and
To Meleri

1

Some fundamental concepts and basic methods

1.1 INTRODUCTION

The aim of this chapter is threefold: to revise some basic mathematics, to introduce some (perhaps) new concepts, but most of all to 'demystify the jargon'. Mathematical English is a precise and efficient language, but unintelligible to the non-native speaker, Indeed, one of the main barriers to improving the mathematical and computational ability of non-mathematicians lies in the unfamiliarity of the vocabulary of the subject. This is not meant to imply that mathematics is easy or that mathematical skills can be acquired without expenditure of effort, simply that the basic concepts often appear to the casual reader to be cloaked in unnecessarily sophisticated and intimidating jargon.

A useful introductory mathematics text is by Francis (1984); for the more advanced reader Perrin (1970) is recommended. There are a very large number of books on aspects of numerical analysis, many of them rather specialized. Recommended amongst the more basic are Churchhouse (1978), Hosking *et al.* (1978) and Ramajaraman (1971); more comprehensive and advanced books worth looking at are Churchhouse (1981), Burden *et al.* (1981) and Cohen *et al.* (1973). For a text with a strong chemistry bias the reader should consult Johnson (1980), which contains many examples and relevant references. For a review of computer aids to chemists, including discussions of commercially available software packages, the reader is directed to Vernin and Chanon (1986).

Chapra and Canale's (1985) book on numerical methods for engineers concentrates on the application of microcomputers and includes many programs.

1.2 APPROXIMATION OF FUNCTIONS

1.2.1 Basic definitions

The term 'function' is most generally and accurately, but perhaps not most helpfully, defined as a relationship between numbers. To say that a quantity represented by the variable y is a function of another variable x, and which is written $y = f(x)$ (or $y = g(x)$ or $y = \alpha(x)$ or any other symbol you care to use), simply means that for any value of x within a defined range, a value of y may be derived by means of some regular procedure. The variable x is termed the independent variable, and y the dependent variable. A common way of representing a function is as a graph of the curve made up of the points (x,y) defined by $y = f(x)$ (Fig. 1.1).

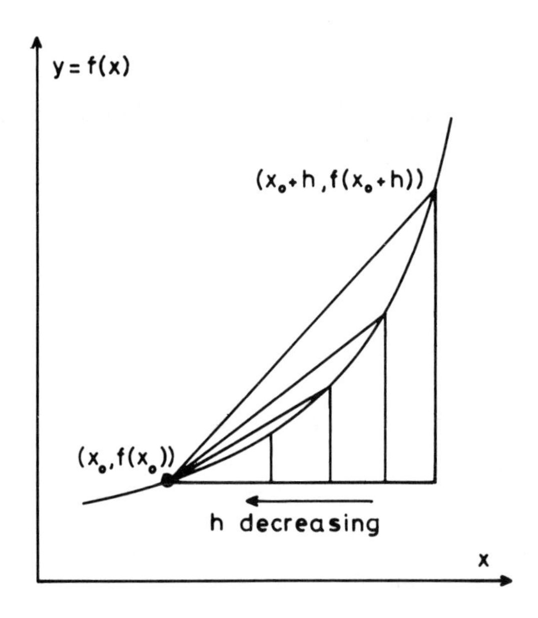

Fig. 1.1—Plot of a general function $y = f(x)$ showing the method of calculating the derivative or tangent at (x_0, y_o) as the limit of a sequence of slopes of straight-line chords.

Some functions are easy to evaluate using basic arithmetic operations, the most well-known being the polynomial

$$f(x) = a_0 + a_1x + a_2x^2 + \ldots + a_nx^n \tag{1.1}$$

Functions which cannot be constructed using basic operations are termed transcendental, examples being $\exp(x)$, $\log(x)$ and $\sin(x)$. Others may be defined in terms of an integral or differential equation, for which there may be no analytical solution, but which nevertheless may be solved numerically to provide a value of y for a given value of x. An example is the error function

$$\mathrm{erf}(x) = (2/\pi^{1/2}) \int_0^x \exp(-z^2)\,\mathrm{d}z$$

It may be useful to think of a function as a 'black box' containing an algorithm (set of rules), possibly very complex, and communicating with the outside world via an input channel and an output channel. For each valid x which enters the input, a corresponding value (or values) of y appears at the output. Such a visualization is analogous to a computer subroutine which returns a function value every time it is CALLed.

A function $y=f(x)$ is said to be differentiable at some point x_0 in its range of definition if the quantity $(f(x)-f(x_0))/(x-x_0)$ approaches a limiting value as x gets closer to x_0. This limiting value is called the derivative of $f(x)$ at x_0 and is written $f'(x_0)$ or $(\mathrm{d}y/\mathrm{d}x)_{x_0}$. In mathematical notation, and replacing $(x-x_0)$ by h

$$\left(\frac{\mathrm{d}y}{\mathrm{d}x}\right)_{x_0}=f'(x_0)=\lim_{h\to 0}\,(f(x_0+h)-f(x_0))/h \tag{1.2a}$$

is the slope of the tangent to the curve $y=f(x)$ at $x=x_0$ (Fig. 1.1), and is also referred to as the rate of change of y with respect to x. For example if $f(x)=x^2$ then

$$\begin{aligned}f'(x_0)&=\lim_{h\to 0}\,((x_0+h)^2-x_0^2)/h\\&=\lim_{h\to 0}\,(x_0^2+2hx_0+h^2-x_0^2)/h\\&=\lim_{h\to 0}\,(2x_0+h)\\&=2x_0\end{aligned}$$

A more symmetrical formula often used to compute a derivative is

$$\left(\frac{\mathrm{d}y}{\mathrm{d}x}\right)_{x_0}=\lim_{h\to 0}\,(f(x_0+h)-f(x_0-h))/2h \tag{1.2b}$$

and this is used in program P1.1 to illustrate how the value of this expression gets closer and closer (that is, converges) to a limiting value as h is progressively reduced (Table 1.1). The function chosen, $F(x)=1/x$, is simple enough for us to differentiate directly ($f'(x)=-1/x^2$) so that we know the correct value $f'(1)=-1$.

The extension to higher derivatives is straightforward; the second derivative $f''(x_0)$ is thus defined

$$f''(x_0)=\lim\,(f'(x_0+h)-f'(x_0))/h$$

and so on.

Table 1.1 — Illustration of the use of equation (1.2b) to approximate the value of $f'(x)$ at $x=1$, for the reduction $f(x)=1/x$. As the step size h in (1.2b) is reduced, the approximation approaches the 'true' value of $f'(1) = -1$

h	$(f(1+h)-f(1-h))/2h$
0.75	−2.286
0.70	−1.961
0.65	−1.732
0.60	−1.563
0.55	−1.434
0.50	−1.333
0.45	−1.254
0.40	−1.190
0.35	−1.140
0.30	−1.099
0.25	−1.067
0.20	−1.042
0.15	−1.023
0.10	−1.010
0.05	−1.003

We have used the terms $\mathrm{d}y/\mathrm{d}x$ and $f'(x)$ interchangeably above. Strictly speaking, we define the differential of the function, written $\mathrm{d}y$, as the response of the function to a small increment $\mathrm{d}x$ in the independent variable; it is given by (see, for example, Phillips (1960))

$$\mathrm{d}y = f'(x)\ \mathrm{d}x \tag{1.3}$$

and tells us how y changes for a small change in x. This may appear a trivial restatement in this case, but it takes on a more complex and useful form in the next section.

Functions often exhibit turning points, defined as points in the neighbourhood of which the first derivative (slope) changes sign — at the turning point $f'(x)$ must consequently be identically zero (Fig. 1.2). Turning points may either be (local) maxima or minima (Fig. 1.2), and are classified mathematically depending on whether the second derivative at the turning point is negative or positive respectively. If both $f'(x)=0$ and $f''(x)=0$ the point is termed a point of inflection (Fig. 1.2).

The inverse of differentiation is called integration, and written

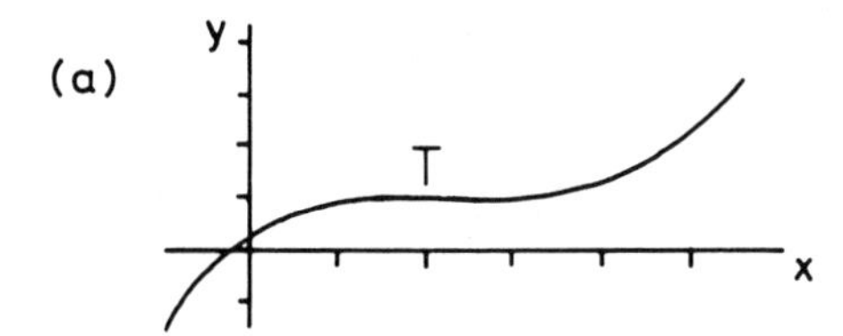

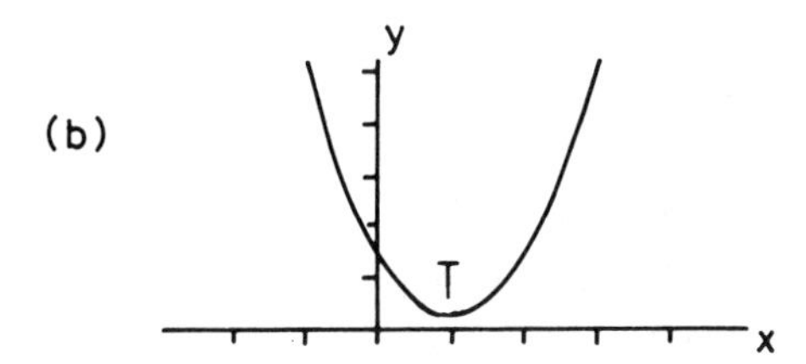

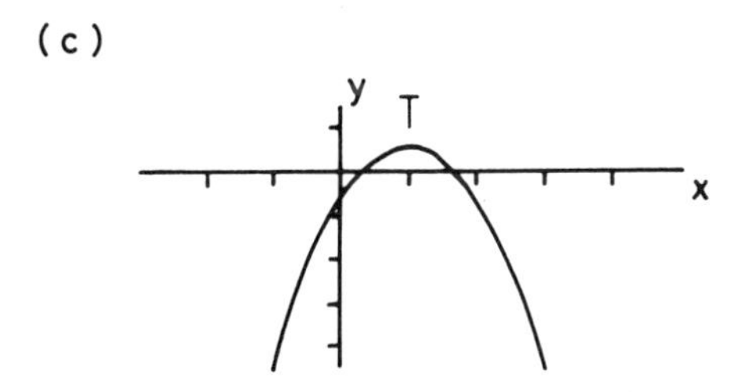

Fig. 1.2 — The three kinds of turning points T, defined as points where $f'(x)=0$: (a) point of inflection (where also $f''(x)=0$), (b) local minimum ($f''(x)>0$) and (c) local maximum ($f''(x)<0$).

$$f(x) = \int f'(x)\ \mathrm{d}x + C$$

where C is a constant which we do not know without further information; remember that d(constant)/d$x = 0$. Thus if $f'(x) = x^3$ then

$$f(x) = \int x^3\ \mathrm{d}x + C = x^4/4 + C$$

and if, for example, we are told that $f(x) = 1$ when $x = 1$ we can calculate $C = 3/4$ and thus $f(x) = x^4/4 + 3/4$. A definite integral is written with limits and the constant of integration disappears:

$$I = \int_a^b x^3 \, dx = [x^4/4]_a^b$$

$$= b^4/4 - a^4/4$$

The definite integral may be given a geometrical interpretation as the area under the curve $y = f(x)$; further discussion is deferred until section 1.6.3.

1.2.2 Functions of more than one variable

The value of the dependent variable may be determined by more than one independent variable, for example position and time; in that case we write $y = f(x,t)$. The basic concept remains the same: we input a value for x and for t and output a value for y, but the graphical representation is not as simple — contour plots (Chapter 6) or pseudo-3-D plots must be used. Our definition of derivatives needs to be modified and we now deal with partial derivatives defined by

$$\left(\frac{\partial f}{\partial x}\right)_{x_0} = \lim_{h \to 0} (f(x_0 + h,t) - f(x_0,t))/h$$

$$\left(\frac{\partial f}{\partial t}\right)_{t_o} = \lim_{h \to 0} (f(x,t_0 + h) - f(x,t_0))/h$$

Thus, if $f(x,t) = x^2 t + t^2 x$ then

$$\left(\frac{\partial f}{\partial x}\right)_{x_0} = \lim_{h \to 0} ((x_0 + h)^2 t + t^2(x_0 + h) - x_0^2 t - t^2 x_0)/h$$

$$= \lim_{h \to 0} (x_0^2 t + 2hx_0 t + h^2 t + t^2 x_0 + t^2 h - x_0^2 t - t^2 x_0)/h$$

$$= \lim_{h \to 0} (2x_0 t + ht + t^2)$$

$$= 2x_0 t + t^2$$

In practice, we differentiate the function 'normally' with respect to the particular variable, assuming the other independent variables to be constant. Naturally, the definitions extend to functions of any number of variables, but simple graphical visualization must be abandoned.

The total differential of the function, dy, is defined in terms of increments dx and dt in the independent variables (Phillips, 1960)

$$\mathrm{d}y = \left(\frac{\partial f}{\partial x}\right) \mathrm{d}x + \left(\frac{\partial f}{\partial t}\right) \mathrm{d}t \tag{1.4}$$

and is the change in y produced by small changes in x and t. This leads to the conditions for a turning point (maximum or minimum) in $f(x,t)$ where we must have d$y = 0$, since momentarily the function is insensitive to changes in either x or t. If d$y = 0$ for non-zero dx and dt then we must have from (1.4)

$$\frac{\partial f}{\partial x} = \frac{\partial f}{\partial t} = 0$$

This result is used extensively in Chapters 5 and 6.

Integration of functions of more than one variable leads to the idea of multiple integration such as

$$\iint f(x,y)\ \mathrm{d}x\ \mathrm{d}y$$

which means that we integrate firstly with respect to x, and then integrate the result with respect to y. For example, if $f(x,y) = xy$ then

$$\iint f(x,y)\ \mathrm{d}x\ \mathrm{d}y = \int x^2 y/2 \mathrm{d}y + A$$

$$= x^2\ y^2/4 + Ay + B$$

By analogy with partial differentiation we are assuming that y is constant whilst we integrate with respect to x, and vice versa. Note that the integrations are performed from the inside outwards, and in the case of a definite integral the limits are applied successively at each stage, not all together at the end of the calculation. Examples of multiple integrals are given in section (1.6.3) and Chapter 3.

1.2.3 Polynomial approximation

It often proves worthwhile to express a 'difficult' function $f(x)$ as a weighted sum of other, more easily manipulated, functions, $g_i(x)$

$$f(x) = a_0 g_0(x) + a_1 g_1(x) + a_2 g_2(x) + \ldots + a_n g_n(x) \tag{1.5}$$

where the coefficients a_i have to be determined. The most popular choice is $g_i(x) = x^i$ (that is, integer powers of x) so that (1.5) becomes a polynomial expression of degree n

$$f(x) = a_0 + a_1x + a_2x^2 + a_3x^3 + \ldots + a_nx^n \tag{1.6}$$

which has the advantages of being easy to differentiate and integrate and to combine with other polynomials. Furthermore, the Weierstrass Approximation Theorem states that any 'well-behaved' function can be approximated as closely as we like by a polynomial (Burden *et al.*, 1981). If we write down the first three derivatives of $f(x)$ from (1.6) we obtain

$$\begin{aligned} f'(x) &= a_1 + 2a_2x + 3a_3x^2 + \ldots \\ f''(x) &= \quad 2a_2 + 6a_3x + \ldots \\ f'''(x) &= \qquad 6a_3 + \ldots \end{aligned}$$

Putting $x = 0$ in these equations and (1.6) we find $a_0 = f(0)$, $a_1 = f'(0)$, $a_2 = (1/2)f''(0)$ and $a_3 = (1/6)f'''(0)$. In general

$$a_n = (1/n!)f^{(n)}(0) \tag{1.7}$$

where $n!$ pronounced n factorial $= n(n-1)(n-2)\ldots1$ and $f^{(n)}(0)$ is the nth derivative of $f(x)$ evaluated at $x = 0$. A polynomial of degree n with these coefficients is known as the n-degree Taylor polynomial representation of the function $f(x)$, or the Taylor polynomial expansion of $f(x)$ about $x = 0$. If the series (1.6) converges as $n \to \infty$ (that is, the partial sums get closer and closer to some limiting value for a given x), it is called a Taylor Series approximation to $f(x)$. The errors involved in truncating the series too early, that is taking too few terms in the expansion, are discussed in section 1.3 and Chapter 2. Fig. 1.3 demonstrates this by showing the effect of increasing the number of terms used from the Taylor series approximation of $\sin(x)$

$$\sin(x) = x - x^3/3! + x^5/5! - \ldots$$

The truncated approximation becomes less accurate the further we move away from the point $x = 0$, but the inaccuracy can be reduced by incorporating more terms of the series. A more mathematically rigid derivation of the Taylor Series approximation to a function is given in Chapter 2.

In some situations, $f(x)$ is better expanded in terms of functions $g_i(x)$ other than simple powers of x; an example is given in Chapter 5. One of the problems with the Taylor Series expansion is that it is only good close to the expansion point since all the information derives from that point (Burden *et al.*, 1981, and Fig. 1.3).

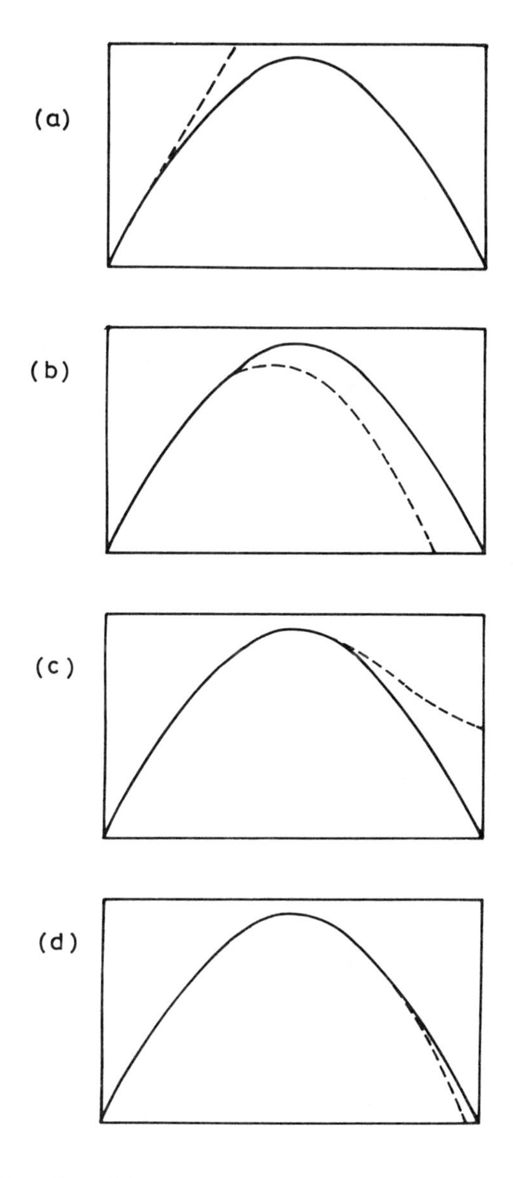

Fig. 1.3 — The effect of increasing the number of terms used in the Taylor polynomial representation of the function $\sin(x)$. The solid line is a plot of $\sin(x)$ (for $0<x<\pi$) and the dotted lines the approximation using (a) 1 term, (b) 3 terms, (c) 5 terms and (d) 7 terms.

1.2.4 Horner's nesting scheme

This method evaluates a polynomial in the most efficient manner (that is, with fewest arithmetic operations and thus least accumulated error) by rearranging the computation. For example, $ax^3 + bx^3 + cx + d$ is evaluated as $(((ax + b)x + c)x + d)$. The procedure is described mathematically by the formula

$$b_k = a_k + b_{k+1}x_0 \qquad k = n-1,..,0 \tag{1.8}$$

where x_0 is the point at which the polynomial (1.6) is to be evaluated, $b_n = a_n$ and the result $f(x_0)$ is given (after repeated application of (1.8)) by b_0. It is advantageous to store all the values b_k since the polynomial

$$F(x) = b_n x^{n-1} + b_{n-1}x^{n-2} + \ldots + b_2 x + b_1$$

has the property that $F(x_0) = f'(x_0)$ which is useful if we are using a procedure which requires first derivatives (such as the Newton–Raphson method, section 1.5). $F(x_0)$ is evaluated by using Horner's method a second time. A program to implement Horner's scheme is giving in listing P1.2.

Horner's scheme may be used with a procedure known as synthetic division to find all the zeros of a polynomial, that is, all the roots of $f(x) = 0$ (Burden *et al.*, 1981; Southworth and Deleeuw, 1965).

1.3 ERRORS

At least as far as the 'practitioner in the street' is concerned, this is probably one of the most neglected and misunderstood areas in numerical analysis; it is often regarded as a 'difficult' subject. We cannot hope to do more in the limited space available in this book than to stress the importance of gaining an appreciation of the basic factors involved, and to indicate where further detailed discussions may be found. We must emphasize straight away that we use the term 'errors' to mean something other than 'mistakes'; error analysis usually consists of asking why the result of a competition is different from the 'correct' answer, by how much, and how we may reduce its magnitude. As such, some of the factors involved in a particular numerical method are peculiar to that method and discussion will be deferred until the appropriate chapter.

If y^* is the 'true', and usually unknown, value of a quantity and y is an approximate estimate of it, then the absolute error E_a is defined by

$$E_a = |y^* - y| \tag{1.9}$$

and the relative error, E_r, by

$$E_r = E_a/|y^*| \tag{1.10}$$

Although E_r is often the more useful quantity, since (for example) a quoted error of 1% is just as meaningful if applied to a quantity whose magnitude is 10^{-3} or 10^{10} whereas an absolute error of 0.00001 is not, we must always be on our guard for situations where $y^* = 0$ or is very close to zero. Since the correct value of a quantity is often unknown (it is just that which we are seeking) the denominator in (1.10) is usually approximated by y.

We may classify errors into three basic types: empirical, arithmetic and computational. This list does not include errors arising from simplifications in the model constructed to emulate the physical system under analysis. These are of a different kind: the analysis itself may be precise but the result, however accurately calculated, may not be close to the physically 'correct' value. The only solution is to recognise when this happens and to modify the model so that the resulting discrepancies are within acceptable bounds.

1.3.1 Empirical errors

In a practical situation the analysis may involve experimental data which will be subject to error, arising from the limited precision with which the measurements may be made. This error will influence the error in the derived result (see, for example, Chapter 6), and indeed the magnitude of the experimental errors may determine how accurately it is worth doing the computation. The magnitude of empirical errors may be reduced by improved instrumentation and/or measurement technique — for example, averaging more readings (see Chapter 5) or taking the readings more carefully. Alternatively, it may be feasible to smooth the experimental data (Chapter 5), in order to reduce the scatter or error in the values. Experimental error or noise may often be modelled by a gaussian distribution function (Chapter 5), which may prove useful when analysing the performance of a particular numerical procedure.

1.3.2 Arithmetic errors

Arithmetic or machine errors arise from the fundamental restriction that a practical calculation can only be made using numbers represented (that is, approximated) by a fixed number of digits, whether in the decimal (human) or binary (machine) system. The errors which arise are the most frequently encountered and the most insidious in their effect.

A BASIC interpreter (which simply means the large program in your microcomputer which translates statements written in BASIC into commands the computer can understand and obey) uses the equivalent of, say, k decimal digits; internally it naturally works with binary numbers. Thus Applesoft outputs all three results to 9 significant figures, and to cope with a wide range of numbers uses a floating point notation, in which any number is represented by a sign and a mantissa (a decimal number between 0 and 9.99...9) multiplied by a power of 10.

$$(+/-)\ d_0.d_1\ d_2\ d_3\ \ldots\ d_k \times 10^n$$

A more common notation, and assumed in the following discussion, is termed normalized floating point notation where the mantissa is always less than 1

$$(+/-)\ 0.d_1\ d_2\ d_3\ \ldots\ d_k \times 10^n$$

Thus the number 4/3 is represented in Applesoft by 1.33333333×10^0 and in normalized floating point notation by 0.133333333×10^1 where the same number of significant digits (nine) is assumed.

There are two basic ways of dealing with a number which, in its exact floating point representation, has more than k digits: chopping and rounding. In chopping we simply ignore all digits after the kth, with a maximum resultant error 10^{-k}, that is one unit in the kth decimal place. Rounding on the other hand, reduces the error in the approximation by taking into account the magnitude of the discarded segment. The most commonly used recipe is to add 5 to the first unwanted digit and chop the new number as before; the maximum error is now $5 \times 10^{-(k+1)}$ or half a unit in the kth place. This method may be alternatively stated: if the $(k+1)$th digit is 5 or greater then add one to the kth digit (round up), it it is less than 5 then leave it alone (round down). Strictly speaking, the case when the $(k+1)$th digit is equal to 5 puts us in a dilemma, since rounding up or down leads to the same absolute error. There are several possible ways around the problem, the main rule being consistency: always round up (as above) or down in this situation, round up and down alternately in the computation, or even round up or down depending on the magnitude of a random number generated in the computer (Churchhouse, 1981).

Round-off error (due to either chopping or rounding) is unavoidable, and we can only seek to reduce its effects. This may be achieved by using higher precision arithmetic on those computers which support it, or by the reformulation of the problem. An important consequence of round-off error is that expected arithmetical identities and rules breakdown. Thus if we write a simple Applesoft program to sum the decimal number 0.1 N times, then for $N=100$ we get the exact answer 10; however, for $N=1000$ the result is 99.9999963. The binary representation of 0.1 is 0.000110011001100... where the dots indicate that the alternating pairs of '0's and '1's go on for ever, so that the chopped value used by the computer is only approximate. The round-off error accumulates with each cycle of the summation program until it eventually intrudes upon the result.

1.3.3 Computational errors

By the term 'computational error' we mean those errors either inherent in the numerical method itself, or in the implementation of it. Further, the error is conveniently approximated by the difference between successive estimates to the true value of quantity. The most common source of this kind of error lies in the truncation of a Taylor's series expansion of some quantity in the analysis. Thus we may simplify the problem of finding the minimum value of a complicated function by expanding it in a Taylor series near the estimated minimum point and discarding all but the first two terms (Chapter 6); clearly this may lead to large errors in the derived minimizing values of the independent variables, but by repeating the procedure in a systematic way this error is progressively reduced to an acceptable level. The lower limit of the magnitude of the error will be determined by round-off errors, but may not be attainable for other reasons. An estimate of the size of the truncation error may be obtained from an analysis of the remainder term (Chapter 2).

A numerical procedure often requires the choice of a step size which is used to change the value of a variable by a well-defined increment. The error in the result will in general depend on the step-size, which is usually progressively reduced until the error is reduced below a pre-set limit. If the step size is reduced to a magnitude where round-off errors dominate the calculations, then the results become meaningless (Chapter 4).

1.3.4 Accumulation and propagation of error

The error involved in the single application of a procedure may be quite small. However, many numerical methods rely on repeated applications, and it is thus important to consider how quickly errors (especially round-off errors) may build up. Firstly, we must investigate how errors are propagated through simple arithmetic operations.

If y is a function of $x_1,x_2,\ldots,x_n$ then the total differential of y (section 1.2) may be interpreted as an expression for the error in the computed value of y in terms of the errors in the variables x_i

$$\Delta y = (\partial f/\partial x_1)\Delta x_1 + (\partial f/\partial x_2)\Delta x_2 + \ldots + (\partial f/\partial x_n)\Delta x_n \tag{1.11}$$

where Δ refers to a small change in a variable.

Hence, if y is the sum of or difference between two variables, that is, $y = x_1 + x_2$ or $y = y_1 - x_2$, then

$$\Delta y = |\Delta x_1| + |\Delta x_2| \tag{1.12}$$

that is, the absolute errors are added. For multiplication and division relative errors are combined

$$\Delta y/y = |\Delta x_1/x_1| + |\Delta x_2/x_2| \tag{1.13}$$

For the difference case, $y = x_1 - x_2$, the relative error in y is given by

$$E_r = \Delta y/y = (|\Delta x_1| + |\Delta x_2|)/|(x_1 - x_2)|$$

so that if x_1 and x_2 are nearly equal then E_r may be very large in magnitude. Round-off errors may thus assume inordinate significance when evaluating expressions containing differences between two similar numbers (see, for example, Chapter 5). In general, any expression containing a mixture of very small and very large numbers will give problems, which may sometimes be reduced by rearranging the expression or the order of computation.

If a procedure is iterated many times, involving many arithmetic operations, the accumulation of error may become important. Fortunately, although the maximum possible accumulated error may be large, the most probable accumulated error is much smaller. For instance, the accumulated round-off error after making N additions increases in proportion with $N^{1/2}$ not N (Churchhouse, 1978). By calculating probability distributions for accumulated round-off error, Southworth and Deleeuw (1965) illustrated clearly that the probability of the occurrence of the maximum possible error is very small (but not zero), and also that the most probable error is much less than the maximum possible error.

The method of successive approximations (section 1.4) is a good one as far as accumulation of errors is concerned, since each iteration may be considered as starting the process afresh with a new initial estimate. The error in the calculation does not therefore increase with repetition.

1.3.5 Ill-conditioned problems

An ill-conditioned problem is one where a small change (error) in the initial data leads to a large change in the computed solution. An example is provided in Chapter 5, where a set of simultaneous equations must be solved to find the coefficients of a fitting polynomial; for polynomials of order greater than about 7 small round-off errors in the coefficients of the equations cause the solution to 'blow up'. In this case a fundamental reformulation of the fitting procedure proves successful; in other cases a reordering of the set of equations may help (Ramajaraman, 1971).

We can classify algorithms as unstable (ill-conditioned) or stable depending on their response to an introduced error (Burden *et al.*, 1981). If an error E is introduced at same point then after n operations the stable algorithm exhibits linear error growth ($E_n = CnE$ where C is a constant), whereas the error grows exponentially in the unstable procedure ($E_n = k^n E$ for some $k > 1$).

1.4 METHOD OF SUCCESSIVE APPROXIMATIONS

1.4.1 Iteration, convergence and limits

These three important concepts, which were introduced with little comment in section 1.2 and program P1.1, lie at the heart of many numerical methods. Iteration is the repetitive application of some procedure; if the successive results of the procedure get closer and closer (in some, as yet, undefined way) to some limiting quantity then the iteration is said to converge. The convergence is conveniently monitored via the change in the result at each iteration step — when this change has decreased below a pre-chosen tolerance level then the iterative process is terminated and the current result taken as an acceptable approximation to the 'real' result. The term 'order of convergence' with reference to iterative methods will often be encountered; a method is said to be kth order convergent if the error (the difference between the approximation and the real solution) in any step is proportional to the kth power of the error in the previous step. The higher the order, the faster the rate of convergence. Important considerations in any iterative method are the formulation of the refinement algorithm (which gives a better value of the result (we hope) at each iteration step), the termination criteria and the choice of initial starting guess, which we use to 'seed' the process. In order to illustrate these ideas we will consider the method of successive approximations, variations of which are used widely in fields ranging from solving partial differential equations to calculating square roots. We shall focus on its application to the solution of equations of the type $f(x) = 0$ where it is also referred to as the fixed-point method (Burden *et al.*, 1981).

1.4.2 The procedure

The method proceeds by first recasting the equation in the form $x = g(x)$. This rearrangement, which may appear trivial at first sight, provides for us an iterative improvement formula or rule

$$x_{k+1} = g(x_k) \tag{1.14}$$

where if x_k is the result after the kth iteration step then x_{k+1} is a better approximation to the solution or root of $f(x) = 0$; if x_k were the exact root then $g(x_k) = x_k$ and $x_{k+1} = x_k$. Figs 1.4(a) and (b) give a graphical interpretation of the method. The

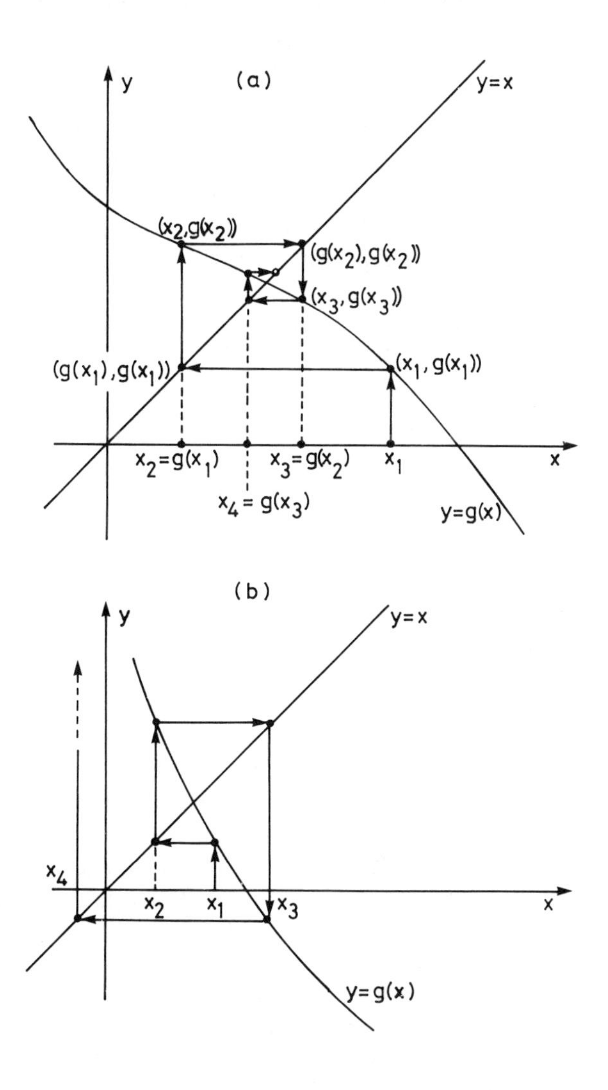

Fig. 1.4 — The method of successive approximations applied to the solution of $f(x) = 0$ recast as $x = g(x)$. The solution (value of x satisfying the equality) is the point of intersection of the graphs of $y = x$ and $y = g(x)$. In example (a) we home in successfully on the solution from an initial guess x_1, whereas in (b) we diverge catastrophically, for reasons explained in the text.

root(s) of $f(x)=0$ are the value(s) of x at the intersection of the curve $y=g(x)$ and the line $y=x$, and this forms the basis of a well-known pencil-and-paper method taught at school for solving equations like $f(x)=0$. The method of successive approximations works by 'bouncing' between the curve and the line, homing in on the intersection point (Fig. 1.4(a)); it fails if the slope of the curve $g(x)$ (more strictly $|g'(x)|$) ever becomes greater than unity during the procedure. This is illustrated in Fig. 1.4(b); a rigorous proof is given by, for example, Ramajaraman (1971). The method is first-order convergent, the ratio of successive errors being equal to the magnitude of $g'(x)$ at the root; convergence may thus be very slow if this value is close to unity.

As an example, consider the chemical reaction $A+B \rightleftharpoons C$, with equilibrium constant $K=10$. Our task is to calculate the concentration of C at equilibrium given initial concentrations of A, B and C equal to 1, 2 and 0 respectively. If we designate by c the desired concentration then simple theory tells us that $K=10=c/(1-c)(2-c)$. Hence, rearranging, $f(c)=10c^2-31c+20=0$. Now this equation could be solved using a calculator and the well-known schoolboy formula, but it serves to illustrate how much more complicated equations may be tackled. We rewrite the equation to give the iterative formula

$$c_{k+1}=10(2+c_k^2)/31 \tag{1.15a}$$

A short program to implement this algorithm might be

```
100  CO = 0
110  CN = 10*(2 + CO*CO)/31
120  IF ABS(CN − CO) < LI THEN END
130  CO = CN
140  GOTO 110
```

Rather than store all the intermediate results c_k in a 1-dimensional array C(K), using up memory, we only need the two variables CN and CO to represent the new (i.e. improved) and old (i.e. previous) values of c. After each updating (line 110) the two values are compared (line 120) to see whether the change in value is less than our previously defined tolerance LI. If it is not then CO is replaced by the value of CN (line 130), another value for CN calculated (line 110) and so on. To start the process c_0 is chosen to be its actual initial value, 0 (line 100). The procedure converges rapidly (Table 1.2) to the correct solution $c=0.916$ which is the physically feasible of the two roots of $f(c)=0$, calculated according to the classical formula. A great advantage of iterative methods is that they are self-correcting and do not lead to a build-up of rounding error since each step may be regarded as starting the process afresh with a new initial guess. Hence, even a gross user error during a manual working of the method need not prevent convergence, but simply delay it (unless a lucky accident placed the new value nearer the root).

It is frequently possible to rearrange the equation $f(x)=0$ in several different ways to give alternative forms for $g(x)$, not all of which will be successful. Furthermore, some will be 'better' than others in their convergence behaviour. The example just discussed, for instance, could be rewritten as

Table 1.2 — Fixed-point method (method of successive approximations) applied to the solution of $10c^2 - 31c + 20 = 0$. The two iterative formulae are derived from different rearrangements of this equation, and converge to different roots

Cycle	$c_{k+1} = 10(2 + c_k^2)/31$		$c_{k+1} = 3.1 - 2/c_k$
	$c_0 = 0$	$c_0 = 0.5$	$c_0 = 0.5$
1	0.64516	0.72581	−0.90000
2	0.77943	0.81510	5.32222
3	0.84113	0.85948	2.72422
4	0.87339	0.88345	2.36584
5	0.89123	0.89693	2.25464
6	0.90138	0.90467	2.21293
7	0.90726	0.90917	2.19622
8	0.91068	0.91180	2.18935
9	0.91269	0.91335	2.18649
10	0.91387	0.91426	2.18529
⋮	⋮	⋮	⋮
20	0.91556	0.91556	2.18443

$$c_{k+1} = 3.1 - 2.0/c_k \tag{1.15b}$$

which fails immediately if we choose a starting value $c_0 = 0$ as above, due to a devision of zero, but also gives 'problems' if we choose a better starting value $c = 0.5$, converging to the other of the two roots of $f(c) = 0$, $c = 2.184$ (Table 1.2). Physically this root is unreasonable since the maximum value c could take if there was no back-reaction (i.e. no equilibrium) is 1 unit. However, what we have just dismissed as an undesirable reformulation would be a positive boon to a mathematician who was searching for all the roots of $f(c) = 0$.

Burden *et al.* (1981) and Churchhouse (1981) have discussed the factors involved. The former authors have highlighted the problem with the example $f(x) = x^3 + 4x^2 - 10 = 0$ which they recast in five different ways

(1) $x = x - x^3 - 4x^2 + 10$
(2) $x = ((10/x) - 4x)^{1/2}$
(3) $x = 0.5(10 - x^3)^{1/2}$
(4) $x = (10/(4 + x))^{1/2}$
(5) $x = x - (x^3 + 4x^2 - 10)/(3x^2 + 8x)$

The first two of these rearrangements fail whilst the efficiencies of the other three vary greatly. The reader is urged strongly to adapt the program segment just introduced for the chemical equilibrium example, and investigate the behaviour of these formulations. The reasons for success and failure mainly revolve around the magnitude of $|g'(x)|$ near the root.

A general expression which may be used for $g(x)$ is $g(x) = (1/m)(f(x) + mx)$ where m is chosen so as to make $|g'(x)|$ small and thus hasten the convergence (Churchhouse, 1978). We could experiment by choosing m manually, and the reader should try this to see the effect on the convergence. Now, we have seen that the smaller $|g'(x)|$ near the root, the faster the convergence of the method. Thus, differentiating, $g'(x) = (1/m)(f'(x) + m)$ which will be small if we choose $m = -f'(x)$ where ideally $f'(x)$ is evaluated at the root x_r. Since we naturally do not know x_r then we need to approximate it, and the obvious candidate for the approximation is the current estimate of the root, x_k, which leads to the idea of an adaptive method of successive approximations.

$$x_{k+1} = (1/m_k)(f(x_k) + m_k x_k) \tag{1.16}$$

with $m_k = -f'(x_k)$. Indeed, this algorithm shows second order convergence — it is in fact the Newton–Raphson method which we discuss in more detail later (section 1.5.3).

The BASIC interpreter on a microcomputer usually incorporates a square root function which is based on an iterative algorithm. At first sight the method of successive approximations could provide the algorithm — the task of finding the square root of a number c may be restated as finding the root of $f(x) = x^2 - c = 0$. This in turn may be rearranged to give the iterative formula $x_{k+1} = c/x_k$ which fails miserably — if x_0 is the initial guess then the result of applying the formula oscillates between x_0 and c/x_0; note that $|g'(x)| > 1$ for $x < c$. All is not lost, however, and a little intuition produces a workable formula: if the guess x_k is smaller than the true root then c/x_k is larger by an approximately similar amount so that the mean of the two values ought to be a better estimate, leading to the improvement formula

$$x_{k+1} = (x_k + c/x_k)/2 \tag{1.17}$$

This result may be derived rigorously using the Newton–Raphson method (section 1.5.3); its performance is considered later.

1.4.3 Extended successive approximations

A modification to the method of successive approximations due to Wegstein (see Ramajaraman, 1971) gives a procedure which converges even when $|g'(x)| > 1$. We start by generalizing (1.14) to

$$x_{k+1} = x_k + \alpha(g(x_k) - x_k) \tag{1.18}$$

where the simple method of successive approximations has $\alpha = 1$. The parameter α is chosen so as to speed up the convergence (a procedure known as over-relaxation) and prevent divergence when $|g'(x)| > 1$. Clearly we reach the root x_r in one step if we choose $\alpha = (x_r - x_k)/(g(x_k) - x_k)$ but this requires foreknowledge of the value of the root. An approximation to this optimum value of α may be derived (Ramajaraman, 1971).

$$\alpha = 1/(1 - \beta) \tag{1.19}$$

where

$$\beta = (g(x_k) - g(x_{k-1}))/(x_k - x_{k-1})$$

Note that β is an approximation to the slope of $g(x)$ at x_k. If $|g'(x)| > 1$ then $\alpha < 0$ which serves to constrain the iteration to converge towards the root. Listing P1.3 is a program to apply Wegstein's method. It requires two starting values since the iterative formula (1.18) with α given by (1.19) uses the current and previous values x_k and x_{k-1} to derive the updated value x_{k+1}. One of the starting values is provided by the user and the second obtained by a single application of the simple successive approximations formula (1.14). Wegstein's method copes quite happily with the square root example, the first starting value in this case being conveniently provided by the number itself. Table 1.3 (p. 35) compares Wegstein's methods with formula (1.17) in calculating $2^{1/2}$. The latter is seen to converge more rapidly to a given accuracy than the former — the Newton–Raphson method (of which (1.17) is an example) is second-order convergent whereas Wegstein's method has an order (*c.* 1.6) intermediate between the first-order convergent method of successive approximations and the Newton–Raphson method (Hosking *et al.*, 1978).

In fact, Wegstein's method is the Secant method (section 1.5) in disguise, which in turn may be thought of as an approximate version of Newton–Raphson method (section 1.5) where the derivative of the function $f(x)$ is estimated numerically.

A modified fixed-point method with an enhanced rate of convergence and improved stability has been proposed by Kantaris and Howden (1983). Their so-called 'Universal Equation Solver' starts with the trivial reformulation $x = x \pm f(x)$ but then introduces a 'stability operator' $Q = 2^q$ and an 'attenuating function' $\sinh^{-1}$ (see Perrin, 1970) to give

$$x_{k+1} = x_k \pm 2^q \sinh^{-1}(f(x_k))$$

The attenuating function helps to avoid too large a correction between steps, and so reduces the risk of divergence. Similarly, the parameter q is varied throughout the convergence, adapting to the result of each computation. It was found useful to write

$$q = (p/3) - r - (1/3)$$

where r is incremented when $f(x_k)$ changes sign, thus halving the 'step size' when we near the root (which we must be if $f(x_k)$ has changed sign). Similar adaptive approaches prove beneficial in numerical integration (Chapter 3) and the solution of differential equations (Chapter 4). The parameter p is decremented or incremented depending on the rate of convergence. For a discussion of the convergence properties and stability of this algorithm, and its application to a wide range of problems in numerical analysis, we refer the reader to the original text (Kantaris and Howden, 1983).

1.5 SOLUTION OF NON-LINEAR EQUATIONS

In section 1.4 we met the method of successive approximations and applied it to the solution of equations of the form $f(x)=0$, after a suitable rearrangement. In this section we discuss methods which we can apply directly to the original function $f(x)$.

1.5.1 Bisection method and Regula Falsi

We consider these methods together since they both require that we start knowing two values of x, x_1 and x_2 such that $f(x_1)<0$ and $f(x_2)>0$. Clearly x_1 and x_2 straddle the root since $f(x)$, providing it is continuous, must pass through zero as it changes sign between x_1 and x_2 (Fig. 1.5).

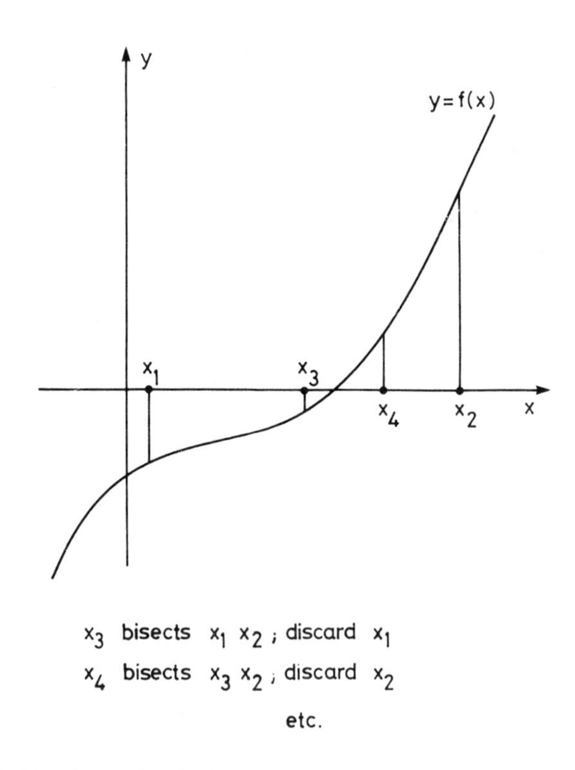

Fig. 1.5 — The bisection method for finding the roots of $f(x)=0$, that is finding the value(s) of x where the graph of $f(x)$ intersects the x-axis. We start with initial estimates x_1 and x_2 which straddle the root, and proceed so as always to straddle the root as we home in on it.

In the bisection method (Balzano's method), we consider the function value $f(x_3)$ at the point whose x-coordinate $x_3=(x_1+x_2)/2$ is midway between the current end-points (Fig. 1.5), and discard the end-point which has a function value of the same sign. Thus we maintain two points which straddle the root, but which we hope are now closer to the root. The procedure is repeated (Fig. 1.5) until a satisfactory accuracy has been achieved — this might be conveniently measured by the difference in x-values between the two straddle-points. The method has the great advantage that it is certain to converge, but may do so rather slowly.

The method of false position or Regula Falsi proceeds in a very similar fashion to the bisection method, except that the new intermediate point each cycle is defined by the intersection of the line joining the end-points with the x-axis (Fig. 1.6), obtained

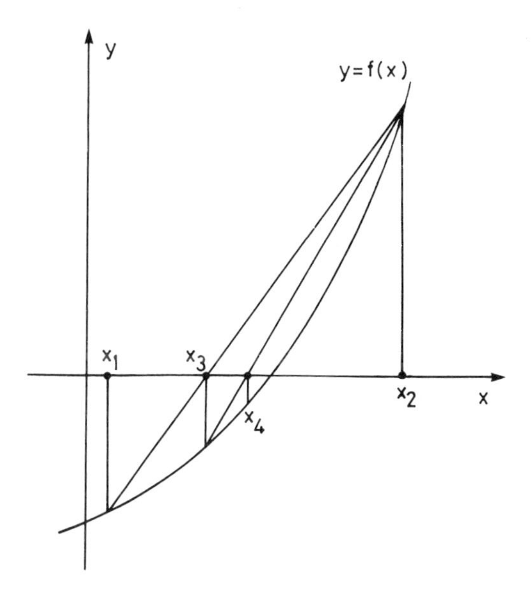

Fig. 1.6 — Method of Regula Falsi. It incorporates a more complex procedure for choosing the next approximation to the root of $f(x) = 0$ than does the bisection method (see text).

by linear interpolation between the end-points. The discarded end-point is chosen as before. Should a particular end-point persist for many cycles, then the convergence may be very slow. One suggested solution is to replace $f(x_i)$ by $f(x_i)/2$ if x_i is used repetitively (Wilkes, 1966). That the method of false position, like many others for the solution of equations, bears more than a passing relationship to the Newton–Raphson method is seen if we consider the expression for the intersection point

$$x_3 = (x_1 f(x_2) - x_2 f(x_1))/(f(x_2) - f(x_1)) \qquad (1.2)$$

which may be rearranged to give

$$x_3 = x_1 - f(x_1)(x_2 - x_1)/(f(x_2) - f(x_1)) \qquad (1.21)$$

The term $(f(x_2)-f(x_1))/(x_2-x_1)$ is an approximation to the first derivative $f'(x_1)$ at x_1 (section 1.2), and we may thus write

$$x_3 \sim x_1 - f(x_1)/f'(x_1)$$

which should be compared with (1.23).

The choice of a suitable pair of starting points which enclose the root may be helped by evaluating the function at a number of regularly spaced points distributed over a wide range of x. Once a sub-interval has been isolated this process may be repeated until the final sub-interval is deemed small enough on which to apply the bisection method or method of false position. A similar initial crude search may indeed be found useful as a precursor to any of the methods discussed in this and the previous section.

Both of these methods show only first-order convergence, but this must be balanced against the reliability of the convergence. A program for the method of false position is given in listing P1.4; its modification to perform the method of bisection is straightforward and is left as an exercise to the reader.

The method of false position may be improved by performing a quadratic rather than linear interpolation between the end-points of the current interval (see, for example, Kantaris and Howden, 1983 and Chapter 2). We construct the parabola passing through the points $(x_1, f(x_1))$, $(x_2, f(x_2))$ and the mid-point $((x_1+x_2)/2, f((x_1+x_2)/2))$. The root of this curve is taken to be the next approximation to the root of $f(x)=0$.

1.5.2 Secant method

Also known as Lin's method, the Secant method resembles the method of false position, but the constraint that the root must always be straddled by successive results is lifted; the penalty paid is that convergence is no longer guaranteed. Once again we must start with two values of x, x_0 and x_1, and proceed according to the iterative formula.

$$x_{k+1} = x_k - f(x_k)(x_k - x_{k-1})/(f(x_k) - f(x_{k-1})) \tag{1.22}$$

which is illustrated graphically in Fig. 1.7. Equation (1.22) should be compared with (1.21) and the Newton–Raphson formula (1.23) — once again we have an approximation (a better one this time) to the Newton–Raphson procedure. The equivalence of the Secant method and Wegstein's modified successive approximations approach (section 1.4) is apparent if we replace $f(x)$ by $x-g(x)$ in (1.22); this is left as an exercise for the reader.

1.5.3 Newton–Raphson method

The basis of this powerful, second-order convergent method lies in the expansion of $f(x)$ as a Taylor polynomial (section 1.2) near the root. We assume that we have an estimate of the root x_k which is close to the true value, x_r differing from it by an amount h. The Taylor expansion of $f(x_k+h)$ is given by

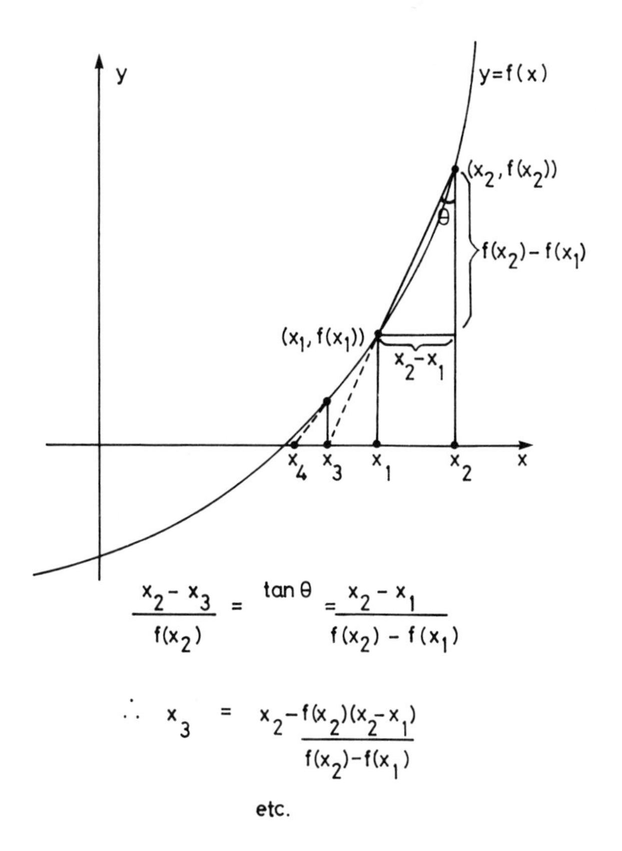

Fig. 1.7 — The Secant method applied to the solution of $f(x) = 0$, starting with estimates x_1 and x_2. It is an approximation to the Newton–Raphson procedure (see later).

$$f(x_k + h) = f(x_k) + hf'(x_k) + R$$

where R is the remainder term (that is, the rest of the series except the first two terms). Ignoring R and recalling that $f(x_k + h) = 0$, since $x_k + h$ is the root, we have $h = -f(x_k)/f'(x_k)$ and thus $x_k + h$ should be a better estimate of the root (it would equal the root if the analysis were exact) and we have the Newton–Raphson iterative improvement formula

$$x_{k+1} = x_k - f(x_k)/f'(x_k) \tag{1.23}$$

A graphical representation of the method is shown in Fig. 1.8. The formula (1.23) may be interpreted as the x-coordinate of the intersection of the tangent at x_k with the x-axis; the Newton–Raphson method is thus seen to be the limiting case of the Secant method.

The method has fast convergence, but convergence is not assured (which it was in

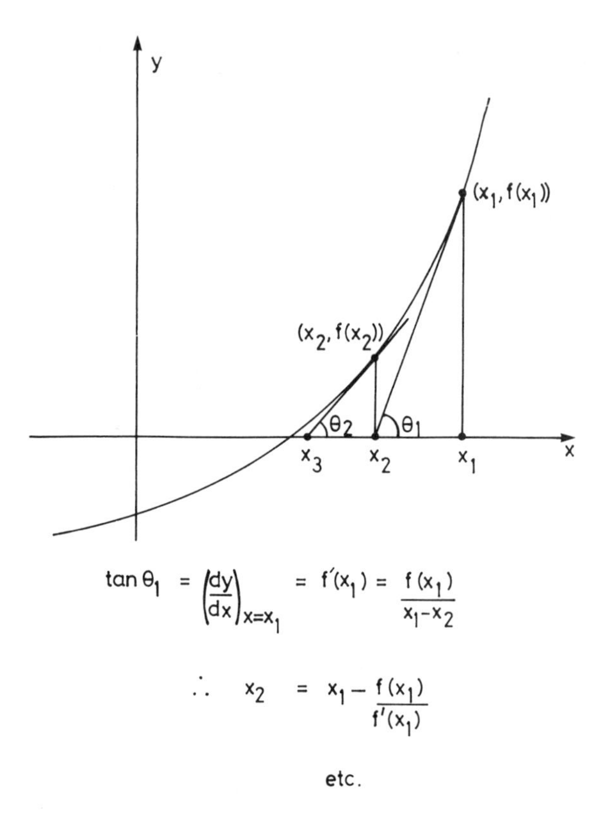

Fig. 1.8 — The Newton–Raphson method. It may be regarded as the limiting case of the Secant method (Fig. 1.7) as $x_2 \to x_1$, and the chords become tangents.

the bisection and Regula Falsi methods) especially if one starts too far from the root. It fails catastrophically if $f'(x)$ becomes zero at any stage, and this possibility should be monitored in any computer program which implements the method (see P1.5); it may be possible to restart the method from a different point or perhaps to change methods for a few cycles.

If $f(x) = x^2 - c = 0$ then $f'(x) = 2x$ and the updating formula is $x_{k+1} = x_k - (x_k^2 - c)/2x_k = (x_k + c/x_k)/2$ which is the formula for obtaining a value for c 'derived' earlier (section 1.4). A comparison of the performance of the Newton–Raphson method and the Secant method (in its Wegstein guise) is provided in Table 1.3.

One drawback of the Newton–Raphson method is that it tends to make a large initial jump from the starting guess x_k, and this may result in divergence of the solution. This may be controlled by use of the attenuating function of Kantaris and Howden (1983) which we encountered in section 1.4.3, leading to

$$x_{k+1} = x_k - \sinh^{-1}\,(f(x_k)/f'(x_k))$$

Table 1.3 — Comparison of Wegstein's method and the Newton–Raphson procedure applied to the computation of $\sqrt{2}$ via the solution of the equation $x^2-2=0$. Both methods start with an estimate of 1.5 for the root

Cycle	Wegstein	Newton–Raphson
1	1.50000	1.50000
2	1.37500	1.41667
3	1.42361	1.41422
4	1.41009	1.41421
5	1.41522	
6	1.41377	
7	1.41432	
8	1.41417	
9	1.41422	

Stability is improved at the expense of a reduced 'capture range' — that is, its ability to reach a root from a distant starting estimate may be impaired.

If we include the term in h^2 in the truncated Taylor series expansion of $f(x+h)$ we have

$$f(x+h)=f(x)+hf'(x)+h^2f''(x)/2+R$$

which leads to the third-order convergent Newton–Raphson formula (Kantaris and Howden, 1983)

$$x_{k+1}=x_k+(f'(x_k)/f''(x_k))(-1\pm(1-2f(x_k)f''(x_k)/(f'(x_k))^2)^{1/2} \qquad (1.24)$$

Although convergence has increased to third order, more criteria have to be satisfied for the method not to fail, and we now have to calculate both first and second derivatives. Note that if $f(x)=ax^2+bx+c$ then this expression reduces to the well-known schoolboy formula for the solution of a quadratic equation (the reader should confirm this). Thus, if $f(x)$ is a second order polynomial then the root is found in one iteration.

1.5.4 Steffensen's method

One major disadvantage with the Newton–Raphson method is the need to calculate the first derivative $f'(x)$ (and $f''(x)$ if we use the higher order method), which may not be easy or convenient. Steffensen's method comes to the rescue by retaining quadratic convergence, whilst avoiding derivative calculations. The iterative formula is (Churchhouse, 1981).

$$x_{k+1} = x_k - f^2(x_k)/(f(x_k + f(x_k)) - f(x_k)) \quad (1.25)$$

and has the further desirable property that it only requires two function evaluations. It is left to the reader to write a suitable program for this method.

1.5.5 Halley's method

Halley's formula is

$$x_{k+1} = x_k - f_k/(f'_k - (f_k f''_k/(2fd'_k)) \quad (1.26)$$

where the subscript k means the function and its derivatives are evaluated at x_k. It exhibits third-order or cubic convergence (Churchhouse, 1981) but unfortunately requires the calculation of both first and second derivatives, which may prove a disadvantage in some situations. In the special case of $f(x) = x^2 - c = 0$ we obtain a simple, efficient formula for determining the square root of a number c (Churchhouse, 1981).

$$x_{k+1} = x_k(x_k^2 + 3c)/(3x_k^2 + c)$$

Halley's method (1.26) and the higher order Newton–Raphson method (1.24) share the property of third-order convergence, but the former proves more stable under a variety of conditions and exhibits a larger capture range. Thus Halley's method finds an accurate solution of $x^3 - 27 = 0$ in 76 steps starting from $x_0 = 50$, whereas (1.24) fails as the term under the square root sign becomes negative.

1.5.6 Some examples

In Table 1.4 we compare the results of applying the simple method of successive approximations, the Newton–Raphson method and Halley's formula to the chemical equilibrium example of section 1.4, that is with $f(c) = 10c^2 - 31c + 20 = 0$ and using the fixed-point rearrangement $c = 10\ (2 + c^2)/31$. The relative rates of convergence are clearly demonstrated, and also how the root to which the method converges, if at all, depends on the starting estimate x_0.

A more challenging problem is that of computing pH–volume curve for a general acid–base titration. Consider a volume V_A of acid, of concentration c_A, in a beaker. A volume V_B of base of concentration c_B is added from a burette and we wish to compute c, the concentration of hydrogen ions in the resultant solution of volume V $(= V_A + V_B)$. If K_A and K_B are the dissociation constants for the acid and base respectively, and K_W is the ionic product for water, then c is one of the roots of the equation

$$a_4c^4 + a_3c^3 + a_2c^2 + a_1c + a_0 = 0 \quad (1.27)$$

where

Table 1.4 — Demonstration of the relative rates of convergence of the methods of Newton–Raphson and Halley and the fixed-point method. The test problem is the solution of $10c^2 - 31c + 20 = 0$ recast for the fixed-point method as $c = 10(2 + c^2)/31$. The starting estimate for the root is $c = 0$

Cycle	Fixed-point	Newton–Raphson	Halley
1	0.6452	0.6452	0.8147
2	0.7794	0.8752	0.9151
3	0.8411	0.9144	0.9156
4	0.8734	0.9156	0.9156
5	0.8912	0.9156	
6	0.9014		
7	0.9073		
8	0.9107		
9	0.9127		
10	0.9139		
⋮	⋮		
18	0.9155		
19	0.9156		

$$a_4 = K_B V$$

$$a_3 = V(K_A K_B + K_W) + K_B c_B V_B$$

$$a_2 = -(VK_W(K_B - K_A) + K_A K_B(c_A V_A - c_B V_B))$$

$$a_1 = -(K_W V(K_A K_B + K_W) + K_W K_A c_A V_A)$$

$$a_0 = -K_A K_W^2 V$$

Program P1.6 solves this equation using the Newton–Raphson method. The procedure is applied repeatedly, incrementing V_B by a small amount each cycle, thus generating a set of (c, V_B) values. Plotting $-\log(c)$ ($=$ pH) against V_B then gives the conventional pH–volume curve. Some examples are shown in Fig. 1.9.

The reader may have wondered how we decide which of the four possible roots of the equation (1.27) is the correct one. It transpires that provided we use an initial guess for c which is greater than the sought-after root, then the process converges to the requisite answer. Thus, to ensure that P1.6 can be left running for a complete range of V_B values without failing, we must provide it with an initial guess which is greater than any value of c likely to be encountered.

Another approach to the solution of (1.27) is to rearrange it to express V_B in terms of c, increment c (or pH) over a suitable range of values and calculate directly corresponding values of V_B. The shape of the pH–volume curve may still be plotted, but if one requires the value of c at a particular value of V_B not equal to one of these

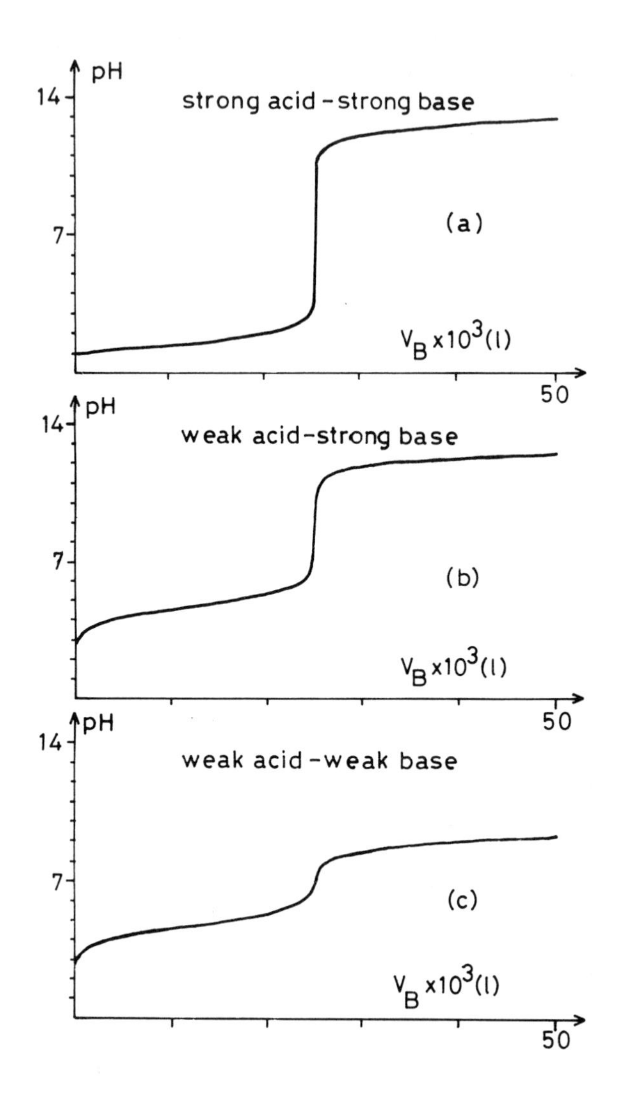

Fig. 1.9 — pH–volume curves computed using (1.27) using the Newton–Raphson method. In each case we are adding 0.1 M base to 25 ml of 0.1 M acid. For the strong acid and strong base we assume $k_A = K_B = 10^6$. We take acetic acid as the weak acid with $k_A = 1.75 \times 10^{-5}$, and ammonium hydroxide as the weak base with $k_B = 1.78 \times 10^{-5}$.

calculated values, then one must resort to a procedure known as interpolation (Chapter 2).

1.6 MONTE CARLO METHODS

This exotic appellation was coined during World War II by John von Neumann as a code-name for the mathematical approach developed to investigate the problem of neutron diffusion through shielding materials (McCracken, 1955). Although the physicists knew a great deal about the collision between a neutron and an atomic nucleus, such as the average distance between collisions and the energy and momentum transfer involved in the scattering, an analytic solution of the problem (that is, a practical predictive formula) proved impossible. The approach of von Neumann and Ulam (see McCracken, 1955) was to simulate the problem on the atomic scale by following the individual wanderings of neutrons through a lattice of atomic nuclei. Random numbers were used to make decisions affecting the fate of each neutron after each collision: how much energy it lost, in which direction it was scattered, how far it travelled to the next collision, etc. By repeating this process for a large number of neutrons a reasonably accurate distribution for the neutrons within the material could be computed for a chosen diffusion time; in particular, the number which escape through a certain thickness of shielding could be estimated. Such a model, where a particle moves in a stepwise fashion, its next move at any point in time depending essentially on a random choice between options, is frequently referred to as a 'random walk' simulation (section 1.6.1). The success of such methods rests to a large extent on the fact that many physical phenomena, such as Brownian motion (Hersh and Griego, 1969), particle diffusion (Atkins, 1982) and reactions between gas molecules and solid surfaces (Dumont and Dufour, 1986), have a fundamentally random origin at the microscopic level, which is hidden in the derived analogue (usually partial differential) equations. In these cases the Monte Carlo or random walk approach may be regarded as simply using Nature's model (Marshall, in Meyer (1956)), or as Yakowitz (1977) puts it 'a return to reality'. For a general discussion of the application of Monte Carlo methods the reader should consult Meyer (1956) and Yakowitz (1977).

There are several apparently non-probabilistic mathematical problems where Monte Carlo methods have also proved useful, including matrix inversion and numerical integration. In this more general context a Monte Carlo method may be defined as a procedure which involves the use of sampling procedures based on probabilities (that is, random sampling) to approximate the solution of mathematical or physical problems. This rather obscure definition will become clearer later.

The Monte Carlo method naturally demands access to a source of random (which in practice means pseudo-random) numbers, either from tables of random numbers or more usually these days from a so-called random number generator, which is simply an algorithm (computer subroutine) which every time it is called outputs a random number. An important question to answer is 'how random is pseudo-random' and as a consequence the performance and design of random number generators has received much attention (section 1.6.4).

1.6.1 A simple random walk

One of the simplest examples where a random walk model is used to represent a physical phenomenon is that of one-dimensional diffusion. The practical example we will use is that of a solvent in which is placed a plate coated with a solute. At time $t = 0$ the so-called initial condition is that all the N_0 solute particles are concentrated in the yz-plane at $x = 0$; we wish to compute the spatial distribution of particles, that is how many are at any particualr distance x from the plate after any particular diffusion time, t. The reason why this is an example of one- rather than two- or three-dimensional diffusion is that there is no concentration gradient (and hence no net diffusion) in any direction perpendicular to the x-axis.

The diffusion process is modelled by considering each particle in turn and allowing it to take M steps or jumps of length d and duration Δt (where $M\Delta t = t$). At each step the particle is assumed to have an equal chance of moving to the left or to the right (except that the presence of the plate means that x can never be negative), the actual choice being determined by the output of a random number generator (see P1.7). In Applesoft BASIC the command takes the form

```
X = RND(1)
```

which loads the variable X with a random number between 0 and 1. The walk is repeated for all the particles and the number $N(x_k)$ at each attainable discrete distance $x_k = kd$ ($k = 0,\ldots,M$) is totalled. Clearly, most particles remain close to the plate (intuitively one expects the left and right steps to cancel out on average), but there is a finite probability that any attainable point will be reached, this probability decreasing with distance from the plate. Typical distributions for various diffusion times, computed from such a simulation (using P1.7) are shown in Fig. 1.10. The characteristic half-bell-shape is part of what is called the normal or gaussian curve and it results from a solution of the partial differential equation describing diffusion

$$\partial C(x,t)/\partial t = D\ (\partial^2 C(x,t)/\partial x^2) \tag{1.28}$$

where D is the diffusion coefficient, and $C(x,t)$ is the concentration of material at distance x and time t. For the solute-plate example taken here Atkins (1982) gives the solution

$$C(x,t) = (N_0/A(\pi\ Dt)^{1/2})\ \exp(-x^2/4Dt) \tag{1.29}$$

where A is the area of the plate. In order to compare this with the random walk solution we need a relationship between D and the step parameters d and Δt. This is the Einstein–Smoluchowski equation (Atkins, 1982)

$$D = d^2/2\Delta t$$

Substituting this in (1.29), and multiplying each side by Ad so that the left-hand side is now the number $N(x,t)$ of particles in a slab of area A and thickness d at a distance x from the solute plate we obtain

$$N(x,t) = N_0(2/\pi M)^{1/2}\ \exp(-L^2/2M) \tag{1.30}$$

where $M = t/\Delta t$ and $L = x/d$. This solution is compared with the Monte Carlo result in Fig. 1.10; the agreement is seen to be excellent.

P1.7 illustrates one of the great advantages of random walk or Monte Carlo

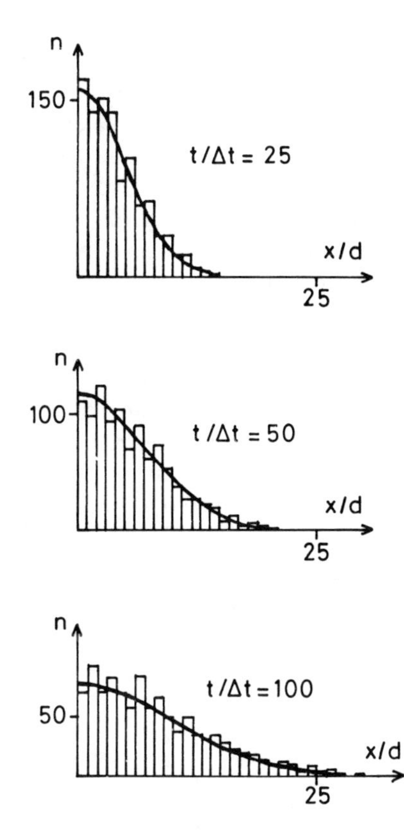

Fig. 1.10 — One-dimensional diffusion modelled by a Monte Carlo approach (histograms), and compared with the exact mathematical solution of the diffusion equation (solid line). The meaning of the quantities $t/\Delta t$ and x/d is discussed in the text.

methods — the programs are usually relatively short, in length if not in time of execution. Providing the decision-making rules are correctly formulated then we simply start the program running and watch as the system of particles evolves, in much the same way that we monitor the evolution of a real experiment.

1.6.2 A not-so-simple random walk

There is a classical problem in mathematics with the rather intimidating name of the Dirichlet boundary problem. In two dimensions it may be stated thus: given a close boundary (perimeter curve) C and a function $f(x,y)$ defined at all points on the closed curve C then there exists a unique function $u(x,y)$ which is equal $f(x,y)$ on C, and in the interior of C satisfies Laplace's equation (such functions are termed harmonic functions.) Laplace's equation in two dimensions is

$$\partial^2\phi/\partial x^2 + \partial^2\phi/\partial y^2 = 0 \tag{1.31}$$

and is an example of a very important class of equations known as elliptic partial differential equations. The quantity $\phi(x,y)$ is specified on a closed boundary and the solution of (1.31) gives the magnitude of $\phi(x,y)$ at every point (x,y) within this boundary. Laplace's equation has application in diffusion problems where $\phi(x,y)$ represents concentration $c(x,y)$ and the solution gives the concentration within the boundary at equilibrium (that is when $\partial c(x,y)/\partial t = 0$), and also describes the variation of potential within an electrolytic medium (Chapter 4). Indeed, in the 1940s electrolytic tanks were used as analogue computers to provide solutions of Laplace's equation (Birkhoff in Metropolis *et al.*, 1980), until digital techniques (of the kind discussed in Chapter 4) asserted their superiority. A major difference between the Monte Carlo approach and the methods of Chapter 4 is that Laplace's equation is solved at chosen isolated points rather than for the whole region.

For the purposes of this example we will restrict ourselves to the problem of calculating the equilibrium concentration of a substance at regular grid points within a square boundary (Fig. 1.11). The boundary conditions along the sides of the square

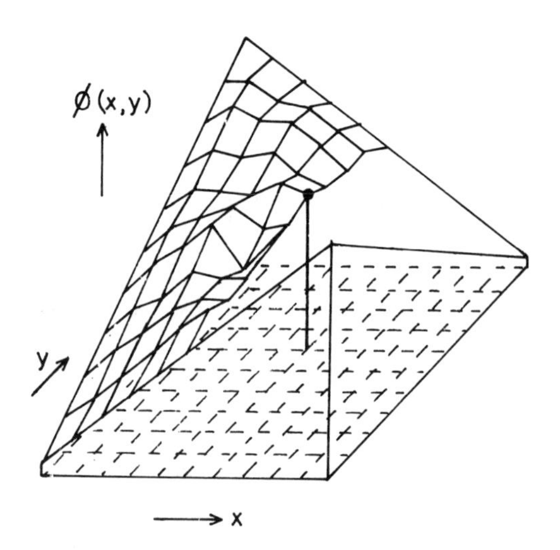

Fig. 1.11 — Region and boundary conditions used for the Monte Carlo solution of Laplace's equation. $\phi(x,y)$ represents the particle concentration in the two-dimensional region (square), and some computed values are shown.

are that one pair of opposite corners have unit concentration and the other pair a maximum value, the concentration varying linearly along the sides between these points (Fig. 1.11). Although this may appear a rather unrealistic situation, it serves as well as any to illustrate the method of solution, and has the merit of simplicity of programming (P1.8). The procedure is very straightforward:

(1) Choose a starting point (x_0,y_0), the point at which we would like to know the concentration.

(2) Get a random number and on the basis of its magnitude choose an increment in the x-direction of $-1, 0$, or 1, that is choose whether to move left one grid point, stay where you are or move right one grid point.
(3) Similarly, compute a y-increment of $-1, 0$, or 1.
(4) Move to the new grid point defined by (2) and (3); in geographical terms you either stay where you are or move to a point N, NE, E, SE, S, SW, W, or NW of your current position.
(5) Repeat (2)–(4), thus taking a random walk within the grid, until you hit the boundary. Record the magnitude of the concentation at this boundary point.
(6) Repeat (1)–(5) a pre-chosen number of times. The average of all the boundary values 'hit' is then an estimate of the concentration at the starting point, in other words the solution of Laplace's equation at that point.

The statement in step (6) will undoubtedly appear to many readers as some unconvincing piece of mathematical magic or trickery. In fact it is simply a demonstration of the point we made earlier, that we observe many physical phenomena as the average effect of essentially random processes. Thus each random walk (steps (4) and (5)) simulates the diffusion of a particle (or ion in the electrolytic example) from (x_0, y_0) to the boundary. Since we are at equilibrium, there must be no net flow between the boundary and any point within the grid — there must be as many particles making the journey from (x_0, y_0) to the boundary as are making the return trip. Since the flow is governed by the concentration gradient, the average gradient between the boundary and (x_0,y_0) must be zero and thus the best estimate of the concentration at (x_0,y_0) is as stated in (6). Note that we cannot simply take the mean of all the boundary values since in general not all boundary point are equally accessible by a diffusion (random walk) process. For example, if we start at the grid point (2,2) in Fig. 1.11 then we will make many more hits at boundary points in that corner than in the diagonally opposite corner.

A rigorous justification of the Monte Carlo approach is given by Ralston and Wilf (1960) for the more general equation

$$\partial^2\phi/\partial x^2 + \partial^2\phi/\partial y^2 = F(x,y)$$

(known as Poisson's equation) of which Laplace's equation is a special case, with $F(x,y) = 0$. These authors also give an expression for the variance (error) of the estimate of the solution, and this is incorporated into program P1.8. Some computed values at various grid points are shown in Fig. 1.11.

A clear, non-mathematical explanation of the relationship between phenomena such as Brownian motion and the solving of Laplace's equation is provided by Hersh and Griego (1969).

1.6.3 Monte Carlo integration

For integrating 'well-behaved' functions, the standard methods we discuss in Chapter 3 prove both accurate and efficient. However, there are certain situations where Monte Carlo methods offer distinct advantages, most notably in multidimensional integration (where we are integrating with respect to several variables) and integration over irregular regions (this statement will become clear later.) In such cases even if the Monte Carlo approach is not theoretically the best, it may be the most attractive on the grounds of its simplicity. As with the random walk example, this simplicity is generally obtained at the expense of lengthy execution times.

The most primitive weapon in the Monte Carlo armoury is the hit-or-miss method, which might be termed the Dartboard method. The simple definite integral

$$I = \int_a^b f(x)\, dx$$

is represented by the shaded area bounded by the curve $y = f(x)$, the ordinates $x = a$ and $x = b$ and the x-axis (Fig. 1.12). The basis of the hit-or-miss approach is to fire N

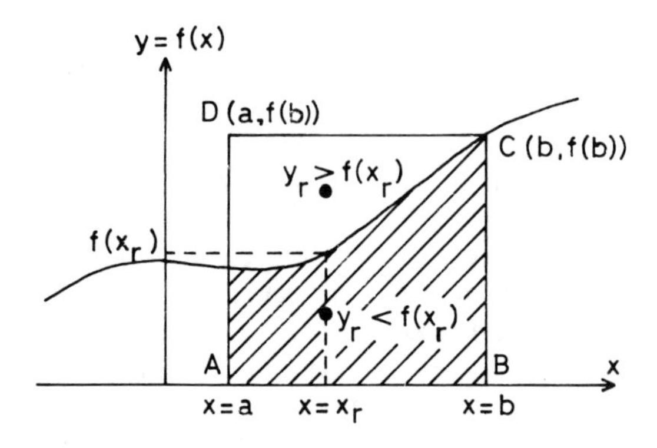

Fig. 1.12 — The hit-or-miss method of Monte Carlo integration, applied to evaluation of the definite integral $\int_a^b f(x)\, dx$ (shaded area).

hypothetical darts at random at the rectangle ABCD in Fig. 1.12, and to count the number of hits n_h, a hit being defined as a dart landing in the shaded area. Assuming purely random throwing, the fraction of darts which hit is equal to the fraction of the target which is shaded; since the area of the target ABCD is $(b - a)f(b)$, the area of the shaded region which is taken as an estimate of the definite integral is

$$I^* = (n_h/N)(b - a)f(b) \tag{1.32}$$

The darts are provided in practice by a random number generator. We generate two random numbers x_r from the interval (a, b) and y_r from the interval $(0, f(b))$, and evaluate $f(x_r)$. If $y_r < f(x_r)$ we register a hit, that is increment a hit counter; after a pre-chosen number of attempts we estimate I from (1.32) using program P1.9.

Although this method is simple to understand and easy to implement, its main drawback is in the number of 'throws' required to obtain any acceptable accuracy.

A more useful approach is termed the 'sample mean method' (Yakowitz, 1977). The basis of this technique is that if we have a set of N random numbers x_i $(i = 1, \ldots, N)$ and corresponding function values $f(x_i)$ then an estimate I^* of the definite integral I is given by the mean or average function value multiplied by the range of integration

$$I^* = (b - a)(1/N) \sum_{i=i}^{n} f(x_i) \tag{1.33}$$

Providing the random number generator is adequate and N is quite large the random numbers x_i $(i = 1, \ldots, N)$ will be evenly distributed along the line segment (a, b) with a spacing $\Delta x = (b - a)/N$. We can thus approximate the shaded area (Fig, 1.12) by the sum total of all the areas of rectangular strips

$$\text{integral} = \text{shaded area} = \sum_{i=1}^{N} f(x_i)\Delta x$$

which is the basis of the fundamental concept of the definite integral as the limit of such a summation; by limit we mean the value of the sum as we let Δx get smaller and smaller (see Francis, 1984). Clearly we could implement this approach to integration simply by dividing up the interval (a, b) into N segments from the beginning, without having to resort to a random number generator. This is no longer true when we graduate to integrals with respect to, say, two variables (that is, over a region) rather than along a simple line segment (a, b), as we shall see later, or if we wish to integrate 'poorly behaved' functions (Yakowitz, 1977). The method is easily implemented and gives a more accurate result than the hit-or-miss procedure.

The accuracy of the Monte Carlo approach to integration may be improved by so-called variance-reduction techniques, in which much effort has been invested. The reader is referred elsewhere (Yakowitz, 1977; Meyer, 1956; Cohen *et al.*, 1973) for details, and to Yakowitz (1977) in particular for a comparison of the performance of many of these methods.

The problem we use to demonstrate Monte Carlo integration again concerns diffusion, but now in two dimensions. Consider a petri dish on which is smeared a thin layer of gel. Into the centre C of the glass is dispensed a drop of dye, which proceeds to diffuse away from C. Our task is to calculate how much dye, after a diffusion time t, is contained in a circle of radius a centred at (u,v) a distance r from C. A practical analogy is an optical densitometer with an aperture of radius a, which is moved in a straight line away from C. The variation of the densitometer output voltage with $r(= (u^2 + v^2)^{1/2})$ then depends on the amount of dye in the sampling circle. The one-dimensional equation (1.29) may be extended to the two-dimensional case to give

$$c(r,t) = (1/4)(\pi Dt)^{-1} \exp(-r^2/4Dt) \tag{1.34}$$

where we assume that unit mass of dye has been injected at C ($r = 0$) at $t = 0$. This is the expression which we need to integrate within a circle of radius a centered at (u,v). Although it is feasible, though far from trivial, to perform the integration using the 'normal' methods of Chapter 3, the Monte Carlo approach is much easier to implement (P1.10). Basically, we use a random number generator to generate the coordinates of points within the integration circle, and compute the average of all the corresponding concentration values; the sample mean method is then invoked to calculate the integral as this mean value multiplied by the area of the circle. Figure 1.13 shows examples of particle distribution profiles calculated for various diffusion times t by repeating the integration for regularly spaced values of r. A convenient measure of how far the dye has spread is the width at half-maximum (Δ), which for curves (a)–(c) compare well with the theoretical (ideal) values computed from (1.34). The effect of using too large a sampling circle is also shown — the profile for a given t is broadened significantly. This sort of broadening besets many different kinds of experimental measurement and is described mathematically by the operation of convolution (Bracewell, 1978).

As in the one-dimensional case we can simulate the diffusion using a random walk model, thus bypassing the integration process. We split the unit mass injected at ($r = 0, t = 0$) into a large number N, say 1000, particles and allow each to perform a random walk for a certain number of steps, dependent on the diffusion time t. At the end of the walks we count the number n of particles which lie within a distance a of the centre (u,v) of the sampling circle — the required integral is then equal to n/N. In order to relate the random walk to the analogue world via the Einstein–Smoluchowski equation (section 1.6.1), we must implement the random walk carefully (P1.11). Since the steps in the x- and y-directions are independent, we can rearrange any particular random walk into a sequence of pure x-steps followed by a sequence of pure y-steps, in other words two successive one-dimensional random walks. Thus the probability of a particle reaching (u,v) is the probability of moving a distance u along the x-axis multiplied by the probability of moving a distance v along the y-axis (in the same way that the chances of throwing two successive sixes at dice is $(1/6) \times (1/6) = 1/36$); this leads to an equation analogous to (1.34) and thence to the Einstein–Smoluchowski relation (Brown, 1968). Since in the normal one-dimensional random walk we did not allow a no-move step, in the two-dimensional case we can not allow a no-move step in either direction. Thus, the allowed moves at any point take the particle to any of the four corners of a square centred on the current point and with side equal to twice the step length (Fig. 1.14). With this restriction, the random talk method gives results in agreement with the Monte Carlo integration, but only after much lengthier execution times.

1.6.4 Random number generators

It must be stressed that all random number generating algorithms only output pseudo-random numbers, since given the same initial conditions they will generate the same sequence of numbers. However, the distribution of numbers in this (perhaps very long) sequence may still satisfy stringent statistical tests of random-

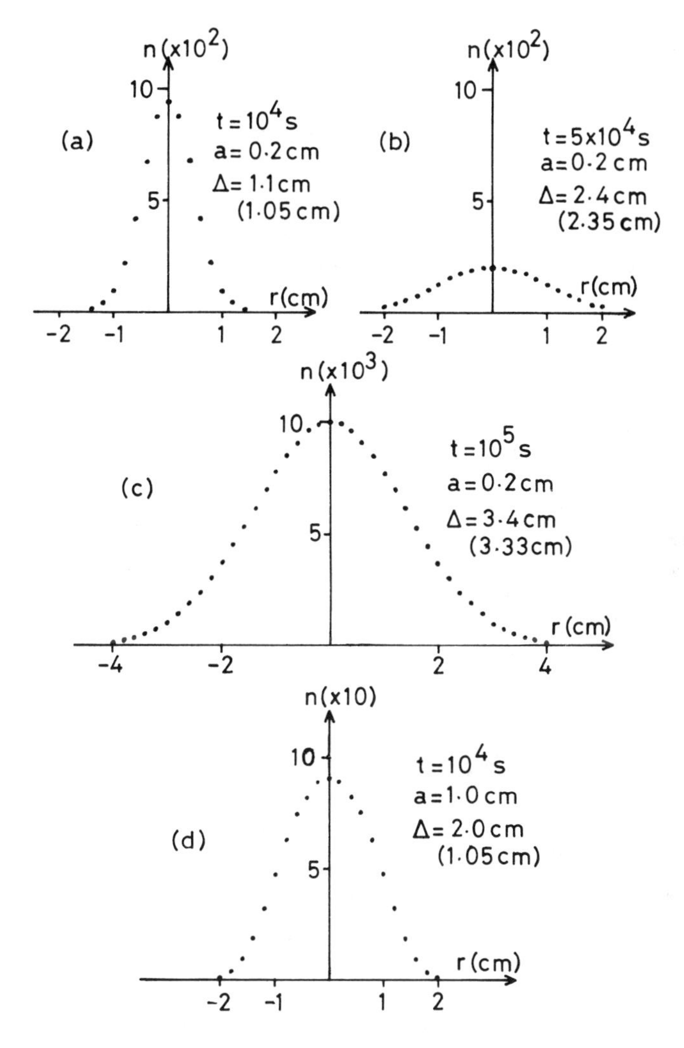

Fig. 1.13 — Two-dimensional diffusion modelled using the Monte Carlo approach. The full-width-at-half-maximum (FWHM) values (Δ) of the calculated profiles are indicated; quantities in brackets are the theoretical values (see text).

ness. The definitive text on the subject is Knuth (1981) which should be consulted by those interested in the theoretical analysis of random number generators (RNG), which is beyond the scope of this chapter; a brief but instructive introduction is provided by Ripley (1983). This latter article appears in the microcomputer publication *Personal Computer World*, which in recent years has been a rich source of RNG algorithms, and more importantly has provided machine code implementations for several microprocessors.

There are two basic types of RNG currently in use: shift register generators and

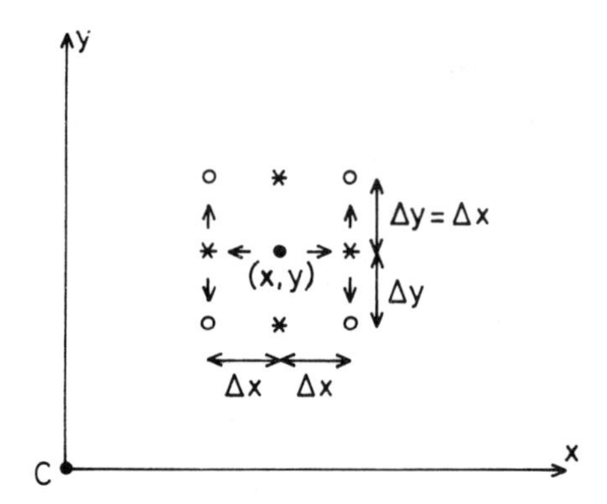

Fig. 1.14 — Allowed moves of a particle in the two-dimensional Monte Carlo simulation. Open circles represent allowed destination points.

congruential generators. The latter kind are by far the most widely used and generate the next random integer in the sequence from the previous one according to.

$$r_{i+1} = (ar_i + c) \bmod M \tag{1.35}$$

where 'mod M' means the remainder left after dividing the preceding quantity by M. Clearly the process must be seeded by some initial choice r_0. The values chosen for the positive integers a and M are important in determining both the randomness and periodicity of the sequence r_i; the integer constant c seems to be much less critical but must either be odd or zero. By periodicity we mean how long before the sequence r_i inevitably repeats itself. Naturally, we desire that the sequence be as randomly distributed and to have as long a period as possible. Certain conditions ensure that particular congruential generators exhibit the maximum possible period (Ripley, 1983), and their randomness may be visually demonstrated by plotting successive (r_i, r_{i+1}) pairs as points on a graph. The pattern produced is a lattice composed of M repeating cells; the optimum randomness occurs when the cell is as square as possible. This procedure, known as the spectral test, may be extended to three and higher dimensions (Ripley, 1983).

There are almost as many different suggestions for random number generators as there are proposers, but the conclusion drawn from the contributors to *Personal Computer World* was that the recommended 16-bit and 32-bit RNG were, respectively

$$r_{i+1} = (1509r_i + 41) \bmod 2^{16}$$

and

$$r_{i+1} = (69069r_i + 41) \bmod 2^{31}$$

A simple RNG involving only the addition of three previously generated (or initially seeded) random numbers, namely

$$r_{i+3} = r_{i+2} + r_{i+1} + r_i$$

(followed by scaling) has been claimed to be both uniform and fast (Swain and Swain, 1980).

1.7 ARRAYS

We often have occasion to make a number of closely related measurements on a sample, for example the concentration of a number of metal ions in a set of water samples, or the absorbance of a sample at a particular wavelength measured every second after the start of a reaction. We deal with such lists of numbers in BASIC, and many other languages, by using arrays. An array may be viewed as a set of 'pigeon-holes' labelled with integer numbers, the whole set being given a name. Thus the contents of any of the pigeon-holes (a particular absorbance measurement in the earlier example for instance) may be accessed by specifying the name of the set (array) and the label of the particular pigeon-hole — it is analogous to describing where someone lives by giving the street name and house number. As an example, we might call an array of absorbance 'A' in a program and refer to the 10th measurement by writing $A(10)$; $A(10)$ is termed an element of the array A. In BASIC we have to tell the computer (more precisely the BASIC interpreter) something about the array before we use it. This is done using the DIM (DIMension) statement which takes the form

```
10 DIM A(100)
```

and results in sufficient memory being reserved for the array A. BASIC usually allows zero to be used as an array index (that is, $A(0)$ is allowed) so that the previous DIM statement reserves space for 101 elements, but in practice it may be more convenient to restrict element labels to positive integers. A short program segment which inputs elements into an array might be

```
10 INPUT "NUMBER OF VALUES = ";NV
20 DIM A(NV)
30 FOR I = 1 TO NV
40 INPUT A(I)
50 NEXT I
```

If we wished to use the zeroth element of A then lines 20 and 30 would read

```
20 DIM A(NV – 1)
30 FOR I = 0 TO NV – 1
```

The maximum size permitted for the array (NV or NV + 1) will naturally depend on the amount of available memory. Now that we have filled the array we can use it by specifying the array name and element label (pigeon-hole); for example, we could go on to find the element with the largest magnitude (using positive indexing)

```
60 HI = A(1)
70 N = 1
80 FOR I = 1 TO NV
90 IF A(I) > HI THEN HI = A(I) : N = I
100 N EXT I
110 PRINT "LARGEST ELEMENT IS NUMBER ";N
120 PRINT "LARGEST VALUE = ";HI
```

We can extend the idea further to arrays with two or more indices (labels) such as $A(P,Q)$, which might represent the absorbance of a solution at time P and wavelength Q, for example. Whereas an array $A(I)$ represents a list, one like $A(P,Q)$ represents a two-variable table of values; P and Q then label the rows and columns of the table respectively. To find the maximum absorbance of this set, $A(P,Q)$, we modify the previous program segments to give

```
10 INPUT "NUMBER OF TIME READINGS = ";NV
20 INPUT "NUMBER OF WAVELENGTH READINGS =";MV
30 DIM A(NV,MV)
40 FOR I = 1 TO NV
50 FOR J = 1 TO MV
60 INPUT A(I,J)
70 NEXT J
80 NEXT I
90 HI = A(1,1)
100 N = 0 : M = 0
110 FOR I = 1 TO NV
120 FOR J = 1 TO MV
130 IF A(I,J) > HI THEN HI = A(I,J) : N = I : M = J
140 NEXT J
154 NEXT I
```

The data-input routine (lines 10–80) is as primitive as is possible, though adequate for the single-index (one-variable) case. There is great potential for confusion in the two-variable case and in practice textual prompts and helpful screen formatting would be used to ensure that the correct values ended up in the correct pigeon-holes.

Three-index arrays are sometimes encountered and may be interpreted as a room-full of pigeon holes as compared with a shelf- or wall-full. Higher-dimensional arrays, using four or more indices, are allowed in some BASIC dialects but are not very easily visualized.

The true beauty of the array concept can be most appreciated when one realizes that a whole branch of mathematics exists for which they are tailor-made — the application of matrices, which gives coherence to such problems as the solution of simultaneous equations, quantum-mechanical calculations and the computer graphics of molecular models.

1.8 MATRICES, VECTORS AND SYSTEMS OF EQUATIONS

1.8.1 Basic concepts

A formal mathematical development of these concepts is beyond the scope of this book, and the interested reader is directed elsewhere (Carley and Morgan, 1989). We shall restrict ourselves here to a less rigorous description, but one which contains all the elements necessary to an understanding of the material contained in the subsequent chapters of this book.

A good place to start at is simply to regard matrices and vectors as a useful mathematical notation for representing systems of simultaneous equations. For example, consider the set of equations

$$\begin{array}{lrcl} \text{(i)} & 2x - 3y - z & = & 5 \\ \text{(ii)} & x + y + z & = & 2 \\ \text{(iii)} & x - y - 2z & = & -3 \end{array} \tag{1.36}$$

The left-hand side involves the unknowns x, y and z and a set of multipliers

$$\begin{array}{ccc} 2 & -3 & -1 \\ 1 & 1 & 1 \\ 1 & -1 & -2 \end{array}$$

whilst the right-hand side consists of three constants 5, 2 and -3. We can regard these as comprising two lists and a table, and invoke the BASIC construction of arrays encountered in the previous section. Mathematically, such ‘tables’ and ‘lists’ are referred to as matrices and column matrices respectively. We shall show this by writing them in square brackets, thus

$$\begin{bmatrix} x \\ y \\ z \end{bmatrix}, \begin{bmatrix} 2 & -3 & -1 \\ 1 & 1 & 1 \\ 1 & -1 & -2 \end{bmatrix} \text{ and } \begin{bmatrix} 5 \\ 2 \\ -3 \end{bmatrix}$$

although alternative notational forms will be encountered by the reader in other texts. Column matrices are frequently referred to as vectors, although there is an important (albeit subtle) distinction between the two quantities (section 1.8.3). For a detailed discussion of this point, the reader should consult Carley and Morgan (1989). However, the purposes of this text, we shall follow common practice and use the terms column matrix and vector interchangeably, except where an explicit distinction is necessary in order to avoid confusion.

The usefulness of the matrix formulation becomes more apparent if we define the operation of matrix–vector multiplication by rewriting (1.36) as

$$\begin{bmatrix} 2 & -3 & -1 \\ 1 & 1 & 1 \\ 1 & -1 & -2 \end{bmatrix} \begin{bmatrix} x \\ y \\ z \end{bmatrix} = \begin{bmatrix} 5 \\ 2 \\ -3 \end{bmatrix} \tag{1.37}$$

By comparing (1.37) with (1.36) we can determine the rule for such a multiplication, which is naturally more complicated than that for multiplying simple numbers.

We proceed to generalize the problem by writing

$$\begin{bmatrix} M_{11} & M_{12} & M_{13} \\ M_{21} & M_{22} & M_{23} \\ M_{31} & M_{32} & M_{33} \end{bmatrix} \begin{bmatrix} x_1 \\ x_2 \\ x_3 \end{bmatrix} = \begin{bmatrix} y_1 \\ y_2 \\ y_3 \end{bmatrix} \tag{1.38}$$

or more compactly as

$$[M]\,[x] = [y] \tag{1.38}$$

where the subscripts of the x and y values indicate the index of the element in the vector. For the matrix, the double subscripts can be thought of as like coordinates: for example, the element M_{32} is that situated at the intersection of row 3 and column 2. Note that the row index is always the first of the two indices. The reader may have also noticed that we have adopted the convention, adhered to throughout this book, of denoting matrices by upper-case letters and column matrices or vectors by lower-case characters.

Matrix multiplication is defined by noting, for example, that

$$y_1 = M_{11}x_1 + M_{12}x_2 + M_{13}x_3$$

or more generally: the ith element of $[y]$ is obtained by taking the ith row of $[M]$, multiplying it element by element by $[x]$ and adding up all the products. Mathematically, for an $N \times N$ matrix and an N-element vector

$$y_i = \sum_{j=1}^{j=N} \underbrace{M_{ij}}_{\text{elements of row } i \text{ of } [M]} \quad \underbrace{x_j}_{\text{elements of } [x]} \tag{1.39}$$

A program segment to perform this would be

```
10 FOR I = 1 TO N
```

```
20 Y(I) = 0
30 FOR J = 1 TO N
40 Y(I) = Y(I) + M(I,J)*X(J)
50 NEXT J
60 NEXT I
```

The inner loop (lines 30–50) does the summation part of (1.39), whilst the outer loop repeats it for each element of the new vector $[y]$. Note the way we accumulate the summation into a variable (line 40) which has been previously set to zero (line 20).

Returning to our original problem, the solution of a set of simultaneous equations is equivalent to solving (1.38) for $[x]$ given $[M]$ and $[y]$. If (1.38) were a simple arithmetic equation like $3x = 1$, then we would divide each side by the multiplier, 3, and obtain $x = 1/3$. Strictly speaking, what we do is to multiply each side by the reciprocal of the multiplier ($1/3 = 3^{-1}$) thus

$$\begin{aligned} 3x &= 1 \\ 3^{-1}\,3x &= 3^{-1} \\ 1x &= 0.333 \end{aligned}$$

and relies on the fact that $3^{-1}\,3 = 1$. This re-statement may appear pedantic but it allows us to consider the analogous process in matrix–vector arithmetic. The 'reciprocal' of a matrix is known as its inverse and is written $[M]^{-1}$. It has the property that

$$[M]^{-1}[M] = [I] = [M][M]^{-1} \tag{1.40}$$

where $[I]$ is the $N \times N$ identity matrix, which performs the same role in matrix algebra as does the number 1 in simple arithmetic. The elements of $[I]$ are zero everywhere except along the diagonal where they are unity. For example, the 3×3 identity matrix is

$$\begin{bmatrix} 1 & 0 & 0 \\ 0 & 1 & 0 \\ 0 & 0 & 1 \end{bmatrix}$$

In (1.40) we have introduced the idea of matrix–matrix multiplication. This is a little more complicated than matrix–vector multiplication, each element of the product matrix arising from the 'product' of a row of the first matrix and a column of the second matrix. Consider the multiplication of two 3×3 matrices $[M]$ and $[N]$ to give a new matrix $[P]$

$$\begin{bmatrix} P_{11} & P_{12} & P_{13} \\ P_{21} & P_{22} & P_{23} \\ P_{31} & \boxed{P_{32}} & P_{33} \end{bmatrix} = \begin{bmatrix} M_{11} & M_{12} & M_{13} \\ M_{21} & M_{22} & M_{23} \\ M_{31} & M_{32} & M_{33} \end{bmatrix} \begin{bmatrix} N_{11} & N_{12} & N_{13} \\ N_{21} & N_{22} & N_{23} \\ N_{31} & N_{32} & N_{33} \end{bmatrix}$$

The boxed element P_{32} is formed from the row and column indicated according to

$$P_{32} = M_{31}N_{12} + M_{32}N_{22} + M_{33}N_{32}$$

Thus, we can extend (1.39) to give the general expression for matrix–matrix multiplication

$$P_{ij} = \sum_{k=1}^{k=N} \underbrace{M_{ik}}_{\text{elements of row } i \text{ of } [M]} \quad \underbrace{N_{kj}}_{\text{elements of column } j \text{ of } [N]} \tag{1.41}$$

The requirement to multiply two matrices together lies at the heart of so many computational procedures that the reader is urged to ensure, before proceeding further, that he or she is totally familiar, and at ease, with the correponding program segment given below. The three nested loops may appear at first sight intimidating, but a closer inspection will reveal a simplifying symmetry

```
10 FOR I = 1 TO N
20 FOR J = 1 TO N
30 P(I,J) = 0
40 FOR K = 1 TO N
50 P(I,J) = P(I,J) + M(I, K)*N(K,J)
60 NEXT K
70 NEXT J
80 NEXT I
```

In some dialects of BASIC this segment can be speeded up significantly by making a few small changes

```
30 P = 0
50 P = P + M(I,K)*N(K,J)
65 P(I,J) = P
```

In Applesoft BASIC this modified routine is 30% faster than the original. The reason is that the BASIC interpreter takes longer to look-up an array variable than it does a simple variable (like *P*) and this adds significantly to the timing of the loop in lines 40–60.

1.8.2 Methods for solving systems of equations

Returning once again to the original problem, assuming we can find $[M]^{-1}$ (and this is not always feasible), we can proceed thus

$$[M]^{-1}[M][x] = [M]^{-1}[y]$$

$$[I][x] = [M]^{-1}[y]$$

and finally

$$[x] = [M]^{-1}[y] \tag{1.42}$$

since multiplying any vector or matrix by $[I]$ leaves it unchanged.

In practice, however, this direct method is not often used. Instead, we employ a formal implementation of the familiar school pupil's method for solving sets of simultaneous equations. We shall recapitulate this approach by considering equations (1.36(i)–(iii)). We proceed by subtracting multiples of equations (i)–(iii) from each other. For example, (ii) minus (iii) gives

(iv) $2y + 3z = 5$

and (i) minus twice (ii) gives

(v) $-5y - 3z = 1$

Then 5 times (iv) plus 2 times (v) results in

(vi) $9z = 27$

and thus $z = 3$. Substituting this value in (v) gives $-5y - 9 = 1$ and thence $y = -2$. Finally, with these values substituted in (i) we find $x = 1$.

In order to automate this process we need to perform the multiplications and subtractions in a systematic manner, rather than the 'let us see which equation is best to use next' approach. We take the matrix formulation of the problem

$$[M][x] = [y]$$

and subtract multiples of the rows of $[M]$ and the elements of $[y]$ from each other, until the resulting matrix equation allows the elements of $[x]$ to be extracted in a simple manner. For example, one method known as Gauss–Jordan elimination performs the subtractions in such a way that we are left with

$$[I][x] = [y]' \tag{1.43}$$

where $[I]$ is the identity matrix and $[y]'$ is the vector left after subjecting the right-hand side to the same operations as were performed on the matrix. Clearly, the solution vector is given directly by $[y]'$ in this case. An alternative approach, called simply Gaussian elimination, reduces the matrix to what is known as upper-triangular form — that is, all the elements below the diagonal or zero. Thus we would obtain

$$\begin{bmatrix} U_{11} & U_{12} & U_{13} \\ 0 & U_{22} & U_{23} \\ 0 & 0 & U_{33} \end{bmatrix} \begin{bmatrix} x_1 \\ x_2 \\ x_3 \end{bmatrix} = \begin{bmatrix} y_1' \\ y_2' \\ y_3' \end{bmatrix} \tag{1.44}$$

which represents the system

$$\begin{aligned} &\text{(i)} \quad U_{11}x_1 + U_{12}x_2 + U_{13}x_3 = y_1' \\ &\text{(ii)} \quad U_{22}x_2 + U_{23}x_3 = y_2' \\ &\text{(iii)} \quad U_{33}x_3 = y_3' \end{aligned} \tag{1.45}$$

These equations can be solved by a process called back-substitution — we obtain x_3 from (iii), substitute in (ii) to obtain x_2 and finally substitute x_3 and x_2 in (i) to get x_1. Of course, to use either of these methods we need to formulate the set of rules (operations to be performed on the original matrix) so that we end up with the appropriate matrix on the left-hand side of the equation.

We consider first the Gauss–Jordan method, which can easily be modified to give also the inverse of the matrix $[M]$. The following program segment assumes the system matrix $[M]$ has been input, together with the vector $[y]$. Since we perform the same operations on $[M]$ and $[y]$ it proves convenient to construct what is called an augmented matrix $[M]'$ which has $N+1$ columns, the extra column being the elements of $[y]$, and to apply all the operations to this now non-square matrix.

```
100 N1 = N+1                : REM number of columns in augmented matrix
110 FOR C = 1 TO N          : REM step through the diagonal elements
120 DD = M(C,C)             : REM divisor = diagonal element
130 FOR J = 1 TO N1         : REM for all elements of row C ...
140 M(C,J) = M(C,J)/DD      : REM ... normalize by dividing by DD ...
150 NEXT J                  : REM ... so that diagonal element now = 1
160 FOR I =1 TO N           : REM step through all the rows ...
170 IF I = C THEN GOTO 220  : REM ... except the current one row (C)
180 MU = M(I,C)             : REM multiplier is the element in row I which is in the same column as M(C,C)
```

```
190 FOR J = 1 TO N1                   : REM for all the elements of row I
                                        ...
200 M(I,J) = M(I,J) - MU*M(C,J)       : REM ... subtract MU times the
                                        corresponding element in row C,
210 NEXT J                            : REM leaving M(I,C) = 0
230 NEXT I                            : REM after this loop, all elements
                                        in column C (except M(C,C)) are
                                        zero
230 NEXT C                            : REM repeat for next diagonal
                                        element
240 RETURN
```

The working of this program will become clearer if we consider a real example (1.36) and print out the intermediate results as the method proceeds. In what follows remember that operations on rows of the augmented matrix are equivalent, in more familiar terminology, to operations on equations. We start with the matrix

$$\begin{bmatrix} 2 & -3 & -1 & 5 \\ 1 & 1 & 1 & 2 \\ 1 & -1 & -2 & -3 \end{bmatrix}$$

CYCLE 1 (C = 1)

$$\begin{matrix} \text{divide} \\ \text{row 1 by} \\ M(1,1)=2 \end{matrix} \rightarrow \begin{bmatrix} 1 & -1.5 & -0.5 & 2.5 \\ 1 & 1 & 1 & 2 \\ 1 & -1 & -2 & -3 \end{bmatrix}$$

$$\begin{matrix} \text{row 2} \\ \text{minus} \\ 1 * \text{row 1} \end{matrix} \rightarrow \begin{bmatrix} 1 & -1.5 & -0.5 & 2.5 \\ 0 & 2.5 & 1.5 & -0.5 \\ 1 & -1 & -2 & -3 \end{bmatrix}$$

$$\begin{matrix} \text{row3} \\ \text{minus} \\ 1 * \text{row 1} \end{matrix} \rightarrow \begin{bmatrix} 1 & -1.5 & -0.5 & 2.5 \\ 0 & 2.5 & 1.5 & -0.5 \\ 0 & 0.5 & -1.5 & -5.5 \end{bmatrix}$$

All the elements in column 1, except for (1,1) are now zero.

$$\begin{matrix}\text{divide} \\ \text{row 2 by} \\ M(2,2) = 2.5\end{matrix} \rightarrow \begin{bmatrix} 1 & -1.5 & -0.5 & 2.5 \\ 0 & 1 & 0.6 & -0.2 \\ 0 & 0.5 & -1.5 & -5.5 \end{bmatrix}$$

$$\begin{matrix}\text{row 1} \\ \text{minus} \\ -1.5 * \text{ row 2}\end{matrix} \rightarrow \begin{bmatrix} 1 & 0 & 0.4 & 2.2 \\ 0 & 1 & 0.6 & -0.2 \\ 0 & 0.5 & -1.5 & -5.5 \end{bmatrix}$$

$$\begin{matrix}\text{row 3} \\ \text{minus} \\ 0.5 * \text{row 2}\end{matrix} \rightarrow \begin{bmatrix} 1 & 0 & 0.4 & 2.2 \\ 0 & 1 & 0.6 & -0.2 \\ 0 & 0 & -1.8 & -5.4 \end{bmatrix}$$

CYCLE 3

$$\begin{matrix}\text{divide} \\ \text{row 3 by} \\ M(3,3) = -1.8\end{matrix} \rightarrow \begin{bmatrix} 1 & 0 & 0.4 & 2.2 \\ 0 & 1 & 0.6 & -0.2 \\ 0 & 0 & 1 & 3 \end{bmatrix}$$

$$\begin{matrix}\text{row 1} \\ \text{minus} \\ 0.4 * \text{row 3}\end{matrix} \rightarrow \begin{bmatrix} 1 & 0 & 0 & 1 \\ 0 & 1 & 0.6 & -0.2 \\ 0 & 0 & 1 & 3 \end{bmatrix}$$

$$\begin{matrix}\text{row 2} \\ \text{minus} \\ 0.6 * \text{row 3}\end{matrix} \rightarrow \begin{bmatrix} 1 & 0 & 0 & 1 \\ 0 & 1 & 0 & -2 \\ 0 & 0 & 1 & 3 \end{bmatrix}$$

The process is now complete and the solution vector is given by the last column of the augmented matrix.

Each of the individual operations above can be represented by a matrix. For example, the last operation in cycle 3 is equivalent to multiplying by the matrix

$$\begin{bmatrix} 1 & 0 & 0 \\ 0 & 1 & -0.6 \\ 0 & 0 & 1 \end{bmatrix}$$

and the whole process is the result of the successive application of many such matrices. The overall matrix describing this method is thus the product of all of these individual matrices, and we shall designate it $[T]$. Since the effect of $[T]$ on $[M]$ is to generate the identity matrix $[I]$, clearly $[T]$ is the inverse of $[M]$, $[M]^{-1}$, and the method turns out to be equivalent to the direct solution mentioned earlier. By 'remembering' all of the matrices making up $[T]$ we can thus compute the inverse matrix as a bonus. The easiest way to do this is to start with the identity matrix $[I]$ and apply all of the operations to it as well as to the augmented matrix; the final result is then $[T]=[M]^{-1}$. The program changes to the Gauss–Jordan routine are minimal, and consist of the insertion of two extra lines

```
145 I(C,J) = I(C,J)/DD
205 I(I,J) = I(I,J) - MU*I(C,J)
```

The matrix $I(i,j)$ must be initialized to the identity matrix before this routine is called.

A problem arises with this method if any of the diagonal elements are zero, since this results in a 'DIVIDE BY ZERO' error at line 140. For example the system

$$\begin{array}{lrl} \text{(i)} & y+z= & 1 \\ \text{(ii)} & x \quad +z= & 4 \\ \text{(iii)} & 2x+2y-z= & -5 \end{array}$$

has the same solution as (1.36) but also has $M(1,1)=M(2,2)=0$ and the program fails first time round at line 140. In fact, it is desirable that the pivots (the diagonal elements which form the divisors) are not only non-zero, but are in fact as large as possible, since division by small numbers, especially on computers with arithmetic of limited precision can lead to the accumulation of large errors. We can overcome both problems by a process known as pivoting, where we swap rows in the augmented matrix both to avoid zero pivots and also to keep the pivots as large as possible. It is simply equivalent, in non-matrix terms, to changing the order in which the equations are listed, and so clearly does not affect the solution. The necessary changes to the Gauss–Jordan routine to accommodate pivoting are as follows

```
115 GOSUB 800 : REM swap rows if needed
800 FL = C : REM FL contains the row number with the largest element
810 FOR K = C+1 TO N : REM check rows below C
820 IF ABS(M(K,C))>ABS(M(C,C)) THEN FL = K
830 NEXT K
840 IF FL = C THEN RETURN : REM M(C,C) was the largest element
850 REM if not then swap rows
860 FOR K = 1 TO N1
870 T = M(C,K) : M(C,K) = M(FL,K) : M(FL,K) = T
880 T = I(C,K) : I(C,K) = I(FL,K) : I(FL,K) = T
890 NEXT K
900 RETURN
```

Note that it is the absolute value of the matrix element which is important (line 820), and that we must not forget the evolving inverse matrix (line 870) if it is being computed. On those systems which support matrix arithmetic, that is arithmetic where matrices can be manipulated as single entities, lines 860–890 would be much simpler and less clumsy, as indeed would all the matrix routines we have encountered.

A program for the Gauss elimination method, where the system matrix is merely reduced to upper triangular form, is derived from the Gauss–Jordan routine by removing line 170 and changing line 160 to

```
160 FOR I = C+1 TO N
```

thus only affecting elements below the diagonal. We then require a segment to perform the back-substitution. This is quite a tricky exercise, and we firstly consider the solution of the simple case represented by (1.45). We have

$$x_3 = y_3'/U_{33}$$
$$x_2 = (y_2' - U_{23}x_3)/U_{22}$$
$$x = (y_1' - U_{12}x_2 - U_{13}x_3)/U_{11}$$

There is a pattern to these equations which we can generalize to

$$x_j = (y_j' - \sum_{k=j+1}^{N} U_{jk}x_k)/U_{jj} \tag{1.46}$$

and implement as the following program segment

```
240 X(N) = M(N,N1)/M(N,N)
250 FOR J = N-1 TO 1 STEP -1
260 S = 0
270 FOR K = J+1 TO N
280 S = S + M(J,K)*X(K)
290 NEXT K
300 X(J) = (M(J,N1)-S)/M(J,J)
310 NEXT J
320 RETURN
```

Remember that both the upper triangular matrix $[U]$ and the vector $[y]'$ are contained in the augmented matrix. Lines 260–290 perform the summation part of (1.46), and line 300 completes the computation of x_j. Note the downwards counting FOR. . .NEXT loop (lines 250–310). The computational effort involved in the Gauss elimination method with back-substitution is less than that in the Gauss–Jordan method, and is thus to be preferred where this consideration is important. The advantage of the latter procedure is that it can provide the inverse of $[M]$ if this is required. Pivoting is recommended whichever method is used. Although we have only considered pivoting by exchange of rows, full pivoting involves column exchange as well, which is much more tricky (Carley and Morgan, 1989). A procedure for reducing a matrix to upper triangular form which avoids pivoting is Givens' method, which we discuss in Chapter 6.

The more sophisticated methods which exist are based on matrix factorization techniques (Chapter 6). In the same way that a simple number may be decomposed into factors, for example $30 = 5 \times 4 \times 2 \times 1$, a matrix may be decomposed into several matrices whose product reconstruct the original matrix. We came across this earlier in the Gauss–Jordan elimination method, where the inverse matrix $[M]^{-1}$ was generated as the product of many other simpler matrices. In this case, the sequence of simple matrices can be regarded as a factorization of $[M]^{-1}$. An illustration of why this may prove useful in the solution of systems of equations is given by the LU decomposition technique. In this approach the matrix $[M]$ is factorized into an upper triangular and lower triangular matrix (the latter is a matrix with zeros above the diagonal) thus

$$
\begin{aligned}
[M] &= [L][U] \\
&= \begin{bmatrix} L_{11} & 0 & 0 \\ L_{21} & L_{22} & 0 \\ L_{31} & L_{32} & L_{33} \end{bmatrix} \begin{bmatrix} U_{11} & U_{12} & U_{13} \\ 0 & U_{22} & U_{23} \\ 0 & 0 & U_{33} \end{bmatrix}
\end{aligned}
$$

where we have written the matrices out explicitly for the (3×3) case. The system of equations thus becomes

$$[L][U][x] = [y]$$

If we write $[z] = [U][x]$, then

$$[L][z] = [y]$$

which we solve for $[z]$ by forward substitution — this is like backward substitution except that now we first find $z_1 = y_1/L_{11}$, then substitute to get z_2, etc. Having obtained $[z]$ we have

$$[U][x] = [z]$$

which we solve by back-substition as explained earlier. Combined with piovoting, this method is robust and popular. A detailed discussion, including how the matrices $[L]$ and $[U]$ are computed, is to be found elsewhere (Carley and Morgan, 1989).

1.8.3 Vectors, column matrices and products

We noted earlier in passing (section 1.8.1) that a vector is an entity quite distinct from its column matrix representation, although this distinction is often (implicitly) ignored. A vector has an existence independent of any axis system, and it is only when we impose a particular system to describe it that we 'create' its column matrix analogue. For example, in three-dimensional space we usually employ a right-angled x–y–z axis system, and the elements of a column matrix are conveniently regarded as coordinates of a point — the vector is then often visualized as a directional line joining the origin of the axis system to that point (Fig. 1.15(a)). However, the vector is the same (the directional line does not move) if we use another, say oblique, axis system, but the numbers describing it — its column matrix representation — will be different. In order to highlight this distinction we denote a vector by $\mathbf{x}$ and its column matrix description by $[x]$. In practice, the axis system is usually defined (either explicitily or implicitly) and we are not troubled by these subtleties.

The modulus or length of a vector $\mathbf{x}$ is denoted by x, and is given by

$$x^2 = x_1^2 + x_2^2 + x_3^2$$

where x_1, x_2 and x_3 are the elements of the column matrix $[x]$. A vector may be described in terms of unit vectors (vectors of unit length) pointing along the direction of each axis (Fig. 1.15(b)). The elements x_i of $[x]$ may then be thought of as components along each of the axes of the coordinate system, and we can write

$$\mathbf{x} = x_1\mathbf{i} + x_2\mathbf{j} + x_3\mathbf{k} \tag{1.47}$$

We can interpret this as meaning that we get from the origin to the point (vector) $\mathbf{x}$ by moving x_1 units in the direction $\mathbf{i}$, then x_2 units in the direction $\mathbf{j}$ and finally x_3 units in the direction $\mathbf{k}$. The result is the same as moving x units in the direction $\mathbf{x}$ (Fig. 1.15(c)).

We are now in a position to understand an important concept in vector analysis, the scalar or dot product of two vectors. It is written $\mathbf{x}\cdot\mathbf{y}$ and is defined by

$$\mathbf{x}\cdot\mathbf{y} = xy\cos\theta \tag{1.48}$$

where θ is the angle between the vectors (Fig. 1.16). It may be thought of as the product of the length of one of the vectors with the projection of the other on it (Fig. 1.16). A physical example is the work done by the movement of a force, which is given by

$$W = \mathbf{f}\cdot\mathbf{d}$$

where W is the work done, $\mathbf{f}$ is the force and $\mathbf{d}$ is the distance vector. Thus

$$W = fd\cos\theta$$

or

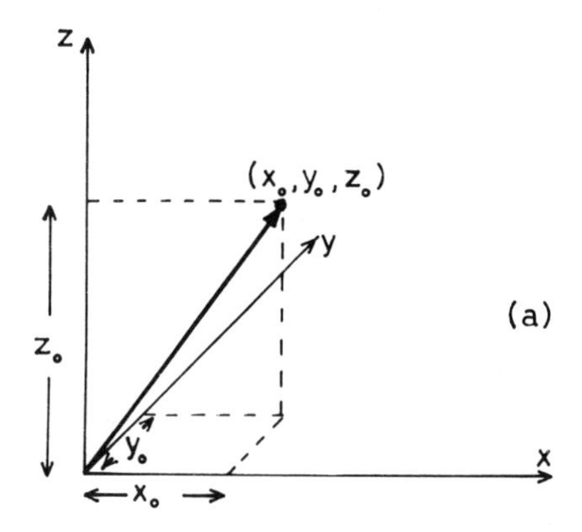

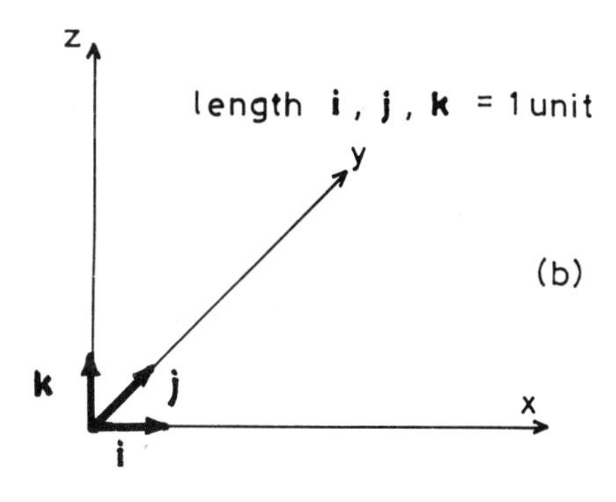

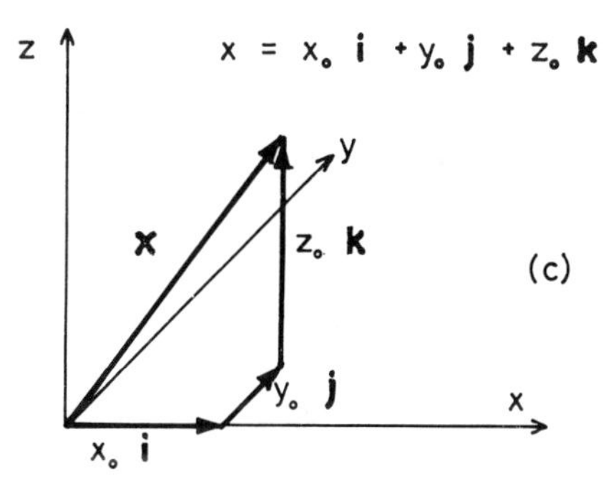

Fig. 1.15 — Graphical representation of a three-element vector [**x**]. (a) The three components may be interpreted as coordinates of a point in a x–y–z axis system. It is common practice to join the origin to this point and regard the directed line as the vector. (b) we can define unit vectors [**i**], [**j**] and [**k**] which point along each axis and are of unit length. The vector [**x**] can then be expressed as the sum of three vector components (c), each component being at right-angles to the other two.

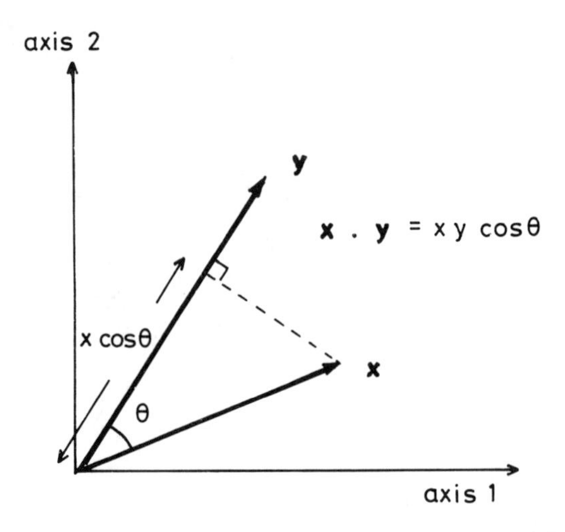

Fig. 1.16 — The dot or scalar product of the two-element vectors [**x**] and [**y**].

work = (force) × (distance moved in direction of force)

or, in other words

work = (component of force in direction of movement) ×
× (distance moved)

An alternative formulation of the dot product arises from the use of (1.47). We have

$$\mathbf{x} \cdot \mathbf{y} = (x_1\mathbf{i} + x_2\mathbf{j} + x_3\mathbf{k}) \cdot (y_1\mathbf{i} + y_2\mathbf{j} + y_3\mathbf{k}) \tag{1.49}$$

which on expansion contains many products of the kind $x_2y_3\ \mathbf{j}\cdot\mathbf{k}$. Since the dot product of any two vectors which are at right angles to each other is zero, most of these terms vanish. Since, further

$$\mathbf{x} \cdot \mathbf{x} = x^2$$

we have $\mathbf{i} \cdot \mathbf{i} = \mathbf{j} \cdot \mathbf{j} = \mathbf{k} \cdot \mathbf{k} = 1$ and (1.49) reduces to

$$\mathbf{x} \cdot \mathbf{y} = x_1y_1 + x_2y_2 + x_3y_3 \tag{1.50}$$

Equation (1.50) should remind us of the expression at the heart of matrix–matrix multiplication — it would result from multiplying a column with the elements y_1, y_2 and y_3 by a row with the elements x_1, x_2 and x_3. In fact, we write the matrix version of the scalar product as $[x]^T[y]$, where the superscript T means the transpose of the vector. The transpose of a matrix is that matrix which results from the interchange of its rows and columns; in the case of a vector, a column vector simply becomes a row vector. Thus

$$\mathbf{x} \cdot \mathbf{y} = [x]^T[y]$$

$$= [x_1 \quad x_2 \quad x_3] \begin{bmatrix} y_1 \\ y_2 \\ y_3 \end{bmatrix}$$

$$= x_1y_1 + x_2y_2 + x_3y_3$$

The angle between two vectors **x** and **y** is now easily computed from

$$\cos(\theta) = [x]^T[y]/xy \tag{1.51}$$

The effect of a matrix on a vector can now be given a graphical representation — what it does is change the length of the vector and rotate it through an angle. In the context of our initial problem the matrix

$$\begin{bmatrix} 2 & -3 & -1 \\ 1 & 1 & 1 \\ 1 & -1 & -2 \end{bmatrix}$$

transforms an as yet unknown vector $[x]$ into the vector $[5 \quad 2 \quad -3]^T$ (Fig. 1.17). The solution of a system of equations may thus be thought of as undoing this scaling and rotation to find the original vector — this is the effect of the inverse matrix.

We have seen that the scalar product of two vectors is computed as the product of a row matrix and a column matrix. What happens if we multiply them in reverse order, that is, given two column matrices $[x]$ and $[y]$ we form $[x][y]^T$?

The result is called the outer product and turns out to be a matrix, with (in this case) elements computed as below.

$$\begin{bmatrix} x_1 \\ x_2 \\ x_3 \end{bmatrix} [y_1 \quad y_2 \quad y_3] = \begin{bmatrix} x_1y_1 & x_1y_2 & x_1y_3 \\ x_2y_1 & x_2y_2 & x_2y_3 \\ x_3y_1 & x_3y_2 & x_3y_3 \end{bmatrix}$$

We are, essentially, multiplying a matrix with columns containing only one element ($[y]^T$) by a matrix with one-element rows ($[x]$).

A more detailed and rigorous treatment of the dot product and the operation of matrices on vectors is found elsewhere (Carley and Morgan, 1989).

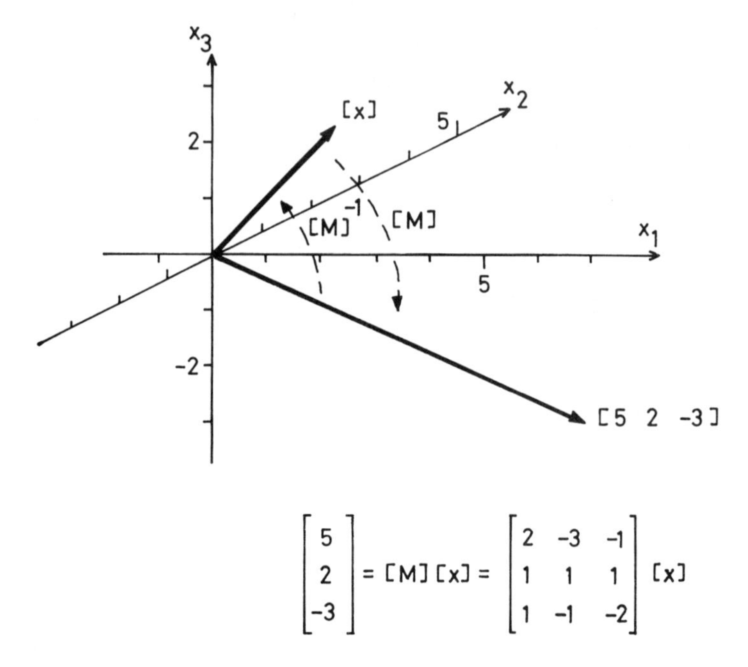

Fig. 1.17 — Graphical representation of the operation of a matrix $[M]$ on a vector represented by $[x]$. In this example $[M]$ is the system matrix for the set of simultaneous equations (1.36) and it scales and rotates an unknown vector $[x]$ into the known vector representing the right-hand side of (1.36). Solving (1.36) is equivalent to finding the matrix $[M]^{-1}$ which 'undoes' this action.

1.8.4 Iterative methods for systems of equations

These are essentially based on the fixed-point method we encountered in section 1.4. Taking the system of equations (1.36) as an example, we restate them as

$$
\begin{aligned}
x &= (5 + 3y + z)/3 \\
y &= 2 - z - x \\
z &= (3 + x - y)/2
\end{aligned}
\tag{1.52}
$$

we then guess values for x, y, and z and use (1.52) to obtain better values for x, y, and z, and so on, repeating the process until the values of x, y, and z do not change more than a user-defined amount between refinement cycles. The method may be implemented in either of two ways. In the Jacobi approach the values of x, y, and z from the previous cycle are inserted in the right-hand side of (1.52), and new values thus computed. The Gauss–Seidel method uses the updated values of x and then y as they are computed within the cycle. Thus at the start of a cycle we use the old values of y and z to compute x, then use this new value of x and the old value of z to give y, and finally these new values of x and y are used to calculate the updated value for z. This may sound like a tricky algorithm to program, but in fact all we are doing is overwriting, at the earliest possible opportunity, the elements of the array $X(I)$ (say) which contains the evolving solution. It is left as an exercise for the reader to write a program for this method.

If we wish to use the termination criterion mentioned above (that the values of x,y, and z change by less than a chosen amount between cycles) with the Gauss–Seidel method, then we must forgo one advantage of the procedure and store both the old and updated values. Alternatively, we can monitor the convergence of the method by the change in modulus (length) of the solution vector $[x\ y\ z]^{\mathrm{T}}$ between cycles.

Returning to (1.52), let us start with an initial guess $[x\ y\ z] = [1\ 1\ 1]$ and demand an accuracy such that the modulus $[x\ y\ z]^{\mathrm{T}}$ changes by less than 0.1% between cycles at convergence. After 44 cycles we obtain the solution $[x\ y\ z] = [1.000\ \ -1.995\ \ 2.998]^{\mathrm{T}}$, expressed to three decinal places.

1.8.5 A practical example

Analytical chemists often use a spectrometer together with a suitable set of standard pure samples to determine the concentrations of the pure materials in a mixture. If the spectra of the standards do not overlap (or at least have some non-overlapped peaks or bands) then we can invoke Beer's Law or an equivalent proportionality and use the absorbance readings from the mixture at suitable wavelengths to get the required concentrations. If, as is more likely, the pure spectra overlap each other badly, then the extraction of this information is not so facile.

For mutually overlapping pure spectra the problem is that the absorbance (say), a_i, at wavelength i is the sum of contributions, c_j, from species $j = 1, 2, \ldots, n$. We may write, for example

$$a_2 = K_{21}c_1 + K_{22}c_2 + K_{23}c_3$$

for three components contributing to the absorbance at the second wavelength. K_{23}, for example, is a Beer's Law-type of constant which determines the contribution of the third species to the absorbance at the second wavelength. In general, for n components and n wavelengths.

$$\begin{aligned} a_1 &= K_{11}c_1 + K_{12}c_2 + \ldots + K_{1n}c_n \\ a_2 &= K_{21}c_1 + K_{22}c_2 + \ldots + K_{2n}c_n \\ &\vdots \\ a_n &= K_{n1}c_1 + K_{n2}c_2 + \ldots + K_{nn}c_n \end{aligned} \tag{1.53}$$

or using summation notation

$$a_i = \sum_{j=1}^{n} K_{ij}\, c_j \tag{1.54}$$

In matrix notation the system of equations (1.53) may be represented as $[a] = [K][c]$, or in full

$$\begin{bmatrix} a_1 \\ a_2 \\ \vdots \\ a_n \end{bmatrix} = \begin{bmatrix} K_{11} & K_{12} & \dots & K_{1n} \\ K_{21} & K_{22} & \dots & K_{2n} \\ \vdots & \vdots & \dots & \vdots \\ K_{n1} & K_{n2} & \dots & K_{nn} \end{bmatrix} \begin{bmatrix} c_1 \\ c_2 \\ \vdots \\ c_n \end{bmatrix} \tag{1.55}$$

Remember that the first index i, labels the rows and the second index, j, the columns.

How can we interpret such sets of equations? We can view the spectrometer as a 'black-box' system with n inputs each corresponding to one of the components (Fig. 1.18). A sample is inserted having concentrations c_j ($j = 1,\dots,n$) of these compo-

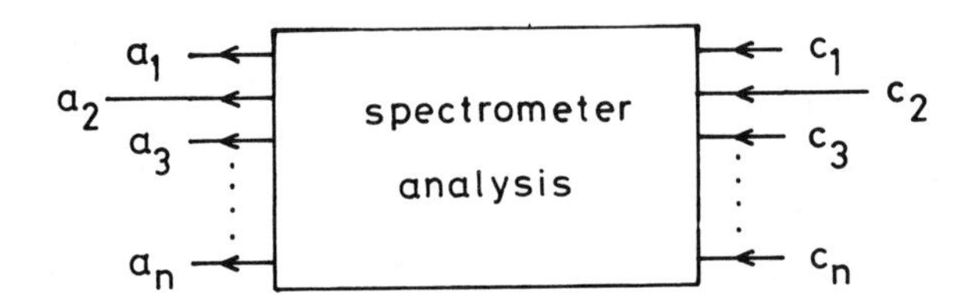

Fig. 1.18 — A spectrometer as a black box — the inputs are the concentrations c_j of the components of a mixture, and the outputs a_i are the measured absorbance values at the ith wavelength.

nents, and out come readings of the absorbance at various wavelengths. The values of K_{ij} are then the recipe or mixing coefficients which determine the contributions of the jth component to the absorbance at the ith wavelength. This is represented graphically in Fig. 1.19 for the case with $n = 3$. The $[K]$ matrix may be regarded as a

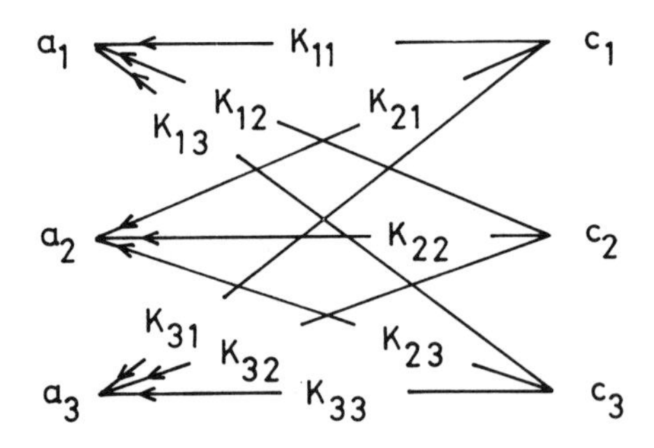

Fig. 1.19 — 'Flow-graph' showing how the elements of $[K]$ in (1.55) may be thought of as mixing coefficients, determining how much each of the concentration inputs in Fig. 1.18 contributes to each of the absorbance measurements.

calibration matrix, each column consisting of the absorbance values obtained from measurements made on a standard amount of each pure component taken in turn, in the absence of the other components. For example if we take $c_1 = 1$ and the other $c_j = 0$ in (1.55), then we obtain $[a] = [K_{11}\ K_{21}\ \ldots\ K_{n1}]^{\mathrm{T}}$.

Fig. 1.20(a) shows four simulated pure component absorption spectra together

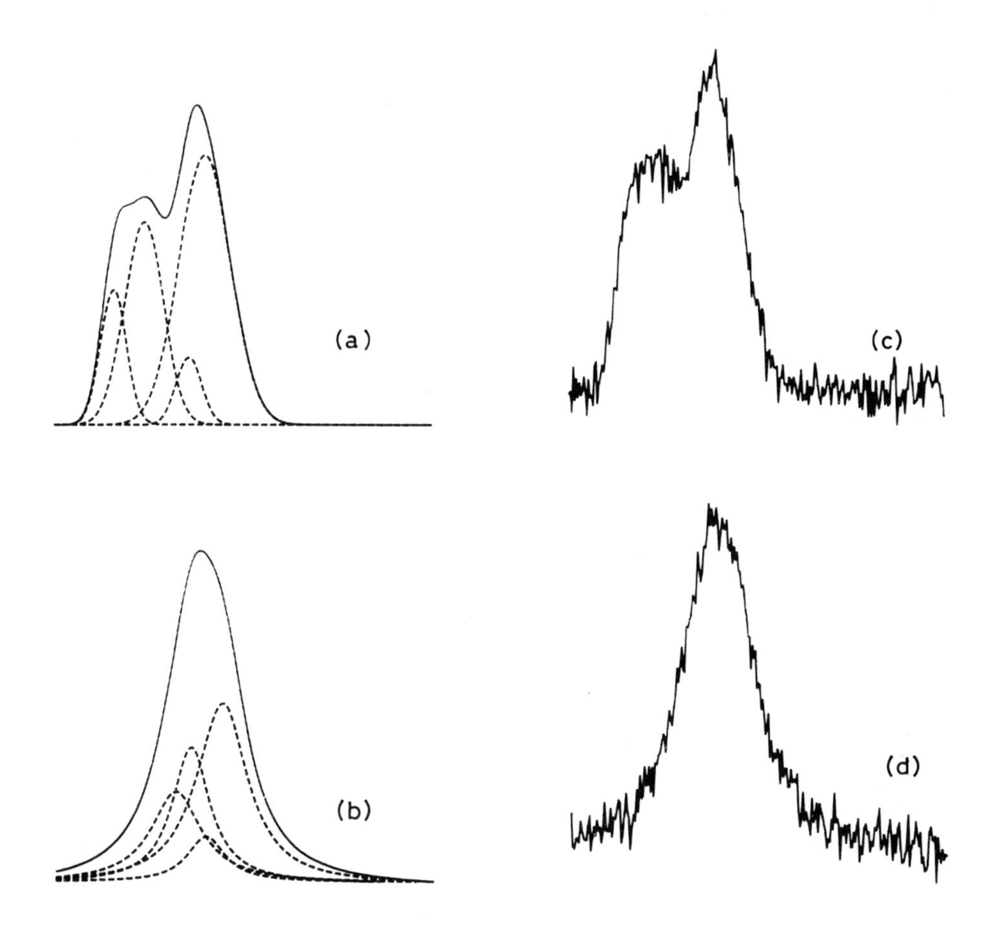

Fig. 1.20 — Synthesized composite spectra used in the example of solving sets of simultaneous equations. The 'pure' components are in the intensity ratio of 1.0:1.5:0.5:2.0. (a) Spectrum generated from four gaussian peaks (Chapter 5). (b) Spectrum constructed from four lorenzian lineshapes (Chapter 3). (c), (d) spectra (a), (b) with noise added (Chapter 5). The signal-to-noise in (c) and (d) is approximately 30:1.

with a simulated spectrum from a mixture, consisting of relative proportions of the pure components of 1:1.5:0.5:2. The wavelengths at which absorbance measurements are made are also chosen to be the positions of the maxima of the pure components. The matrix $[K]$ is calculated from the known pure curves, by loading the columns of $[K]$ with measurements made on these pure spectra at the chosen wavelengths. Fig. 1.20(b) shows a different simulated spectrum, consisting of the same relative proportions of pure components. However, the shape, positions and widths of the pure spectra are now chosen so that the resulting mixture spectrum is severely overlapped compared with Fig. 1.20(a). Indeed, the pure components in Fig. 1.20(b) are such that it would prove difficult to distinguish by eye mixture compositions as varied as 1:1:2:1 and 1:2:1:1 for example. Table 1.5 shows the result

Table 1.5 — Calculated component concentrations for the four synthesized spectra of Fig. 1.20. The method used fails completely in the case of four overlapped lorenzian curves with added noise (Fig. 1.20(d)), although it performs well for the same mixture in the absence of noise (Fig. 1.20(b))

'Spectrum	Components			
	1	2	3	4
Fig. 1.20(a)	1.002	1.501	0.504	1.997
Fig. 1.20(b)	1.003	1.497	0.497	2.003
Fig. 1.20(c)	1.134	1.838	0.690	2.176
Fig. 1.20(d)	0.817	2.400	−0.046	2.374

of applying the Gauss–Jordan elimination procedure to both sets of data. The agreement between the calculated and 'true' concentration values is remarkably good, even for the testing case of Fig. 1.20(b). Pivoting is not required, since the choice of pure component positions as the measurement wavelengths ensures that the diagonal elements of $[K]$ are the largest elements in each column.

The data of Fig. 1.20(a), (b) are very unrealistic, since they do not incorporate any experimental error (noise). Curves (c) and (d) show the same spectra as constructed in (a) and (b) but now with the addition of noise, generated according to an algorithm we shall meet in Chapter 5. If we now solve the systems of equations corresponding to these data, the results for the two cases differ markedly (Table 1.5). The severe overlap in the second case manifests itself in a very inaccurate solution vector, whereas the better-resolved data still gives a reasonable result. The former system of equations is said to be ill-conditioned, and is recognized by this sensitivity of the solution to small changes (noise) in the elements of the matrix or measured vector.

Intuitively, we might expect that such problems could be alleviated if we could incorporate more measurements into the analysis. The difficulty then is that the matrix $[K]$ becomes non-square — in the examples above, if we made measurements at 10 wavelengths then $[K]$ would have 4 columns, each 10 elements long, and $[a]$ would also contain 10 elements. In such a case and with real (noisy) data, it is impossible to find a solution vector $[c]$ whose 4 elements satisfy exactly all 10 simultaneous equations. Fortunately, techniques exist which allow us to calculate a 'best approximation' to the solution. One such method is discussed in Chapter 6, after we have decided on how we measure the 'goodness' of the approximation.

```
    0 REM  P1.1
   10 :
   20 REM compute derivative of a function
   30 :
   40 GOTO 10000
   50 :
  100 REM  subroutine to calculate function
  110 IF ABS(X)<1E-30 THEN PRINT "ZERO X" : END : REM protect against 1/0
  120 F=1/X : RETURN
  130 :
 1000 REM iteration loop
 1010 H=1 : D=1E30 : REM initial value of H, accuracy test fails first time
 1020 :
 1030 X=XX + H : GOSUB 100 : F1=F
 1040 X=XX - H : GOSUB 100 : F2=F
 1050 DN=(F2 - F1)/(2*H) : REM get estimate of derivative value
 1060 PRINT "ESTIMATE OF DERIVATIVE IS ":DN:" AT H = ":H
 1070 IF ABS(DN - D)<AC THEN RETURN : REM check for accuracy
 1080 D=DN : H=H/2
 1090 IF H<1E-20 THEN PRINT "STEP TOO SMALL" : RETURN
 1100 GOTO 1030 : REM save new value and try again
 1110 :
10000 REM main segment
10010 INPUT "X-VALUE = ":XX
10020 INPUT "ABSOLUTE ACCURACY":AC
10030 REM accuracy is ultimately limited by both truncation and round-off error
10040 REM try smaller and smaller values of AC to see effects of round-off
10050 GOSUB 1000
10060 PRINT "VALUE OF DERIVATIVE IS ":DN
10070 REM you can also X=XX:GOSUB 100:F1=F and DN=(F2-F1)/H to compare accuracy
10070 END
```

```
    0 REM  P1.2
   10 :
   20 REM  Horner's scheme for efficient evaluation of a polynomial
   30 :
   40 GOTO 10000
   50 :
 1100 REM evaluate the polynomial at X
 1110 A=A(M)
 1120 FOR I=M-1 TO 0 STEP -1
 1130         A=A(I) + A*X : REM  A is not the same variable as A()
 1140 NEXT I
 1150 RETURN
 1160 :
10000 REM main segment
10010 INPUT "ORDER OF POLYNOMIAL = ";M
10020 PRINT "NOW INPUT THE M+1 POLYNOMIAL COEFFICIENTS"
10030 FOR I=0 TO M
10040         PRINT : PRINT "A(";I;") = "; : INPUT "";A(I)
10050 NEXT I
10060 :
10070 PRINT : INPUT "X = ";X
10080 :
10090 GOSUB 1100
10100 :
10110 PRINT : PRINT "THE ANSWER IS ";A
10120 INPUT "ANOTHER VALUE (Y/N)";Y$
10130 IF Y$="y" OR Y$="Y" THEN 10070
10140 END
```

```
    0 REM  P1.3
   10 :
   20 REM  Compute the square root of a number N using Wegstein's method to
   30 REM  solve the equation X=N\X
   40 :
   50 GOTO 10000
   60 :
  100 REM function subroutine - could use DEF FN G(X)=N/X before GOTO 10000
  110 G=N/X : RETURN
  120 :
 1200 REM subroutine to do iteration - first guess is N itself
 1210 KF=0 : REM initialize iteration counter
 1220 X=XP : GOSUB 100 : XC=G : REM 2nd estimate is simple fixed-point formula
 1230 :
 1240 X=XC : GOSUB 100 : XG=G : REM if using DEF FN write XG=FN G(XC)
 1250 :
 1260 CP=XC - XP : GC=XG - XC : PG=CP - GC
 1270 IF ABS(PG)<1E-20 THEN PRINT "NEAR ZERO DIVISION" : RETURN
 1280 :
 1290 DX=CP*GC/PG
 1300 XP=XC : REM update for next cycle
 1310 XC=XC + DX
 1320 IF ABS(DX)<AC THEN RETURN
 1330 KF=KF + 1 : IF KF>MF THEN PRINT "TOO MANY ITERATIONS" : RETURN
 1340 :
 1350 GOTO 1240
 1360 :
10000 REM main segment
10010 INPUT "NUMBER = ":N
10020 INPUT "FRACTIONAL ACCURACY = ":AC
10030 MF=50 : XP=N : GOSUB 1200 : REM set iteration limit and initial guess
10040 PRINT "ESTIMATE OF ROOT IS ":XC
10050 END
```

```
   0 REM  P1.4
  10 :
  20 REM  Regula Falsi applied to chemical equilibrium example (see text)
  30 :
  40 GOTO 10000
  50 :
 100 REM function subroutine - or use DEF FN F(X)= ... with FN F(..) as P1.3
 110 F=10*X*X - 31*X + 20 : RETURN
 120 :
1300 REM  Regula Falsi routine
1310 KF=0 : REM initialize step counter
1320 REM function values at X1,X2
1330 X=X1 : GOSUB 100 : F1=F
1340 X=X2 : GOSUB 100 : F2=F
1350 :
1360 IF F1*F2>0 THEN RF$="N" : GOTO 1380
1370 INPUT "DO YOU WANT TO USE THE REGULA FALSI METHOD (Y/N)":RF$
1380 IF RF$="N" OR RF$="n" THEN PRINT "NOW USING THE SECANT METHOD"
1390 :
1400 D=(F2 - F1)/(X2 - X1)
1410 IF ABS(D)<1E-6 THEN PRINT "SLOPE OF CURVE IS NEAR ZERO" : RETURN
1420 :
1430 DX=F2/D
1440 X3=X2 - DX : REM ONE less * than (X1*F2-X2*F1)/(F2-F1)
1450 REM also round-off error only affects correction term
1460 PRINT "X1,X2,X3 ":X1,X2,X3 : REM cf. two methods for rate of convergence
1470 IF ABS(DX)<AC THEN RETURN : REM check for convergence
1480 :
1490 KF=KF+1 : IF KF>MF THEN PRINT "TOO MANY TRIALS" : RETURN
1500 :
1510 X=X3 : GOSUB 100 : F3=F : AF=ABS(F3)
1520 IF AF>1E6 THEN PRINT "FUNCTION TENDING TO INFINITY" : RETURN
1530 IF AF<1E-10 THEN PRINT "FUNCTION NEAR 0  CHANGE IN X IS ":DX : RETURN
1540 :
1550 IF ( RF$="Y" OR RF$="y" ) AND F3*F1<0 THEN 1570 : REM discard X1 or X2 ?
1560 X1=X2 : F1=F2 : REM always make X2 the most recent value
1570 X2=X3 : F2=F3
1580 GOTO 1400 : REM retry
1590 :
```

```
10000 REM main segment
10010 INPUT "ABSOLUTE ACCURACY":AC
10020 INPUT "FIRST GUESS AT ROOT = ":X1
10030 PRINT "IF YOU INTEND USING THE REGULA FALSI METHOD THEN"
10040 PRINT "THE VALUE OF THE FUNCTION AT YOUR SECOND GUESS"
10050 PRINT "MUST BE OF OPPOSITE SIGN TO THAT AT YOUR FIRST"
10060 INPUT "SECOND GUESS AT ROOT = ":X2
10070 :
10080 IF X1=X2 THEN 10020
10090 MF=50 : REM set maximum number of new function evaluations
10100 :
10110 GOSUB 1300
10120 PRINT "ESTIMATE OF ROOT IS ";X3
10130 END
```

```
   0 REM  P1.5
  10 :
  20 REM  Newton-Raphson method applied to the chemical equilibrium example
  30 REM  see text
  40 :
  50 GOTO 10000
  60 :
 100 REM subroutine to calculate function
 120 REM for other functions involving /X, EXP(X) etc., protect against
 130 REM very small or very large numbers
 140 F=10*X*X - 31*X + 20 : RETURN
 150 :
 200 REM subroutine to calculate derivative of function
 210 D=20*X - 31 : RETURN
 220 :
1400 REM  Newton-Raphson formula
1410 GOSUB 200
1420 IF ABS(D)<1E-20 THEN PRINT "ZERO DERIVATIVE" : END
1430 :
1440 GOSUB 100
     DX=F/D
```

```
1450 X=X - DX : REM calculate improved estimate of root
1460 :
1470 IF ABS(DX)<AC THEN RETURN
1480 KF=KF + 1 : IF KF>MF THEN PRINT "TOO MANY TRIALS" : RETURN
1490 :
1500 GOTO 1400 : REM set new starting value and retry
1510 :
10000 REM main segment
10010 INPUT "GUESS AT VALUE OF ROOT = ";X
10020 INPUT "ABSOLUTE ACCURACY = ";AC
10025 REM whatever AC, unless root=0 result cannot have any more sig. figs.
10026 REM than machine is using - use GOSUB 11000 to estimate relative precision
10030 MF=50 : GOSUB 1400 : REM set maximum number of trials and start
10040 PRINT : PRINT "ESTIMATE OF ROOT IS ";X
10050 PRINT "LAST CHANGE IN X WAS ";DX
10060 PRINT "FUNCTION VALUE WAS ";F
10070 END
10080 :
11000 REM estimate relative precision of computer
11010 PR=1
11020 PR=PR/2 : REM reduce value of PR
11030 Q=1 + PR
11040 IF Q>1 THEN 11020 : REM if adding PR to 1 makes any difference reduce PR
11050 PRINT "ESTIMATED RELATIVE PRECISION IS ";PR
11060 RETURN
```

```
    0 REM  P1.6
   10 :
   20 REM  Solution of titration equation using Newton-Raphson method
   30 :
   40 VA=0.025 : REM volume of acid in cubic decimetres
   50 KW=1E-14 : REM ionic product for water
   60 CA=.1 : CB=.1 : REM molar concentrations of acid and base
   70 LF=LOG(10)
   80 GOTO 10000
   90 :
  200 REM  Horner's method for evaluating a polynomial
  210 Q=B(M) : FOR I=M-1 TO 0 STEP -1 : Q=B(I) + Q*C : NEXT I : RETURN
  220 :
10000 REM main segment
10010 INPUT "KA = ";KA : REM acid dissociation constant
10020 INPUT "KB = ";KB : REM base dissociation constant
10030 INPUT "VOLUME (IN CUBIC CM.) RANGES FROM ";V1 : INPUT " TO ";V2
10040 INPUT "VOLUME INCREMENT IN CUBIC CM.";VS
10050 :
10060 HGR : HCOLOR=3
10070 X=0 : V1=V1/1000 : V2=V2/1000 : VS=VS/1000 : C=1 : REM initial C value
10080 FOR VB=V1 TO V2 STEP VS
10090 :
10100 V=VA + VB : REM calculate total volume
10110 :
10120 REM calculate coefficients of polynomial equation derived from
10130 REM equilibrium expression for hydrogen ion concentration
10140 A(4)=KB*V : A(3)=V*( KA*KB + KW ) + KB*CB*VB
10150 A(2)= - ( KB*KW*V + KA*KB*CA*VA - KA*KW*V - KA*KB*CB*VB )
10160 A(1)= - ( KW*V*( KA*KB + KW ) + KA*CA*VA*KW )
10170 A(0)= - KA*KW*KW*V
10180 :
10190 FOR I=0 TO 3 : D(I)=(I + 1)*A(I+1) : NEXT I
10200 REM calculate coefficients of derivative of polynomial
10210 :
10220 M=4 : FOR I=0 TO M : B(I)=A(I) : NEXT I
10230 GOSUB 200 : F=Q : REM calculate function value
10240 :
10250 M=3 : FOR I=0 TO M : B(I)=D(I) : NEXT I
```

```
10260 GOSUB 200 : DF=Q : REM calculate derivative value
10270 :
10280 CN=C - F/DF
10290 IF ABS( (CN - C)/C )>1E-6 THEN C=CN : GOTO 10190
10300 REM check on relative error
10310 PH=-LOG(CN)/LF : REM calculate pH - note use of Naperian logs
10320 Y=INT(10*PH + .5) : REM rounded result for plotting
10330 HPLOT X,159 - Y : REM origin at top LHS of screen
10340 X=X + 1
10350 NEXT VB : REM old value of C used as new starting value
10360 Y=70.5 : HPLOT 0,159 - Y TO X - 1,159 - Y : REM axis at pH=7
10370 REM if a starting value of C is chosen which is larger than the root
10380 REM method converges - it will not always converge if converse is true
10390 END
```

```
    0 REM  P1.7
   10 :
   20 REM  Random walk simulation of 1-D diffusion from solute covered plate
   30 :
   40 DIM N(100)
   50 :
   60 GOTO 10000
   70 :
  200 REM plotting routine
  210 ST=INT(250/NS) : SC=150/NP : REM horizontal and vertical scaling
  220 HGR : HCOLOR=3
  230 FOR I=0 TO NS
  240         HPLOT ST*I,159 TO ST*I,159 - N(I)*SC
  250 NEXT I
  260 RETURN
  270 :
10000 REM main segment
10010 INPUT "NUMBER OF STEPS = ";NS
10020 INPUT "NUMBER OF PARTICLES = ";NP
10030 FOR I=1 TO NP
10040         X=0
```

```
10050         FOR J=1 TO NS
10060                 DX=1 : IF RND(1)<.5 THEN DX=-1 : REM DX either 1 or -1
10070                 X=X + DX : REM or DX=2*INT(2*RND(1))-1 but takes longer
10080                 IF X<0 THEN X=0 : REM barrier at X=0
10090         NEXT J
10100         IF X>100 THEN 10120 : REM stop error if DIM N()>100)
10110         N(X)=N(X) + 1 : REM register particle in bin X
10120 NEXT I
10130 :
10140 GOSUB 200 : REM now plot histogram
10150 END
```

```
  0 REM  P1.8
 10 :
 20 REM  Random walk solution of Laplace's equation allowing diagonal steps
 30 :
 40 DIM A(20,20)
 50 :
 60 DATA 8,1,0,0,1,-1,0,0,-1,-1,1,1,1,1,-1,-1,-1
 70 :
 80 GOTO 10000
 90 :
200 FOR I=1 TO 19 : FOR J=1 TO 19 : A(I,J)=-10000 : NEXT J : NEXT I
210 REM set interior points to a signal value
220 FOR I=0 TO 20
230         A(0,I)=I + 1 : A(I,0)=I + 1 : REM set boundary values
240         A(20,I)=21 - I : A(I,20)=21 - I
250 NEXT I
260 :
270 READ NS : REM read number of moves allowed
280 FOR I=0 TO NS-1 : READ DX(I),DY(I) : NEXT I : REM read moves
290 RETURN
300 :
```

```
10000 REM  main segment
10010 GOSUB 200
10020 INPUT "COORDINATES OF STARTING POINT FOR RANDOM WALKS":XS,YS
10030 IF A(XS,YS)<>-10000 THEN SU=A(XS,YS) : GOTO 10210
10040 INPUT "NUMBER OF TRIALS = ":NT
10050 SU=0 : SS=0
10060 FOR CT=1 TO NT
10070         X=XS : Y=YS
10080         W=INT( NS*RND(1) ) : REM choose an allowed move at random
10090         REM some BASICs chop array index others round - safer to use INT
10100         X=X + DX(W) : Y=Y + DY(W)
10110         IF A(X,Y)=-10000 THEN 10080 : REM if point is interior continue
10120         V=A(X,Y)
10130         SU=SU + V : REM sums boundary value into accumulator SU
10140         SS=SS + V*V : REM accumulate sum of squares
10150 NEXT CT
10160 :
10170 REM calculate solution at (XS,YS) and standard deviation for answer
10180 REM see Ralston and Wilf (1960)
10190 SU=SU/NT : SS=(SS - SU*SU*NT)/(NT - 1)
10200 REM reduces to SS/NT-SU^2 for large NT
10210 SS=SQR(SS/NT) : REM SD of mean of NT results
10220 REM only true if SU values for individual walks are distributed Normally
10230 PRINT "SOLUTION IS ":SU:" WITH STANDARD DEVIATION ":SS
10240 END
```

```
  0 REM  P1.9
 10 :
 20 REM Monte Carlo integration using hit or miss method
 30 :
 40 NH=0 : REM initialize hit counter
 50 DEF FN F(X)=EXP(X) - 1 : REM define function to be integrated
 60 :
100 INPUT "LIMITS OF INTEGRATION = ";XL,XR
110 YL=0 : YR=FN F(XR) : REM bottom and top of "box" containing all of
120 REM the graph of the function
130 INPUT "NUMBER OF TRIALS = ";NT
140 :
150 FOR I=1 TO NT
160         RX=XR*RND(D) : RY=YR*RND(D) : REM get random point inside box
170         IF RY<FN F(RX) THEN NH=NH + 1 : REM score a hit
180 NEXT I
190 :
200 IL=XR*YR*NH/NT : REM integral is total area of box times fraction of hits
210 PRINT "ESTIMATE OF INTEGRAL = ";IL : REM if XL=0,XR=1 IL=EXP(1)-2 (e-2)
220 END
```

```
    0 REM  P1.10
   10 :
   20 REM Given an initial (unit) mass of material at (0,0) at T=0, calculates
   30 REM the amout of material within a circle of radius A centred at (U,V)
   40 REM at time T by integrating the concentration expression within a
   50 REM circular boundary (program easily modified to define other regions)
   60 REM If D*T is small C is strongly peaked about (0,0) and random ordinates
   70 REM stand little chance of "hitting" it - number of trials must be huge.
   80 REM Choice of sampling points crucial for such "difficult" functions.
   90 :
  100 PI=4*ATN(1) : REM TAN of PI/4 is 1 hence PI is 4*arctan(1)
  110 :
  120 DEF FN C(X)=K1*EXP(-X/K2) : REM from theory - see text
  130 REM some BASICs need protecting from U <<0 as well as >>0 in EXP(U)
  140 :
  150 GOTO 10000
  160 :
  170 REM a subroutine to find C can be placed here instead of DEF FN
  180 :
10000 REM main segment : REM for D=.1 U=1 V=1 A=1 T=1, integral is about 0.13
10010 INPUT "DIFFUSION COEFFICIENT = ";D
10020 INPUT "COORDINATES OF CENTRE U,V ARE ";U,V
10030 INPUT "RADIUS OF CIRCLE = ";A
10040 :
10050 AR=PI*A*A : REM calculate area of circle
10060 :
10070 INPUT "TIME = ";T
10080 INPUT "NUMBER OF TRIALS = ";NT
10090 IF D*T<1E-10 THEN PRINT "K1 TOO LARGE AND K2 TOO SMALL" : END
10100 K2=4*D*T : K1=1/(PI*K2)
10110 REM constants for concentration dependence on time and radius from (0,0)
10120 :
10130 S=0 : REM initialize accumulator
10140 FOR CT=1 TO NT
10150         REM random -1<X,Y<1 so that (X,Y) in circle radius 1 centre (0,0)
10160         X=2*( RND(1) - .5 ) : Y=2*( RND(1) - .5) : REM pt. in unit square
10170         R=X*X + Y*Y
10180         IF R>1 THEN 10160 : REM reject points outside unit circle
10190         :
```

```
10200         X=A*X + U : Y=A*Y + V : REM scale and shift to point inside
10210         RR=X*X + Y*Y : REM RR is square of distance from origin
10220         S=S + FN C(RR) : REM accumulate function values
10230 NEXT CT
10240 :
10250 IL=AR*S/NT
10260 REM Integral is average of ordinates at the various (X,Y) times area
10270 REM inside boundary of integration. Each random ordinate contributes
10280 REM a "volume" of a cylinder based on the boundary and of height C
10290 PRINT "ESTIMATE OF INTEGRAL IS ";IL
10300 END
```

```
  0 REM  P1.11
 10 :
 20 REM same calculation as in P1.10 but uses random walk simulation of
 30 REM diffusion to estimate material inside sampling circle
 40 :
100 INPUT "DIFFUSION COEFFICIENT = ";D
110 INPUT "COORDINATES U,V OF CENTRE ARE ";U,V
120 INPUT "RADIUS = ";A : A=A*A : REM save defining inside later loop
130 INPUT "NUMBER OF PARTICLES = ";NP : REM each of mass 1/NP
140 INPUT "SPACE STEP = ";ST
150 INPUT "TIME = ";T
160 :
170 REM use Einstein-Smoluchowsky relationship to calculate time step
180 DT=ST*ST/(2*D) : REM if ST too big random walks will be "grainy"
190 NT=INT(T/DT) : REM calculate number of steps
200 :
210 NH=0 : REM initialize counter for number of hits
220 FOR I=1 TO NP
230         X(0)=0 : X(1)=0 : REM initial coordinates of particle
240         FOR CT=1 TO NT
250                 FOR K=0 TO 1
```

```
260                     DS=1 : IF RND(1)<.5 THEN DS=-1
270                     X(K)=X(K) + ST*DS
280               NEXT K
290         NEXT CT
300         :
320         DU=U-X(0) : DV=V-X(1) : LE=DU*DU + DV*DV
330         REM distance squared between (U,V) and end of walk
340         IF LE<A THEN NH=NH + 1 : REM if particle lands inside score hit
350 NEXT I
360 :
370 IL=NH/NP : REM integral is fraction of points that land in circle
380 PRINT "ESTIMATE OF INTEGRAL IS ":IL
390 REM The simulation can easily be modified to define different sampling
400 REM regions or to include drift effects (by adding a fixed increment to
410 REM X() at each time step). Monte Carlo methods are useful when a rough
420 REM estimate is needed and calculation is hard to express analytically.
430 END
```

2

Interpolation

2.1 FILLING IN THE GAPS

Imagine that you are performing an experiment for which you need values of a specific heat capacity at various temperatures. You find a table of accurate heat capacities given at, say, intervals of 5 K covering the appropriate temperature range, but you probably need values between those that the table provide. How do you obtain these values?

One can suppose that Nature has yielded the values in the table from a complicated function that she knows and you do not. If you could divine this function then the problem would be solved — the 'complete' answer would be to construct a physical model of the material under study and to fit a mathematical representation of this to the tabulated data. This could well be a research project in itself, however, and the extra accuracy obtained might not warrant the effort involved, and would probably not even be necessary. Nevertheless, it is essential in any interpolation scheme that the function we choose to 'flesh out' the table of values should be capable of approximating the aforementioned natural function. If there were a phase change taking place in the range of interpolation, for example, then a very smooth function such as a quadratic curve ($u = ax^2 + bx + c$ where a, b, c are constants, u is the specific heat capacity and x is the temperature) is not suited to approximating the data in this region — whatever values we choose for a, b, c, we can never constrain this quadratic function to display the disontinuous behaviour expected for the specific heat capacity near a phase transition.

Fortunately, over narrow enough ranges of data, most physical quantities are smooth enough to be approximated by a polynomial function, that is, a function of the form

$$u = a_0 + a_1x + a_2x^2 + \ . \ . \ . \ + a_nx^n$$

We are assuming that our natural function can be made up, recipe fashion, from a weighted sum of powers of x, and this assumption is very often justified. Moreover, polynomials exhibit some very useful properties: they can be added together by simply combining coefficients, and furthermore multiplication, division, integration and differentiation are relatively straightforward operations. As an example, the latter two operations yield yet another polynomial. Bearing these useful properties in mind, how are we to perform polynomial interpolation?

2.2 DIVIDED DIFFERENCES AND NEWTON'S FORWARD FORMULA

2.2.1 Derivation

In Fig. 2.1 are shown some values u_0, u_1 u_2, and u_3 plotted against the corresponding

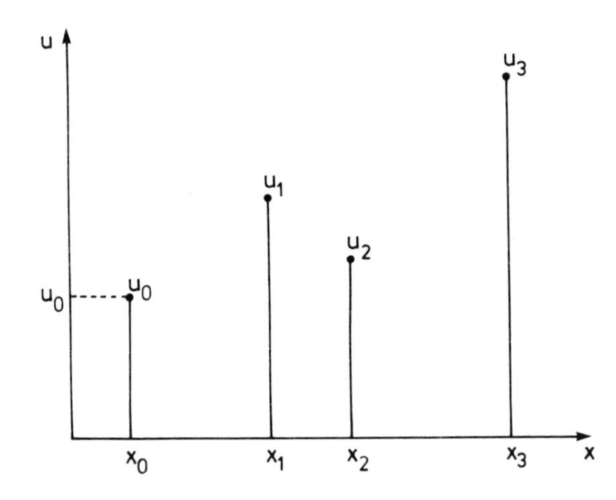

Fig. 2.1 — The example data set used to derive the interpolation formulae.

independent variable values x_0, x_1, x_2, and x_3. If we knew only one value u_0 at x_0, then our safest assumption might be that u is constant in the region of x_0 and our rather trivial expression for u is

$$u(x) = u_0 \tag{2.1}$$

Given another value u_1 at x_1, we can improve on this by adding a term involving x_1 and seeking an expression for u whose graph (Fig. 2.2) is a straight line passing through the points (x_0,u_0) and (x_1,u_1). The dashed line in Fig. 2.2 represents the correction term to be added to (2.1) giving

$$u(x) = u_0 + a_1(x - x_0) \tag{2.2}$$

where a_1 is a constant to be determined. Note that the extra term $a_1(x - x_0)$

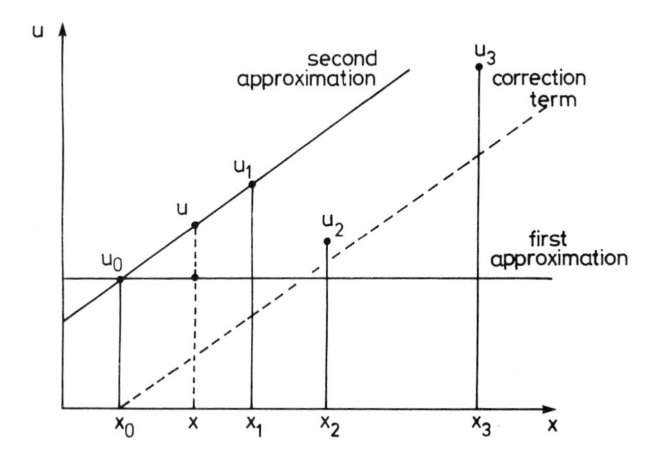

Fig. 2.2 — The first two stages in the derivation of Newton's forward formula for polynomial interpolation.

disappears at $x = x_0$ as it must if it is not to invalidate the approximation at x_0, $u(x_0) = u_0$. From (2.2) at $x = x_1$, $u_1 = u(x_1) = u_0 + a_1(x_1 - x_0)$ and hence

$$a_1 = (u_1 - u_0)/(x_1 - x_0) \tag{2.3}$$

giving for the approximating function

$$u(x) = u_0 + \frac{(u_1 - u_0)}{(x_1 - x_0)}(x - x_0) \tag{2.4}$$

In order to extend the approximation by including another point (x_2, u_2) and another corresponding correction term, we must repeat the procedure. The additional term must be of the form $a_2(x - x_0)(x - x_1)$, a quadratic term which vanishes at the two previously used values x_0 and x_1. Thus at x_2, we have

$$u_2 = u(x_2) = u_0 + \frac{(u_1 - u_0)}{(x_1 - x_0)}(x_2 - x_1) + a_2(x_2 - x_0)(x_2 - x_1)$$

which may be rearranged to give an expression for a_2

$$a_2 = \frac{\dfrac{(u_2 - u_0)}{(x_2 - x_0)} - \dfrac{(u_1 - u_0)}{(x_1 - x_0)}}{(x_2 - x_1)} \tag{2.5}$$

A similar treatment applied to the cubic term $a_3(x - x_0)(x - x_1)(x - x_2)$ at x_3 gives

$$a_3 = \frac{\dfrac{\dfrac{(u_3 - u_0)}{(x_3 - x_0)} - \dfrac{(u_1 - u_0)}{(x_1 - x_0)}}{(x_3 - x_1)} - \dfrac{\dfrac{(u_2 - u_0)}{(x_2 - x_0)} - \dfrac{(u_1 - u_0)}{(x_1 - x_0)}}{(x_2 - x_1)}}{(x_3 - x_2)} \tag{2.6}$$

The coefficients a_0 $(= u_0)$, a_1, a_2, and a_3 are clearly generated by some regular process. They are aptly known as divided differences and are important both practically and theoretically in the understanding of a range of numerical methods. The coefficient a_0 is a zero-order divided difference since it is merely the value of u_0 and no differences or divisions are involved. By extension, a_1 is a first-order divided difference, a_2 is a second-order divided difference (being the divided difference between two divided differences), and a_3 is a third-order divided difference. Since the expressions for a_1, a_2, and a_3 are too complicated repeatedly to represent explicitly in type, we will use the notation $\Delta_{1,0}u$ for a_1, $\Delta_{2,0,1}u$ for a_2, and $\Delta_{3,1,0,2}u$ for a_3, the rationale for which will become apparent later.

Several simple relationships exist between these divided differences. They are formed recursively:

$$\Delta_{0,1}u = \frac{\Delta_0 u - \Delta_1 u}{x_0 - x_1}$$

where we recall that $\Delta_1 \mathrm{u} = \mathrm{u}_1$, a zero-order divided difference,

$$\Delta_{0,1,2}u = \frac{\Delta_{0,1}u - \Delta_{1,2}u}{x_0 - x_2}$$

$$\Delta_{0,1,2,3}u = \frac{\Delta_{0,1,2}u - \Delta_{1,2,3}u}{x_0 - x_3}$$

and in general

$$\Delta_{i,j,\ldots,m,n}u = \frac{\Delta_{i,j,\ldots,m}u - \Delta_{j,k,\ldots,m,n}}{x_i - x_n} \tag{2.7}$$

Furthermore, we find that

$$\Delta_{0,1}u = \frac{u_0 - u_1}{x_0 - x_1} = \frac{u_1 - u_0}{x_1 - x_0} = \Delta_{1,0}u$$

(compare expression (2.3) for a_1) and thus we say that $\Delta_{0,1}u$ is symmetric in the indices 0 and 1, since interchanging them leaves its value unchanged.

We may rewrite this expression for $\Delta_{0,1}u$ as

$$\Delta\!\!\!\!|_{0,1}u = \frac{u_0}{x_0 - x_1} + \frac{u_1}{x_0 - x_1}$$

which seems needlessly complicated until we realize that we can extend it to $\Delta\!\!\!\!|_{0,1,2}u$ given earlier which may be expanded to give

$$\Delta\!\!\!\!|_{0,1,2}u = \frac{u_0}{(x_0 - x_1)(x_0 - x_2)} + \frac{u_1}{(x_1 - x_0)(x_1 - x_2)} + \frac{u_2}{(x_2 - x_0)(x_2 - x_1)}$$

The reader is advised to pause at this point and verify that this last manipulation is indeed correct. The generalization of this procedure to (2.7) gives

$$\begin{aligned}\Delta\!\!\!\!|_{i,j,k,\ldots,m,n}u = &\frac{u_i}{(x_i - x_j)(x_i - x_k)\ \ldots\ (x_i - x_n)} + \\ &+ \frac{u_j}{(x_j - x_i)(x_j - x_k)\ \ldots\ (x_j - x_n)} + \ \ldots\ + \\ &+ \frac{u_n}{(x_n - x_i)(x_n - x_j)\ \ldots\ (x_n - x_m)}\end{aligned} \tag{2.8}$$

If we exchange the indices i and j in (2.8) then we obtain the same expression since all we effectively do is interchange the first two terms; in fact this is true for any pair of indices. This symmetry of expression means that the order of the indices in a divided difference does not affect its value, and the general form of the divided difference $\Delta\!\!\!\!|_{i,j,\ldots,n}\ u$ (2.7) may be replaced by $\Delta\!\!\!\!|_{0,1,\ldots,n-1,n}\ u$ which makes for easier comprehension and more orderly computation.

The symmetry property means that the coefficients a_0, a_1, a_2, and a_3 of the cubic polynomial through the four data points in Fig. 2.1 may be replaced by regularly formed divided differences to give for the polynomial

$$\begin{aligned}u(x) = \Delta\!\!\!\!|_0 u + \Delta\!\!\!\!|_{0,1}u(x - x_0) + \Delta\!\!\!\!|_{0,1,2}u(x - x_0)(x - x_1) + \Delta\!\!\!\!|_{0,1,2,3}u(x - \\ - x_0)(x - x_1)(x - x_2)\end{aligned}$$

which is extended to the $(n+1)$-point case in a straightforward manner

$$\begin{aligned}u(x) = \Delta\!\!\!\!|_0 u + \Delta\!\!\!\!|_{0,1}u(x - x_0) + \Delta\!\!\!\!|_{0,1,2}u(x - x_0)(x - x_1) + \ \ldots\ + \\ + \Delta\!\!\!\!|_{0,1,2,\ldots,n}u(x - x_0)(x - x_1)\ \ldots\ (x - x_{n-1})\end{aligned} \tag{2.9}$$

This celebrated formula is known as the 'Newton forward formula' for the interpolation polynomial since each subsequent term in the expression incorporates the next point in the sequence. The required divided differences are conveniently calculated in tabular form, as is illustrated in Table 2.1 for the case of four data items. An inspection of this table and a comparison with (2.9) reveals that the divided differences we need for Newton's forward formula arise from the uppermost (forward) diagonal.

2.2.2 Computation

If we rewrite Table 2.1 slightly to give Table 2.2, then it is clear that the tabulation is most easily represented, but not necessarily most efficiently, by a two-dimensional array variable (matrix) $DD(I,J)$. The column index J runs from 0 to $N-1$, where N is the number of data points, and represents the order of the divided difference; thus $J=0$ (zero-order) corresponds to the original data u. The row index I runs from J to $N-1$, so that almost half of the available array elements are not used, and $DD(I,J)$ is the Ith divided difference of order J; the boxed element in Table 2.2 is thus represented by $DD(2,1)$.

The recursive formulation for divided differences (section 2.2.1) greatly simplifies the computation, leading to the program in listing P.2.1. The divided difference table calculated by applying this program to the equation $u=x^3-x^2+1$ for various values of x is shown in Table 2.3. One should note that the differences of order 3 are constant and those of order 4 and above are equal to zero. Hence the tabulated values are exactly represented by a third-order polynomial — a trivial conclusion in this case. The procedure is applied to some 'real' data in Table 2.4. The data consists of selected values of the specific heat of air at 20 atmospheres in the temperature range 1000–2800 K (*Handbook of Chemistry and Physics*, 1974); 8 pairs of (T, C_p) values are shown together with derived divided differences.

Newton's forward formula (2.9) is easily implemented as a computer program as is demonstrated in listing P2.2. Since Newton's formula only requires the diagonal elements of the divided difference table, memory savings may be made by storing only these values in the first place. Applying program P2.2 to the data of Table 2.4 yields the interpolation polynomial plotted in Fig. 2.3 together with the original (T, C_p) data points. Also plotted are some data omitted from Table 2.4 but available in the more comprehensive source compilation. The quality ('goodness') of the interpolation may be judged from the excellent agreement between the interpolated curve and the additional experimental data points; thus the interpolated value of C_p at a temperature $T=2100$ K is calculated to be 0.3215 which compares well with the 'actual' value of 0.3214.

A good treatment of the topic of divided differences is given by Bartels *et al.* (1987).

2.3 AITKEN'S METHOD

Although we have assumed so far that the data we are interpolating arise from a function which is capable of approximation by a polynomial, we would like some means of measuring the validity of this assumption. Newton's forward formula (2.9) for $n+1$ points tells us that the polynomial it decribes is of degree n, since the $(n+1)$th term contains the expression $(x-x_0)(x-x_1)\ldots(x-x_{n-1})$ which when expanded contains x^n as its highest power of x. We can therefore always fit our data by taking enough terms. If this data is truly polynomial representable, and we are given a new datum point, then our prediction of this datum value should not differ too much from the actual value. This amounts to saying that the sequence of

Table 2.1 — Method of construction of a divided difference table

	Order of divided difference			
	0	1	2	3
x_0	$\Delta_0 u = u_0$			
		$\Delta_{0,1} u = \dfrac{\Delta_0 u - \Delta_1 u}{x_0 - x_1}$		
x_i	$\Delta_1 u = u_1$	$\Delta_{1,2} u = \dfrac{\Delta_1 u - \Delta_2 u}{x_1 - x_2}$	$\Delta_{0,1,2} u = \dfrac{\Delta_{0,1} u - \Delta_{1,2} u}{x_0 - x_2}$	$\Delta_{0,1,2,3} u = \dfrac{\Delta_{0,1,2} u - \Delta_{1,2,3} u}{x_0 - x_3}$
			$\Delta_{1,2,3} u = \dfrac{\Delta_{1,2} u - \Delta_{2,3} u}{x_1 - x_3}$	
x_2	$\Delta_2 u = u_2$	$\Delta_{2,3} u = \dfrac{\Delta_2 u - \Delta_3 u}{x_2 - x_3}$		
x_3	$\Delta_3 u = u_3$			

Table 2.2 — Divided difference table in a format suited to computer implementation

	Divided difference order			
	0	1	2	3
x_0	$\Delta\!\!\!|_{0}\ u$			
x_1	$\Delta\!\!\!|_{1}\ u$	$\Delta\!\!\!|_{0,1}\ u$		
x_2	$\Delta\!\!\!|_{2}\ u$	$\Delta\!\!\!|_{1,2}\ u$	$\Delta\!\!\!|_{0,1,2}\ u$	
x_3	$\Delta\!\!\!|_{3}\ u$	$\Delta\!\!\!|_{2,3}\ u$	$\Delta\!\!\!|_{1,2,3}\ u$	$\Delta\!\!\!|_{0,1,2,3}\ u$

Table 2.3 — Divided difference table for some values of (x,y) derived from the cubic $y = x^3 - x^2 + 1$

x	y	Divided difference order								
		1	2	3	4	5	6	7	8	9
0	1									
1	1	0								
3	19	9	3							
4	49	30	7	1						
7	295	82	13	1	0					
8	449	154	18	1	0	0				
10	901	226	24	1	0	0	0			
12	1585	342	29	1	0	0	0	0		
15	3151	522	36	1	0	0	0	0	0	
19	6499	837	45	1	0	0	0	0	0	0

interpolation polynomials produced by adding more and more points converges to the true function. A method for calculating the interpolation polynomial and simultaneously checking for this convergence is provided by the ingenious iterated linear interpolation scheme devised by Aitken.

2.3.1 Mathematical description

Referring to Fig. 2.4, a linear interpolation is made from (x_0,u_0) to (x_1,u_1) giving a value u_1' at some point x; u_1 is then replaced by u_1' at x_1. This process is repeated to

Table 2.4 — Divided difference table for some selected specific heat (C_p)–temperature (T) data for air at 20 atmospheres

T (K)	C_p (cal g^{-1} K^{-1}) $\times 10^4$	Divided difference order						
		1	2	3	4	5	6	7
1000	2730							
1200	2814	0.420						
1500	2939	0.417	-6.67×10^{-6}					
1600	2981	0.420	8.33×10^{-6}	2.50×10^{-8}				
1900	3115	0.447	6.67×10^{-5}	8.33×10^{-8}	6.48×10^{-11}			
2300	3332	0.543	1.37×10^{-4}	8.78×10^{-8}	4.06×10^{-12}	-4.67×10^{-14}		
2500	3488	0.780	3.96×10^{-4}	2.88×10^{-7}	2.00×10^{-10}	1.51×10^{-13}	1.32×10^{-16}	
2800	3850	1.21	8.53×10^{-4}	5.08×10^{-7}	1.84×10^{-10}	-1.23×10^{-14}	-1.02×10^{-16}	-1.30×10^{-19}

give u_2' and u_3', and the three new points so generated are indicated by crosses in Fig. 2.4. The procedure is repeated by interpolating from u_1' each time to give two further points $u_2''$and u_3''at x_2 and x_3 respectively A final interpolation is made between $u_2''$and u_3''yielding finally u_3''' at x_3, which is the value which would have been obtained by polynomial interpolation through the original four points. We summarize the calculations in Table 2.5.

The relationship between Aiken's Method and Newton's forward formula is demonstrated by expanding the successive interpolations u_1', u_2'', and u_3'''as shown in Table 2.6. Comparing the upper diagonal entries with our original efforts to obtain an interpolation polynomial (2.9), we see that these entries correspond to the successive partial sums of that polynomial. Thus the Aitken procedure yields the same result as Newton's method, but has the advantage of allowing an estimate of the errors involved in the interpolation to be made.

2.3.2 Computation

Employing a similar notation to that for divided differences, the table of polynomials arising from the use of Aitken's Method may be constructed as shown in Table 2.7. The computation proceeds recursively, each entry in the table being derived from two entries in the previous column: the one at the top of that column, and the one on the same row as the 'target' entry. To be precise

$$I_{0,1,\ldots,n}(x) = (I_{0,1,\ldots,n-1}(x_n - x) - I_{0,1,\ldots,n-2,n}(x_{n-1} - x))/(x_n - x_{n-1}) \tag{2.10}$$

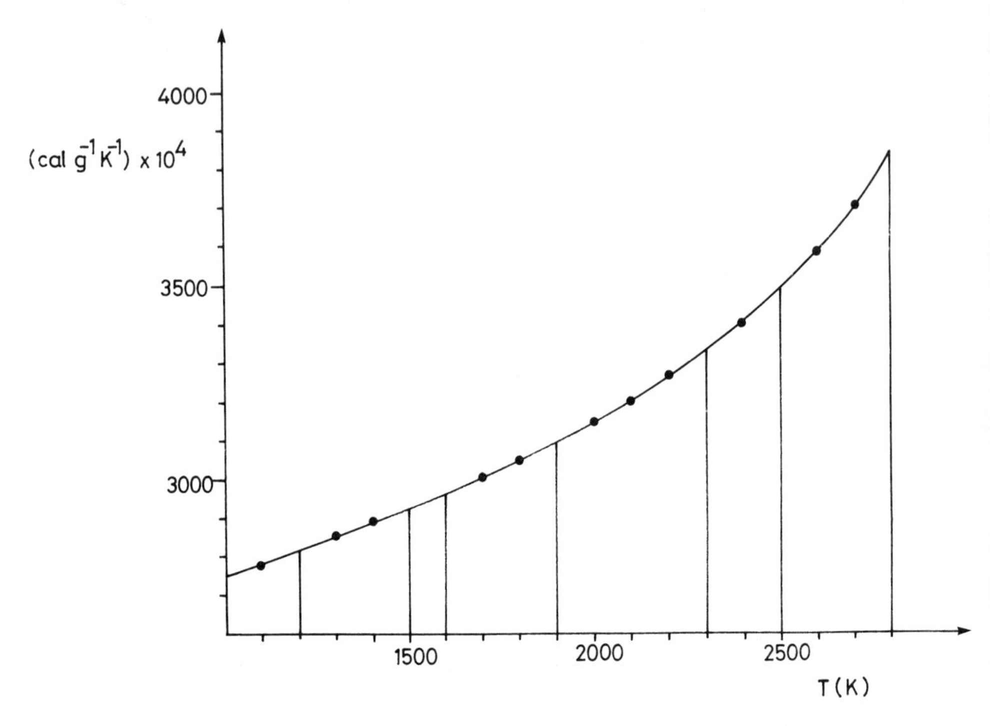

Fig. 2.3 — Plot of the interpolation curve generated by the application of Newton's forward formula to some selected specific heat (C_p)–temperature (T) values for air at 20 atmospheres (Table 2.4). The vertical bars represent the nodes (points) used in the interpolation, and the solid circles are extra experimental points obtained from the source compilation and not used in deriving the interpolation curve.

The computation may be so arranged that a one-dimensional array rather than the expected two-dimensional array suffices for representing the Aitken polynomials, unwanted entries being overwritten 'on the run'; the reader is directed to listing P2.3 for details. Table 2.8 shows the calculated complete Aitken polynomial table for the specific heat example of section 2.2.2, and for $T = 2100$ K. As expected the final interpolated C_p value is identical to that computed using Newton's forward formula.

2.4 OTHER FORMS OF INTERPOLATION POLYNOMIAL

2.4.1 Lagrange's interpolation polynomial

Recalling our derivation of the Newton forward formula, we see that the ordinate values u_i enter the formula in a rather roundabout way via the divided differences, and that this is due to the stepwise addition of new points and terms. An alternative approach is to seek directly a polynomial of degree n passing through $(n + 1)$ points. Let the data values be (x_0,u_0), (x_1,u_1), . . ., (x_n,u_n). If we could find a polynomial which took the value u_0 at x_0, and zero at the other nodes or defined points (but not at the intervening values of course), and thence similar polynomials for x_1,x_2, . . ., x_n,

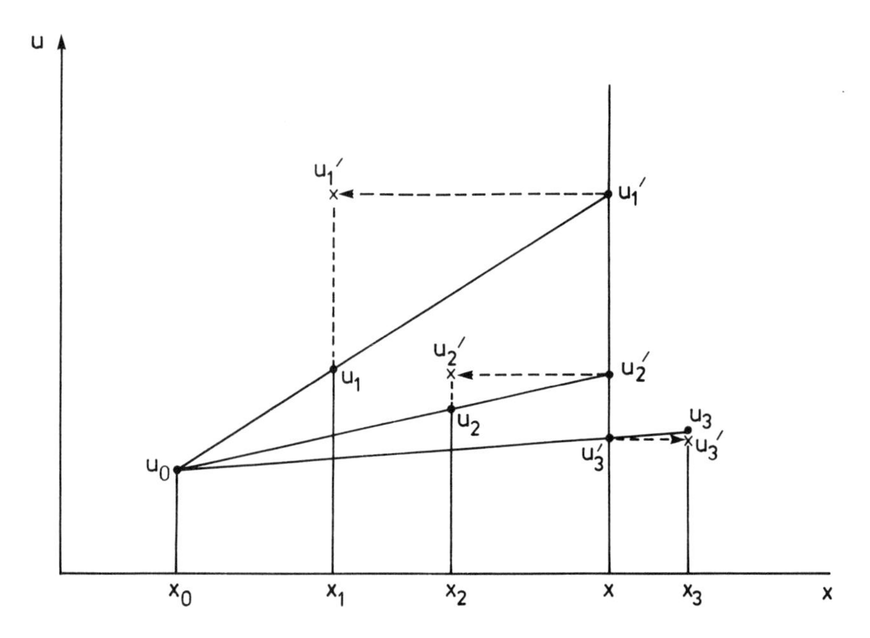

Fig. 2.4 — Graphical representation of Aitken's Method applied to the data set of Fig. 2.1, in order to determine the interpolated u-value at some point x.

Table 2.5 — Construction of the Aitken polynomial table

	linear interpolation →			
x_0	u_0			
x_1	u_1	u_1'		
x_2	u_2	u_2'	u_2''	
x_3	u_3	u_3'	u_3''	u_3'''

then the sum of all these will take the given values at the given points. Taking a hint from the Newton forward formula we write

$$P_0(x) = (x - x_1)(x - x_2) \; . \; . \; . \; (x - x_n)$$

Table 2.6 — Aitken polynomial table showing each polynomial expanded in terms of the original data values

x	u	Aitken polynomials		
x_0	u_0			
		$u_0+\left(\dfrac{u_1-u_0}{x_1-x_0}\right)(x-x_0)$		
x_1	u_1		$u_0+\left(\dfrac{u_1-u_0}{x_1-x_0}\right)(x-x_0)+\dfrac{\left(\dfrac{u_2-u_0}{x_2-x_0}\right)-\left(\dfrac{u_1-u_0}{x_1-x_0}\right)}{x_2-x_1}(x-x_0)(x-x_1)$	$u_0+\left(\dfrac{u_1-u_0}{x_1-x_0}\right)(x-x_0)+\dfrac{\left(\dfrac{u_2-u_0}{x_2-x_0}\right)-\left(\dfrac{u_1-u_0}{x_1-x_0}\right)}{x_2-x_1}(x-x_0)(x-x_1)$
		$u_0+\left(\dfrac{u_2-u_0}{x_2-x_0}\right)(x-x_0)$		$+$
x_2	u_2		$u_0+\left(\dfrac{u_1-u_0}{x_1-x_0}\right)(x-x_0)+\dfrac{\left(\dfrac{u_3-u_0}{x_3-x_0}\right)-\left(\dfrac{u_1-u_0}{x_1-x_0}\right)}{x_3-x_1}(x-x_0)(x-x_1)$	$\dfrac{\dfrac{\left(\dfrac{u_3-u_0}{x_3-x_0}\right)-\left(\dfrac{u_1-u_0}{x_1-x_0}\right)}{x_3-x_1}-\dfrac{\left(\dfrac{u_2-u_0}{x_2-x_0}\right)-\left(\dfrac{u_1-u_0}{x_1-x_0}\right)}{x_2-x_1}}{x_3-x_2}(x-x_0)(x-x_1)(x-x_2)$
		$u_0+\left(\dfrac{u_3-u_0}{x_3-x_0}\right)(x-x_0)$		
x_3	u_3			

Table 2.7 — Notation used in the implementation of Aitken's Method

x	u	Aitken polynomials		
x_0	u_0			
x_1	u_1	$I_{0,1}$		
x_1	u_2	$I_{0,2}$	$I_{0,1,2}$	
x_3	u_3	$I_{0,3}$	$I_{0,1,3}$	$I_{0,1,2,3}$

Table 2.8 — Aitken polynomial table for the $C_p - T$ data of Table 2.4

T (K)	C_p (cal g^{-1} K^{-1}) $\times 10^4$	Aitken polynomial order						
		1	2	3	4	5	6	7
1000	2730							
1200	2814	3192.0						
1500	2939	3189.8	3185.4					
1600	2981	3190.2	3187.9	3200.3				
1900	3115	3200.6	3203.0	3211.8	3219.5			
2300	3332	3239.4	3230.8	3219.4	3213.9	3216.7		
2500	3488	3285.9	3257.0	3228.4	3215.9	3218.3	3215.2	
2800	3850	3414.4	3317.1	3246.2	3219.4	3219.5	3215.6	3214.5

which is zero at $x_1, x_2, \ldots, x_n$ but not at x_0. The next step is to normalize this polynomial to u_0 at x_0. At x_0

$$P_0(x_0) = (x_0 - x_1)(x_0 - x_2) \ldots (x_0 - x_n)$$

and our required polynomial is thus $u_0 P_0(x)/P_0(x_0)$ and the sum of the n polynomials for $x_0, x_1, \ldots, x_n$ is given by

$$u(x) = \frac{u_0 P_0(x)}{P_0(x_0)} + \frac{u_1 P_1(x)}{P_1(x_1)} + \ldots + \frac{u_n P_n(x)}{P_n(x_n)}$$

The terms $P_i(x_i)$ are constants depending on the ordinate spacings. The result may be written as

$$u(x) = \sum_{i=0}^{i=n} u_i \frac{(x - x_0)(x - x_1) \ldots (x - x_{i-1})(x - x_{i+1}) \ldots (x - x_n)}{(x_i - x_0)(x_i - x_1) \ldots (x_i - x_{i-1})(x_i - x_{i+1}) \ldots (x_i - x_n)} \quad (2.11)$$

or even more concisely, albeit more cryptically, as

$$u(x) = \sum_{i=0}^{i=n} u_i \left\{ \prod_{\substack{i \neq j \\ d=1}}^{j=n} \frac{(x - x_j)}{(x_i - x_j)} \right\} \tag{2.12}$$

where the pi symbol stands for taking the product of several terms. This rather elegant formula is the Lagrange form of the interpolation polynomial, and the ordinate values enter it explicitly. It is not very useful in practice because the incorporation of an extra point necessitates the recalculation of every term, whereas the Newton forward scheme requires only the calculation of one extra term to update the approximation.

2.4.2 Generalized divided difference formulae

Very often, functions that are easily approximated by polynomials are quite 'smooth'. This means that the introduction of higher degree terms has an increasingly smaller effect on the approximation, so that the partial sums of the Newton forward polynomial approach a limiting value, as shown previously. It is clearly important to choose an approximation which converges as quickly as possible, and intuition tells us that this ought to involve choosing the ordinates close to the desired region of interpolation. When we derived the first four terms of Newton's forward polynomial we introduced the ordinates in the order 0, 1, 2, 3, but there is no reason why we could not have chosen a different order. We could, for example, have started with x_3 and then included x_2 and so on to x_0. Our polynomial would then have been

$$u(x) = u_3 + \Delta_{3,2}u(x - x_3) + \Delta_{3,2,1}u(x - x_3)(x - x_2) + \Delta_{3,2,1,0}u(x - x_3)(x - x_2)(x - x_1)$$

This equation corresponds to Newton's backward formula. In fact there are 2^n possible polynomials of this type, each formed by choosing a different path from the left-hand column to the right-hand vertex of the divided difference table (see Table 2.9). The rule for the formation of the general polynomial is given by

Table 2.9 — Graphical illustration of Newton's forward and backward formulae

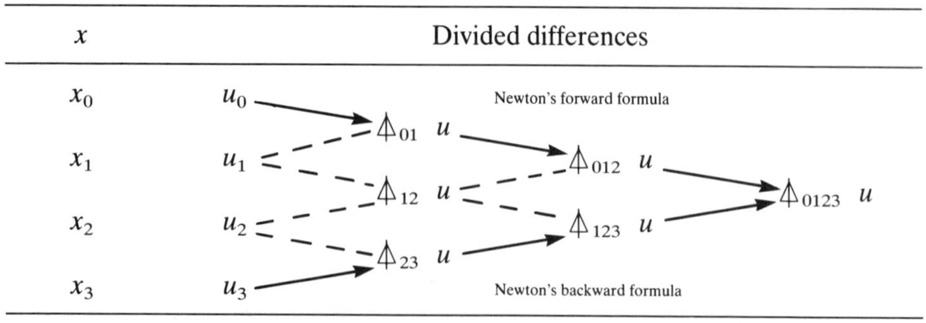

x	Divided differences				
x_0	u_0		Newton's forward formula		
		$\Delta_{01}u$			
x_1	u_1		$\Delta_{012}u$		
		$\Delta_{12}u$		$\Delta_{0123}u$	
x_2	u_2		$\Delta_{123}u$		
		$\Delta_{23}u$			
x_3	u_3		Newton's backward formula		

$$u(x) = u_i + \Delta\!\!\!|_{i,j}u(x - x_i) + \Delta\!\!\!|_{i,j,k}u(x - x_i)(x - x_j) + \ldots +$$
$$+ \Delta\!\!\!|_{i,j,k,\ldots,m,n}u(x - x_i)(x - x_j) \ldots (x - x_m) \tag{2.13}$$

Two common forms of this are Gauss's backward and forward formulae which derive from a zigzag horizontal path through the table of divided differences (see Table 2.10).

Table 2.10 — Graphical illustration of Gauss's forward and backward formulae

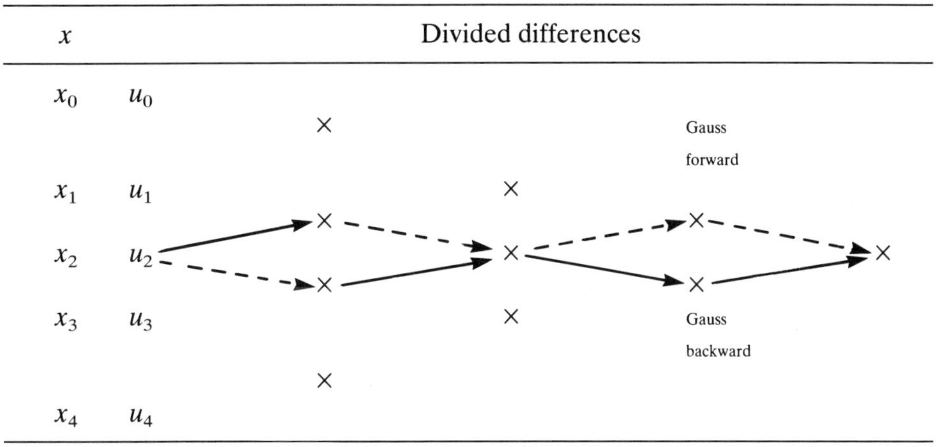

x	Divided differences				
x_0	u_0				
		×		Gauss forward	
x_1	u_1		×		
		×		×	
x_2	u_2		×		×
		×		×	
x_3	u_3		×	Gauss backward	
		×			
x_4	u_4				

The main rule to follow in forming the best polynomial is to introduce the ordinates in order of increasing $|x - x_i|$ where x is the point of interpolation. However, if we cover the entire path from left to right, we obtain exactly the same polynomial!

In order to prove this, let us imagine that we could have two different polynomials of degree n passing through $(n + 1)$ points and consider the particular case with $n = 2$ (Fig. 2.5) and label the two polynomials $P_1(x)$ and $P_2(x)$. Since $P_1(x)$ and $P_2(x)$ vanish (by constraint) at x_0, x_1, and x_2 then so does their difference $P_1(x) - P_2(x)$. If a_0, a_1, and a_2 are the coefficients of this difference polynomial then we may write

$$a_0 + a_1x_0 + a_2x_0^2 = 0$$
$$a_0 + a_1x_1 + a_2x_1^2 = 0$$
$$a_0 + a_1x_2 + a_2x_2^2 = 0$$

and hence

$$a_1(x_0 - x_1) + a_2(x_0^2 - x_1^2) = 0$$

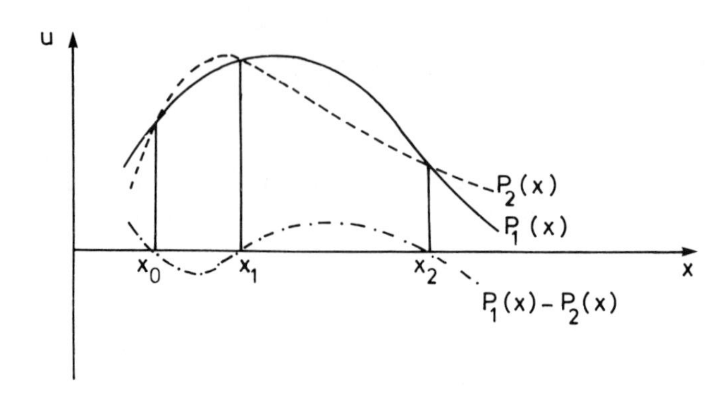

Fig. 2.5 — Plot of two hypothetical quadratics (solid and dashed lines) which pass through the same three data points. Also shown is the curve representing the difference between the quadratics (dot-dashed line).

$$a_1(x_1 - x_2) + a_2(x_1^2 - x_2^2) = 0$$

Cancelling $(x_0 - x_1)$ and $(x_1 - x_2)$ gives

$$a_1 + a_2(x_0 + x_1) = 0$$

$$a_1 + a_2(x_1 + x_2) = 0$$

and subtracting these two equations gives

$$a_2(x_0 - x_2) = 0$$

so that $a_2 = 0$. It follows that $a_0 = a_1 = a_2 = 0$ and thus $P_1(x) - P_2(x) = 0$. Therefore there is only one polynomial of degree 2 through 3 points. The argument is easily extended and in general the polynomial of degree n through $(n+1)$ points is unique.

As a consequence, it is only if we are neglecting some of the higher terms in one of these polynomials that there is any point in choosing one of the forms over another. We will discuss the errors involved in ignoring any of the terms in section 2.6.

2.5 EQUISPACED ORDINATES

Considerable simplifications result from choosing equispaced ordinates, and fortunately many tables of data are conveniently obtained for equally spaced values of x.

Let $h = (x_0 - x_1) = (x_1 - x_2)$, etc. The divided differences are easily represented in terms of ordinary forward differences

$$\Delta\!\!\!\!|_{0,1}u = \frac{u_0 - u_1}{x_0 - x_1} = \frac{-\Delta_0 u}{-h} = \frac{\Delta_0 u}{h}$$

$$\Delta\!\!\!\!|\,_{0,1,2}u = \frac{\Delta\!\!\!\!|\,_{0,1}u - \Delta\!\!\!\!|\,_{1,2}u}{x_0 - x_2} = \frac{\dfrac{\Delta_0 u}{h} - \dfrac{\Delta_1 u}{h}}{2h} = \Delta\frac{\Delta\!\!\!\!|\,_0^2 u}{2h^2}$$

$$\Delta\!\!\!\!|\,_{0,1,2,3}\mathrm{u} = \frac{\dfrac{\Delta_0^2 u}{2h^2} - \dfrac{\Delta_1^2 u}{2h^2}}{3h} = \frac{\Delta_0^3 u}{6h^3}$$

and in general

$$\Delta\!\!\!\!|\,_{0,1,2,\ldots,n}u = \frac{\Delta_0^n u}{n!h^n} \tag{2.14}$$

where Δ denotes a forward difference and Δ^n is a forward difference of order n, defined by (for example) $\Delta_0^n = \Delta_0^{n-1} - \Delta_1^{n-1}$ with $\Delta_0 u = u_1 - u_0$.

There is much less work involved in setting up a table of forward differences than there was in deriving a divided difference table. Newton's forward formula for equispaced data then becomes

$$u(x) = u_0 + \frac{\Delta_0 u}{h}(x - x_0) + \frac{\Delta_0^2}{2!h^2}(x - x_0)(x - x_0 - h) + \\ + \frac{\Delta_0^3}{3!h^3}u(x - x_0)(x + x_0 - h)(x - x_0 - 2h) + \ \ldots \tag{2.15}$$

and again, the differences come from the upper diagonal of the ordinary difference table.

We may remark that, for example,

$$\Delta_0^3 = \Delta_1^2 - \Delta_0^2 = (\Delta_2 - \Delta_1) - (\Delta_1 - \Delta_0) = \Delta_2 - 2\Delta_1 + \Delta_0$$

and hence

$$\Delta_0^3 = (u_3 - u_2) - 2(u_2 - u_1) + (u_1 - u_0) = u_3 - 3u_2 + 3u_1 - u_0$$

and make the general comment that whenever we expand a higher order difference in terms of its lower order differences, we find the same pattern of coefficients as appear as binomial coefficients in the expansion of $(a + b)^m$ where m is an integer.

This form of Newton's forward formula (2.15) is useful for interpolation at the beginning of the data table, whereas the equispaced backward formula

$$u(x) = u_0 + \frac{\Delta_{-1}}{h}u(x - x_0) + \frac{\Delta_{-2}}{2!h^2}u(x - x_0)(x - x_0 + h) + \ \ldots$$

is most suited to interpolation at the end of the table. Similarly, near the centre of the table Gauss's forward and backward formulae are best employed. The choice of a particular formulation only becomes important when we truncate the polynomial, and in this event the best approximation will be given by the truncated polynomial using the tabulated values near the point of interpolation.

The Lagrange interpolation polynomial may also be used to good effect with equispaced ordinates, as the following example demonstrates. We imagine, for instance, that we need to estimate the value of a function at a point mid-way between two of its tabulated values. For the sake of example, we consider the case of four given ordinates u_0, u_1, u_2, and u_3 spaced a distance h apart where the point of interpolation lies half-way between x_1 and x_2. Putting $x_1 = x_0 + h$, $x_2 = x_0 + 2h$, $x_3 = x_0 + 3h$ and $x = x_0 + 3h/2$ in the Lagrange form of the interpolation polynomial (2.11) then we obtain after simplification

$$u_{3/2} = (1/16)(-u_0 + 9u_1 + 9u_2 - u_3) \tag{2.16}$$

Note that neither x_0 nor h enter (2.16) which expresses $u_{3/2}$ in terms of a weighted sum of ordinates.

2.6 ERROR IN INTERPOLATION

Bearing in mind that all we are provided with in an interpolation problem are the tabulated values, we cannot, unless we know the function from which these values derive, reasonably ask for a precise quantitative idea of the error involved in interpolation. We can, however, obtain some qualitative confirmation of the fact that interpolation is best performed with function values as close as possible to the point of interpolation and also be warned of the dangers of polynomial extrapolation.

Let us suppose that the (unknown) function we are interpolating is $f(x)$, and we seek information about the error $f(x) - P_n(x)$ where $P_n(x)$ is the interpolation polynomial of degree n. For the sake of argument we take the case where $n = 2$ (Fig. 2.6). We wish to estimate the error involved in interpolating the data derived from $f(x)$ by passing a quadratic $P_2(x)$ through the points x_0, x_1, and x_2. Let us consider the error at a point x_3 (Fig. 2.6). We construct a new polynomial $P_3(x)$ which passes through $x_0 x_1$,x_2, and x_3 by adding to $P_2(x)$ an extra term according to Newton's forward scheme

$$P_3(x) = P_2(x) + \triangle_{0,1,2,3}f(x)(x - x_0)(x - x_1)(x - x_2)$$

and of course $P_3(x_3) = f(x_3)$. Therefore

$$f(x_3) - P_2(x_3) = P_3(x_3) - P_2(x_3) = \triangle_{0,1,2,3}f(x)(x_3 - x_0)(x_3 - x_1)(x_3 - x_2) \tag{2.17}$$

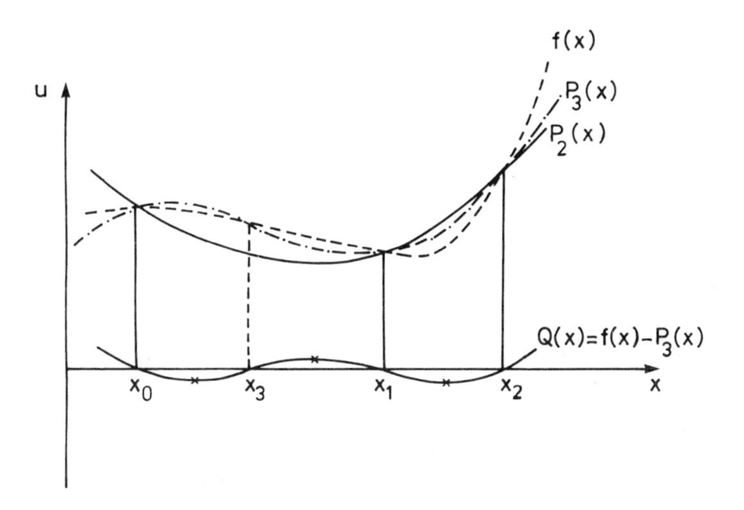

Fig. 2.6 — Plots of the various functions used to derive an expression for the error term in polynomial interpolation (see text).

and is the error at x_3. If we knew the value of $\Delta_{0,1,2,3} f(x)$, our task would be over. Consider the difference $Q(x)$ between $P_3(x)$ and $f(x)$

$$Q(x) = f(x) - P_2(x) - \Delta_{0,1,2,3} f(x)(x - x_0)(x - x_1)(x - x_2) \qquad (2.18)$$

This function is sketched in Fig. 2.6; the points in the interval (x_0, x_2) where the curve is horizontal are marked with crosses. If this function $Q(x)$ is differentiated then the resulting function will have zeros at these points. Since the original function $Q(x)$ has at least four zeros in the interval (x_0, x_2) then its derivative will have at least three zeroes. Similar reasoning shows that differentiating twice will lead to at least two zeros and a further differentiation to at least one. After three differentiations (2.18) becomes

$$Q'''(x) = f'''(x) - \Delta_{0,1,2,3}\ f(x)3!$$

Note that $P_2(x)$ has disappeared, as a quadratic must after three differentiations, the polynomial term $(x - x_0)(x - x_1)(x - x_2)$ has become 3! and $\Delta_{0,1,2,3} f(x)$, being a constant, remains unchanged. Since the third derivative has at least one zero, there exists at least one value of x, say $x = \varepsilon$, where

$$Q'''(\varepsilon) = 0 = f'''(\varepsilon) - \Delta_{0,1,2,3}\ f(x)3!$$

and therefore $\Delta_{0,1,2,3}\ f(x) = f'''(\varepsilon)/3!$. Inserting this in (2.17) gives

$$f(x_3) - P_2(x_3) = f'''(\varepsilon)(x_3 - x_0)(x_3 - x_1)(x_3 - x_2)/3!$$

and replacing x_3 by x we can generalize this to an nth degree polynomial

$$f(x) - P_n(x) = f^{(n+1)}(\varepsilon)(x - x_0)(x - x_1) \ldots (x - x_n)/(n+1)! \tag{2.19}$$

This formula for the error has two major drawbacks: we do not know ε, save that it lies in the interval of interpolation, and we do not know $f(x)$. Even if we did know $f(x)$, it might not be differentiable $(n+1)$ times. A safe though pessimistic way to circumvent the first difficulty would be to replace $f^{(n+1)}(\varepsilon)$ by the maximum value of $f^{(n+1)}(x)$ in the interval. The one feature of the error formula (2.19) that we have under our control is the factor $(x - x_0)(x - x_1) \ldots (x - x_n)$, which we can attempt to make as small as possible by choosing $x_0, x_1, \ldots, x_n$ to be as close as possible to the point of interpolation x.

To simplify the discussion we will assume that the data is equispaced at unit interval and form a Newton forward interpolation polynomial using the three points $x_0 = 0$, $x_1 = 1$ and $x_2 = 2$. Accordingly, the error term (2.19) will contain the factor $x(x-1)(x-2)$ the graph of which is shown in Fig. 2.7. Naturally, the factor $f^{(3)}(\varepsilon)/3!$ which also occurs in the error term will vary with the value of x in the interval (0,2), but Fig. 2.7 indicates that the error is likely to be greater in the interval (1,2) than in the interval (0,1). The error is seen to increase without bound outside the range of interpolation, showing that polynomial extrapolation is a very dangerous procedure indeed.

In view of the useful properties exhibited by polynomials, how are we to anticipate the type of data to which these methods will be applicable? In short, the data must derive from a 'smooth' function; the presence of 'abnormalities' such as rapid changes of slope, discontinuities and other non-polynomial features will clearly obviate the use of a polynomial approximation. Even with apparently suitable functions problems may arise with their representation as a polynomial. Some values of the function $u = 1/(1 + x^2)$ are plotted in Fig. 2.8(a), together with the result of a polynomial interpolation. The oscillatory behaviour of the polynomial about the true curve is known as the Runge phenomenon, after its original investigator. We could be forgiven for believing that by incorporating extra intermediate data points we will cure this interpolation 'disease'. That this is not the case is illustrated in Fig. 2.8(b). It may be shown that by carefully choosing particular unequally spaced ordinates, we may alleviate this problem to some extent, but we are not often in a position to do this. The function $u = 1/(1 + x^2)$ becomes infinite at $x = \pm\mathrm{i}$ $(\mathrm{i} = \sqrt{(-1)})$, and although these singularities occur at imaginary values, they are the hidden rocks on which the interpolation founders. Apart from singularities and discontinuities such as occur in functions like $u = \tan(x)$ and $u = |x|$ (where the slope changes from -1 to $+1$ at $x = 0$) most functions of physical interest behave reasonably over sufficiently narrow ranges of x and polynomial interpolation is sufficient.

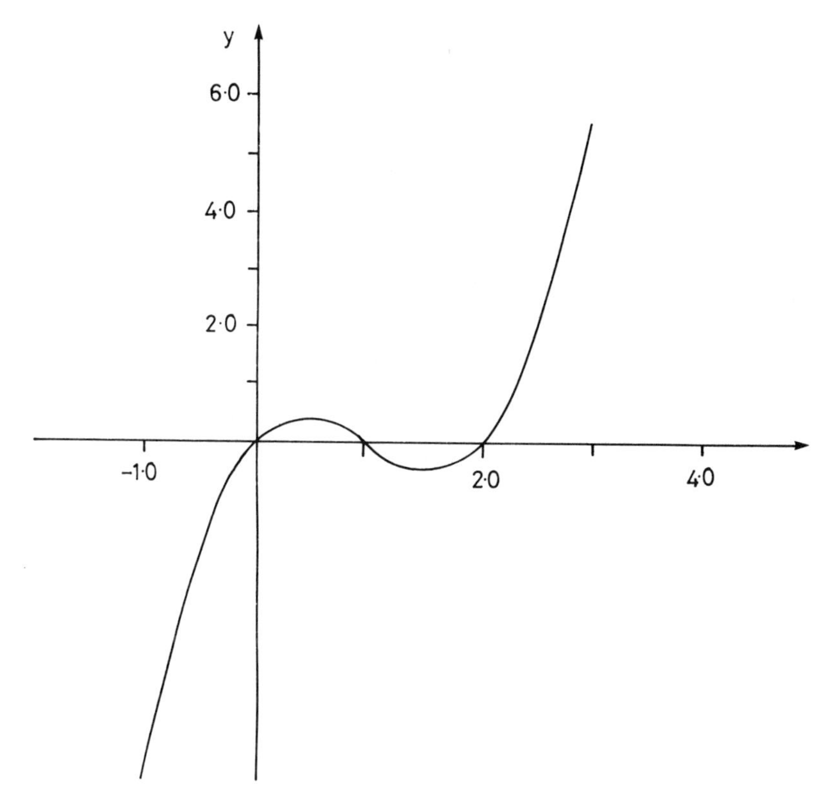

Fig. 2.7 — Plot of the polynomial $y = x(x-1)(x-2)$ which occurs in the error term for quadratic interpolation.

2.7 OTHER FORMS OF INTERPOLATION FUNCTION

Data may often be interpolated using functions other than polynomials. We might employ, for example, sums of sine and cosine functions of varying frequencies known as Fourier Series (Perrin, 1970), and such interpolation possesses special properties, not least for data which is periodic in nature.

The discussion of the problems which we encountered in attempting a polynomial interpolation of data derived from the function $1/(1+x^2)$ (section 2.6) should lead us to consider the use of more flexible interpolation functions such as rational functions $u = P_n(x)/Q_m(x)$, where $P_n(x)$ and $Q_m(x)$ are polynomials of degree n and m respectively, to open the way to the interpolation of data arising from more general types of curve. In practice, it is often found that rational functions give more accurate approximations to data than power series, for a given number of independent constants. Furthermore, rational functions may possess singularities (when $Q_m(x) \to 0$) and so should be capable of representing functions like $u = \tan(x)$. The major problem with rational functions is that they are more difficult to differentiate and integrate than polynomials, and although the derivative of a rational function will still be a rational function, the indefinite integral will generally be much more complicated. Despite this, because of a growing interest in cooperative phenomena

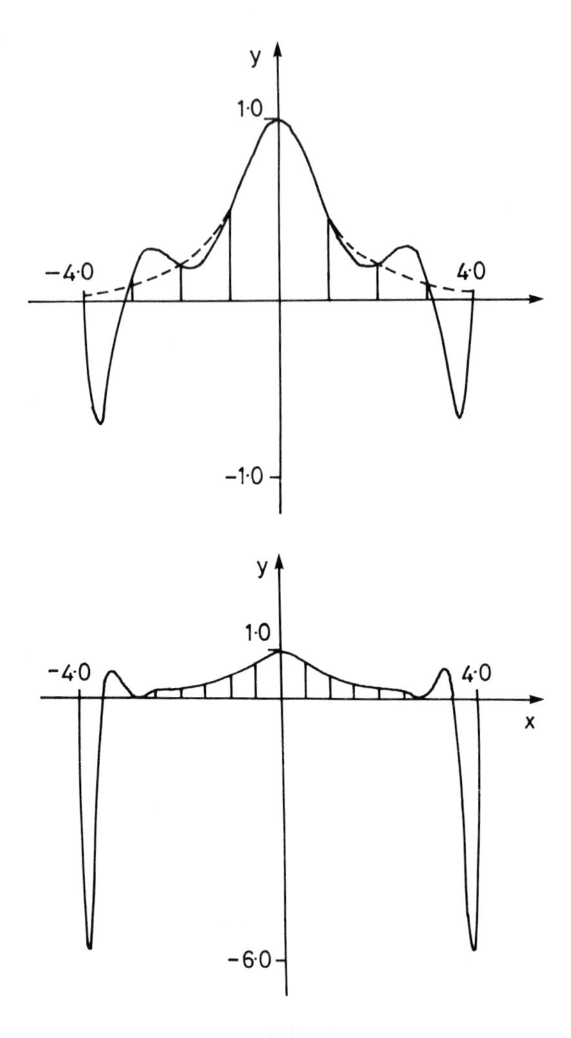

Fig. 2.8 — Interpolation curve for points generated by the function $y = 1/(1+x^2)$ in the range $-4 < x < 4$, using (a) 9 nodes and (b) 17 nodes. As the number of nodes increases the curve does not settle down but exhibits wilder variations, known as Runge's oscillations.

and systems near critical points, a great deal of effort is currently being expended on the use of rational functions to represent functions near singularities.

In order to understand how we interpolate using rational functions, we must return briefly to the theme of polynomial interpolation. Rather than developing the concept of divided differences which led to Newton's forward equation, we could have formulated the interpolation in terms of the system of equations

$$
\begin{aligned}
&a_0 + a_1x_0 + a_2x_0^2 + \ldots + a_nx_0^n = u_0 \\
&a_0 + a_1x_1 + a_2x_1^2 + \ldots + a_nx_1^n = u_1 \\
&\vdots \\
&a_0 + a_1x_n + a_2x_n^2 + \ldots + a_nx_n^n = u_n
\end{aligned}
$$

and solved for $a_0,a_1,a_2, \ldots, a_n$ according to one of the methods discussed in Chapter 1. However, such systems have been shied away from traditionally as being difficult to solve since the number of operations required in the solution increases as n^3. This difficulty has been largely removed with the advent of easily accessible computers, although in this example the approach is still inferior to the divided difference method since extra points are easily incorporated into the latter scheme, whereas the 'sledgehammer' formulation requires a complete recalculation each time it is updated.

In the case of rational functions, however, the direct method is the most straightforward to implement. Let us take the example $u = (a_0 + a_1x)/(1 + b_1x + b_2x^2)$ and four data points. Thus we can write

$$(a_0 + a_1x_0)/(1 + b_1x_0 + b_2x_0^2) = u_0$$

and so on for $x = x_1,x_2,x_3$. These expressions may be rewritten as a set of linear equations

$$
\begin{aligned}
&a_0 + a_1x_0 - b_1x_0u_0 - b_2x_0^2u_0 = u_0 \\
&a_0 + a_1x_1 - b_1x_1u_1 - b_2x_1^2u_1 = u_1 \\
&a_0 + a_1x_2 - b_1x_2u_2 - b_2x_2^2u_2 = u_2 \\
&a_0 + a_1x_3 - b_1x_3u_3 - b_2x_3^2u_3 = u_3
\end{aligned}
$$

Substitution of the values x_i,u_i $(i = 0,1,2,3)$ into these equations is followed by solving for the unknown constants a_0,a_1,b_1, and b_2.

In Fig. 2.9 are plotted some vapour pressure–temperature data for hexachloroethane. Since the vapour pressure undergoes a rapid increase with temperature as the boiling point is approached, we might expect polynomial approximation to be less accurate than a rational approximimation which can cope with such rapid changes; this is illustrated in Fig. 2.9, where the curves resulting from applying both methods are compared. The rational interpolation was made according to the function

$$u(x) = (a_0x^2 + a_1x^3 + a_2x^4)/(1 + b_1x^2)$$

the form of which was determined essentially by trial-and-error — several different rational functions were considered and the one chosen was that which gave the best agreement with the 'extra' data points (Fig. 2.9) not used in the interpolation.

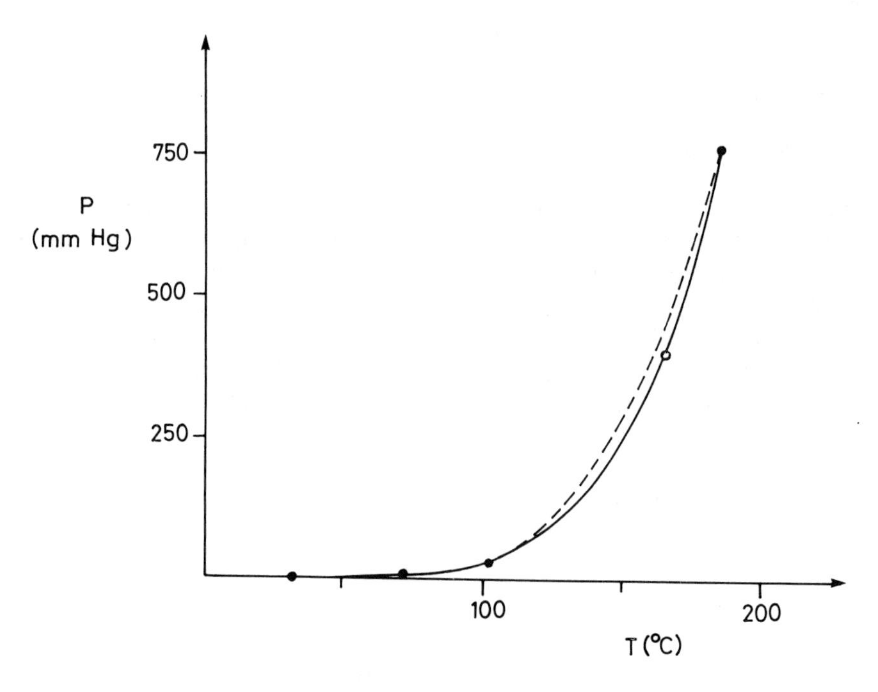

Fig. 2.9 — Interpolation curves for some vapour pressure (p)–temperature (T) data for hexachloroethane using a rational (solid line) and polynomial (dashed line) interpolation function. The solid circles are the nodes employed and the open circle an extra experimental point used to test the 'goodness' of the interpolations.

However, having decided on the best form of the function, it should then be applicable directly to other data of the same type (in this case vapour pressure–temperature data).

Procedures for interpolation by rational functions in a manner analogous to the method of divided differences do exist, and for a discussion of these the reader is referred to Handscomb (1966) for an outline and to Milne (1949) for a more detailed analysis. As a more general guide, a non-polynomial approximation should be kept in reserve for those cases where polynomial representation is clearly unsuitable.

2.8 SPLINE INTERPOLATION

Given a set of graphed points (x_i,u_i), assumed accurate, and set the task of providing an interpolating curve, the simplest solution is either to join up successive points with straight lines, or to use a 'flexicurve' and persuade it to pass through all the points before pencilling around the resulting shape. There exists a systematic mathematical technique called spline interpolation for computing such curves without recourse to physical aids. The method is widely used in computer-graphics applications (Bartels *et al.*, 1987) where smooth shapes need to be defined using the minimum number of data points. It differs from the interpolation methods we have discussed previously in that the interpolation curve consists of a series of connected simple polynomial functions rather than a single high order polynomial. As a consequence it provides greater flexibility in that it can cope with more wildly fluctuating data not easily represented by a single polynomial function.

2.8.1 Mathematical development

A spline function is best illustrated by considering firstly a set of data points and joining them up with consecutive straight line segments. Fig. 2.10 shows such an

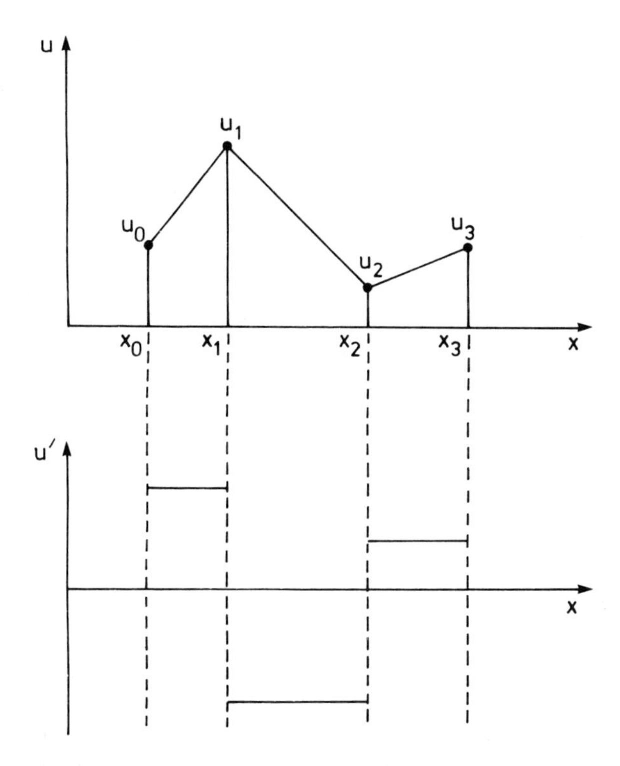

Fig. 2.10 — Interpolation using straight line segments between the nodes. The interpolation 'curve' u and its first derivative u' are shown.

interpolation scheme together with its first derivatives. It is easily seen that although the straight line interpolation ('joining up the dots') is continuous, its first derivative is not, since the slope changes abruptly as each x_i is traversed (the x_i are known as knots, nodes or joints in spline terminology). This discontinuous behaviour arises because each interval (x_{i-1},x_i) is treated in isolation from its neighbours. If we could make the first derivative continuous as well as the original function, then this should improve the smoothness of the curve. Let us explore how this might be achieved by referring to Fig. 2.11 where given values of the first derivative at the nodes, we can

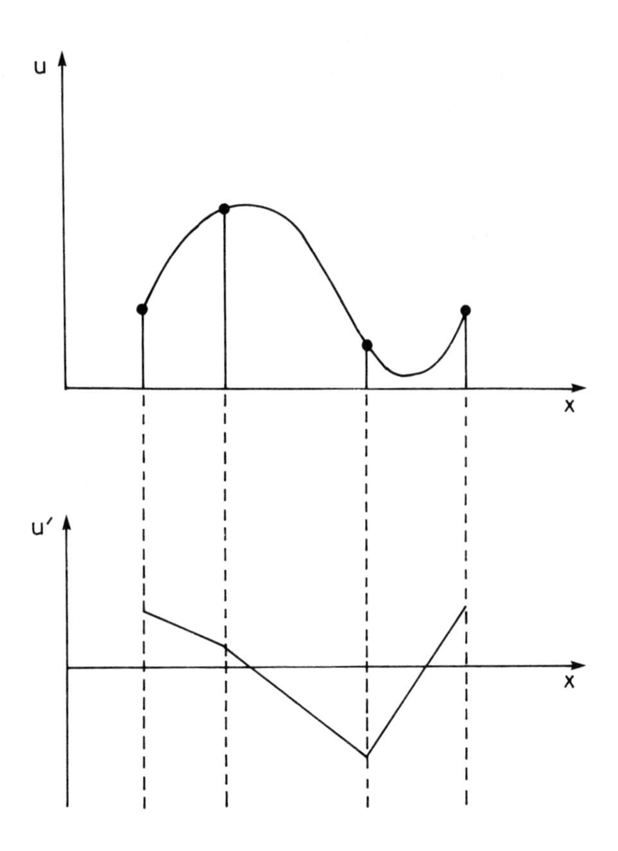

Fig. 2.11 — Interpolation using quadratic segments, showing u and u'.

make up a piecewise linear curve passing through the points and then integrate it to give the second curve. This integrated curve clearly consists of piecewise connected quadratic curves, which by its very manner of construction must be itself continuous and possess a continuous first derivative. The problem is that, in practice, we are provided with values of u and not u' and all that Fig. 2.11 does is to show that we require segments of quadratic curves in order to obtain a continuous first derivative.

In fact, piecewise cubic curves are most often used and we shall concentrate on such cubic splines. A spline function is formally defined as a piecewise connected set of polynomial arcs of degree m, the derivative curves of which are continuous up to order $(m-1)$. A cubic spline, for example, has continuous first and second derivatives.

Our task may thus be stated: given a set of $(n+1)$ data points (x_0,u_0), (x_1,u_1), . . ., (x_n,u_n) determine the cubic curves in each of the n intervals (x_{i-1},x_i) which have continuous first and second derivatives at x_{i-1} and x_i. Within each interval the cubic curve is of the form

$$u(x) = A_i x^3 + B_i x^2 + C_i x + D_i$$

where A_i, B_i, C_i, and D_i are constants, which are different for each interval. Differentiating this function twice gives

$$u''(x) = 6A_i x + 2B_i$$

and knowing the values of $u''(x_{i-1}), u''(x_i), u(x_{i-1})$, and $u(x_i)$ leads to four equations in the four unknowns A_i, B_i, C_i, and D_i. In other words, a knowledge of u''_i, u_i, and x_i would suffice for the complete determination of the cubic spline. However, if we were to determine the values of A_i, B_i, C_i, and D_i in each of the n intervals we would have to store a total of $4n$ values as opposed to the $(2n+2)$ values of u_i and u''_i values.

We now need to find a method for calculating the u''_i values. We do this by first considering the behaviour of the divided difference cubic polynomial

$$u(x) = u_0 + \Delta_{0,1}u(x - x_0) + \Delta_{0,1,2}u(x - x_0)(x - x_1) + \Delta_{0,1,2,3}u(x - x_0)(x - x_1)(x - x_2) \tag{2.20}$$

through the points x_0, x_1, x_2, and x_3 as we let $x_2 \rightarrow x_1 \rightarrow x_0$. The divided differences exhibit the following limiting behaviour

$$\Delta_{0,1}u = \frac{u_0 - u_1}{x_0 - x_1} \rightarrow (\mathrm{d}u/\mathrm{d}x)_0 = u'_0 \qquad \text{(Chapter 1)}$$

$$\Delta_{0,1,2}u \rightarrow (1/2)(\mathrm{d}^2u/\mathrm{d}x^2)_0 = (1/2)u''_0$$

(In general $\Delta_{0,1,2,\ldots,n}u \rightarrow (1/n!)(\mathrm{d}^n/\mathrm{d}x^n)_0$ as $x_n \rightarrow x_{n-1} \rightarrow \ldots \rightarrow x_0$.)

Rewriting (2.20) as

$$u(x) = u_0 + \Delta_{0,1}u(x - x_0) + \Delta_{0,1,2}u(x - x_0)(x - x_1) + \left\{\frac{\Delta_{0,1,2}u - \dfrac{\Delta_{1,2}u - \Delta_{2,3}u}{x_1 - x_3}}{x_0 - x_3}\right\}(x - x_0)(x - x_1)(x - x_2)$$

then as $x_2 \rightarrow x_1 \rightarrow x_0$ we can replace divided differences by derivatives to give

$$u(x) = u_0 + u'_0(x - x_0) + (u''_0/2)(x - x_0)^2 + u''_0/2h_0 - u'_0/h_0^2 + ((u_3 - u_0)/h_0^3)(x - x_0)^3$$

where $h_0 = x_3 - x_0$. If we differentiate this expression twice with respect to x then we can find u''_3

$$u''_3 h_0 = -2u''_0 h_0 - 6u'_0 + 6(u_3 - u_0)/h_0$$

If we identify x_0 with the one end of the spline interval and x_3 with the other end, then we have a relationship between the function values u_0, u_3, second derivative values u''_0, u''_3 and the first derivative u'_0. Replacing the subscripts 0 and 3 by i and $i+1$, and i and $i-1$, we obtain two equations

$$u''_{i+1}h_i = +2u''_i h_i - 6u'_i + 6(u_{i+1} - u_i)/h_i$$

$$u''_{i-1}h_i = -2u''_i h_{i-1} + 6u'_i + 6(u_{i-1} - u_i)/h_{i-1}$$

where $h_i = x_{i+1} - x_i$ and $h_{i-1} = x_i - x_{i-1}$, which yield, upon eliminating u'_i, the 'continuity equation'

$$\begin{aligned} &u''_{i+1}h_i + u''_{i-1}h_{i-1} + 2u''_i(h_i + h_{i-1}) = \\ &= (6/h_i)(u_{i+1} - u_i) + (6/h_{i-1})(u_{i-1} - u_i) \end{aligned} \tag{2.21}$$

This equation expresses the relationship between the function and second derivative values in adjacent intervals, given the continuity of function values and first and second derivatives.

We can write a similar equation for each interior knot (that is, excluding x_0 and x_n) giving $(n-1)$ equations but $(n+1)$ unknowns $(u''_0, u''_1, \ldots, u''_n)$. We must assume some behaviour at the end-points, the so-called boundary conditions, in order to obtain the two extra equations we need. There are a number of possibilities, specifying u''_0 and u''_n, or u'_0 and u'_n, or some combination of these. If we have a periodic function, for example, then we can specify $u'_0 = u'_n$ and $u''_0 = u''_n$. In order to illustrate the practical implementation of the spline method, we shall consider the 'natural spline' where $u''_0 = u''_n = 0$. Our set of continuity equations (2.21) then becomes

$$\begin{array}{llllll}
i=1 & 2(h_0+h_1)u''_1 + & h_1u''_2 & & & \\
i=2 & h_1u''_1 + & 2(h_1+h_2)u''_2 + & h_2u''_3 & & \\
i=3 & & h_2u''_2 + & 2(h_2+h_3)u''_3 + & & h_3u''_4 = r_3 \\
\cdots & \cdots & \cdots & \cdots & \cdots & \cdots \\
i=n-2 & & & 2(h_{n-3}+h_{n-2})u''_{n-2} + & & h_{n-2}u''_{n-1} = r_{n-1} \\
i=n-1 & & & h_{n-2}u''_{n-2} + & & 2(h_{n-2}+h_{n-1})u''_{n-1} = r_n
\end{array}$$

where $r_i = (6/h_i)(u_{i+1} - u_i) + (6/h_{i-1})(u_{i-1} - u_i)$. This system may be written in matrix form as

$$\begin{bmatrix}
2(h_0+h_1) & h_1 & 0 & 0 & \cdots \\
h_1 & 2(h_1+h_2) & 0 & 0 & \cdots \\
0 & h_2 & 2(h_2+h_3) & h_3 & \cdots \\
\vdots & \vdots & \vdots & \vdots & \\
0 & 0 & \cdots & h_{n-2} & 2(h_{n-2}+h_{n-1})
\end{bmatrix}
\begin{bmatrix} u''_1 \\ u''_2 \\ u''_3 \\ \vdots \\ u''_{n-1} \end{bmatrix}
=
\begin{bmatrix} r_1 \\ r_2 \\ r_3 \\ \vdots \\ r_{n-1} \end{bmatrix} \tag{2.22}$$

or

$$[M][u''] = [r] \tag{2.23}$$

This set of $(n-1)$ equations can be solved by one of the special methods for systems of equations having tri-diagonal matrices. Such systems often occur when physical situations are modelled, or approximated by a discrete mesh of points, where each point is only affected by its immediate neighbours.

2.8.2 Computation

The implementation of the spline method in a computer program is best divided into two stages (P2.4). The first step is to calculate the u''_i values knowing the x_i and u_i values; this involves calculating the elements of the matrix $[M]$ and the vector $[r]$ according to (2.22) and then solving (2.23) for the u''_i values, recalling that we chose $u''_0 = u''_n = 0$. The method used is gaussian elimination followed by back substitution (listing P2.4). Given a value of x, the interpolation point, the interpolated ordinate value u is calculated. The basis of this last step is to write

$$u = A_i(x-x_i)^3 + B_i(x-x_i)^2 + C_i(x-x_i) + D_i \qquad (2.24)$$

which is differentiated twice to give

$$u'' = 6A_i(x-x_i) + 2B_i$$

Substituting for $x = x_i$ and $x = x_{i+1}$yields the following expressions for u_i, u_{i+1}, u''_i, and u''_{i+1}

$$u_i = D_i$$

$$u_{i+1} = A_ih_i^3 + B_ih_i^2 + C_ih_i + D_i$$

$$u''_i = 2B_i$$

$$u''_i = 6A_ih_i + 2B_i$$

This set of equations may be solved easily to give A_i, B_i, C_i, and D_i which are then substituted into (2.24) to give, finally, the interpolated value of u.

It transpires that the natural spline ($u''_0 = u''_n = 0$) used in this example is the one which minimizes the total curvature of the approximation as measured by

$$\int_{x_0}^{x_n} (u''(x))^2 \, dx$$

and makes this approximation the 'smoothest' one. Although the analogy is not exact, this is rather similar to a thin beam adopting a configuration of least potential energy where it bends as little as possible.

2.9 INTERPOLATION WITH MORE THAN ONE VARIABLE

There are many areas of chemistry where functions of more than one variable are important. They arise naturally in subjects such as crystallography where diffraction methods give us electron density as a function of the distances along the crystallographic axes. In chemical engineering, concentration and temperature are often a function of position and time. As a consequence attempts have been made to extend the methods of this chapter to interpolate such data. In particular, the recent emphasis on the display of data and theoretical models by computer graphics has led to a great deal of interest in multivariate interpolation.

Not surprisingly, multivariate interpolation poses much greater difficulties than its univariate counterpart. By extending the approach which led to the one-variable Lagrange interpolation polynomial (section 2.4.1) we may derive its bivariate analogue

$$u = \sum_{i=0}^{1=n} u_i \left\{ \prod_{\substack{i \neq j \\ j=0}}^{j=n} \frac{(x - x_j)(y - y_j)}{(x_i - x_j)(y_i - y_j)} \right\} \tag{2.25}$$

where u_i is the datum value at (x_i, y_i). Despite the fearsome notation, we see that each of the terms in the summation vanishes for each pair of the values (x,y) except one, namely (x_i,y_i). There are a number of difficulties with this formulation, the most important of which is the very high degree of the resulting polynomial. For example, using three irregularly-spaced data points the polynomial is of second degree in both x and y, containing as its highest term x^2y^2. However, we can fit a plane whose equation is of the form $u = ax + by + c$ through three points and so our polynomial violates any principle of economy.

Consider now an irregularly spaced set of thermocouples on a plane. Given their positions and readings and the task of interpolating the temperature between them, we must decide the form of the surface to fit through the points and how many points to use at a time. If for argument's sake we use five points and a polynomial of second degree in x and y, then we can choose terms from ax^2y^2, bx^2, cy^2, dxy, ex, fy, and g, where $a, b, \ldots, g$ are constants. Which five of these terms we choose determines the form of the interpolation surface, and we may obtain the constants a, b, etc., by substituting the values of x and y and solving the five resulting linear equations.

Fortunately, real problems, perhaps by design, often deal with regular arrangements of data values; particular simplifications arise from the use of a rectangular array. Some features of two-variable interpolation are illustrated in the example of Fig. 2.12. where we are given four points at the corners of a rectangle, and wish to interpolate the value at the interior point marked X.

The simplest practical interpolation is linear interpolation, and this may be accomplished by fitting a plane through three of the four points. We thus have a choice of four possible triangles; common sense and a suspicion of extrapolation constrain us to choose a triangle which 'contains' the point X (Fig. 2.12)). Writing the equation of the plane in the form

$$u = a(x - x_0) + b(y - y_0) + c$$

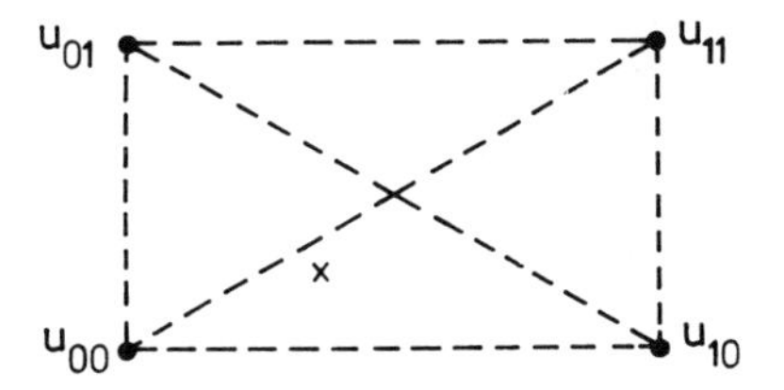

Fig. 2.12 — Bivariate interpolation (see text); we seek the value u at x.

which shifts and origin to the point (x_0,y_0), and substituting the values (x_0,y_0), (x_1,y_0) and (x_0,y_1) (that is, choosing the points u_{00}, u_{10} and u_{01} to form the interpolating triangle) gives

$$\begin{aligned} c &= u_{00} \\ a(x_1 - x_0) + c &= u_{10} \\ b(y_1 - y_0) + c &= u_{01} \end{aligned} \tag{2.26}$$

Solution of these equations yields

$$\begin{aligned} a &= (u_{00} - u_{10})/(x_0 - x_1) \\ b &= (u_{00} - u_{01})/(y_0 - y_1) \\ c &= u_{00} \end{aligned}$$

and the interpolated value u at (x,y) may be written

$$u = u_{00} + (u_{00} + u_{10})(x - x_0)/(x_0 - x_1) + (u_{00} - u_{01})(y - y_0)/(y_0 - y_1) \tag{2.27}$$

which should be compared with the divided difference polynomial encountered earlier (section 2.2), and is the equation of the interpolating surface. A similar expression would be obtained if we chose the triangle with vertices u_{00}, u_{10}, and u_{11} (Fig. 2.12).

The triviality of the solution of sets of equations such as (2.26) arises from the rectangular arrangement of data points, with the rectangle axes parallel to the x- and y-axes. For the general case of three irregularly placed points the expression for the interpolation function would be far more complicated, both in form and derivation.

In Fig. 2.13 we show the region represented by the rectangular projection of Fig. 2.12, broken down into plane triangular sections. The third diagram shows the result of averaging the interpolated values resulting from the two triangular interpolations discussed above. Such a breaking-up of complicated surfaces into simple basic units has been the subject of intense research since the blossoming of computer-generated

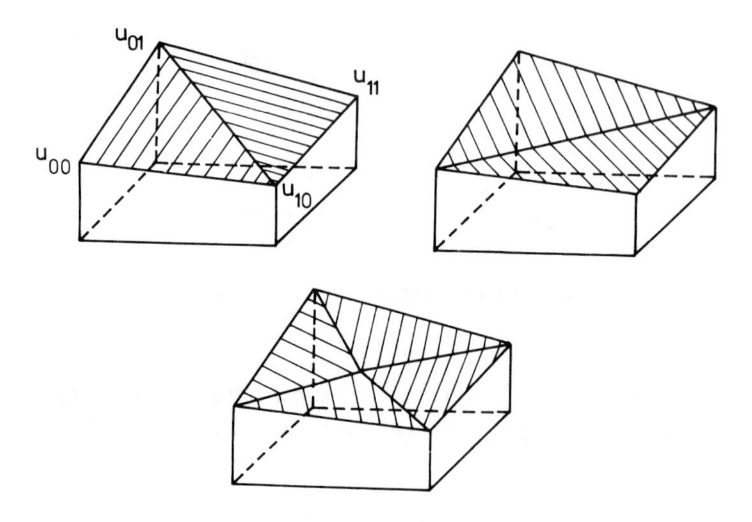

Fig. 2.13 — Three-dimensional representation of bivariate interpolation (see text).

graphic displays (Bartels *et al.*, 1987). However, for applications where we need an expression for the interpolated value of u which may be differentiated, for example, then we must use a smoother surface.

A very general alternative method interpolates each independent variable at a time, as demonstrated in Fig. 2.14. Interpolation in the x-direction along both sides

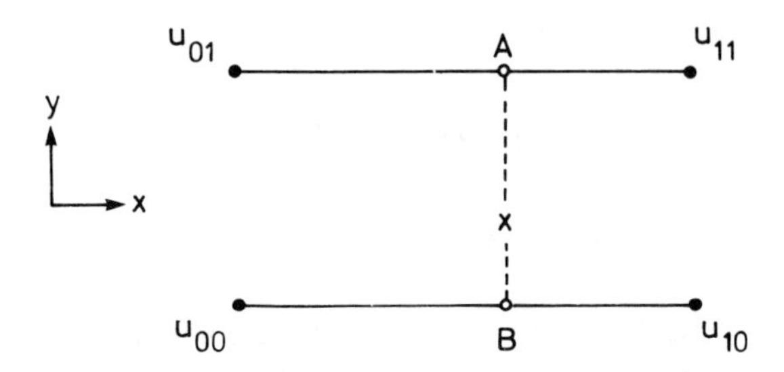

Fig. 2.14 — Bivariate interpolation performed by the interpolation of each variable independently to generate points A and B, followed by interpolation along AB.

of the rectangle yields the points A and B. Further interpolation between these points, in the y-direction, gives the interpolated value of u at x. This procedure is known in general as successive interpolation, and as bilinear interpolation in this special case.

The values at the points A and B are easily written down from our previous discussion as

$$u_A = u_{01} + (u_{01} - u_{11})(x - x_0)/(x_0 - x_1)$$
$$u_B = u_{00} + (u_{00} - u_{10})(x - x_0)/(x_0 - x_1)$$

and the result of the second interpolation is

$$u = u_B + (u_B - u_A)(y - y_0)/(y_0 - y_1)$$

or expanding in full

$$u = u_0 + \frac{u_{00} - u_{10}}{x_0 - x_1}(x - x_0) + \frac{u_{00} - u_{01}}{y_0 - y_1}(y - y_0) + \frac{\dfrac{u_{00} - u_{10}}{x_0 - x_1} - \dfrac{u_{01} - u_{11}}{x_0 - x_1}}{y_0 - y_1}(x - x_0)(y - y_0) \tag{2.28}$$

This expression is linear in x and y, as expected, but non-linear overall and so describes a curved surface of interpolation. In fact, this is a ruled surface made up of infinitely many straight line segments stacked side-by-side, and of the type beloved by designers of curved roofs since they can be made up from straight beams.

```
    0 REM P2.1
   10 :
   20 REM construction of a divided difference table
   30 :
   40 DIM X(50),U(50),DD(50,50)
   50 GOTO 10000
   60 :
  500 REM subroutine to read datafile or input from keyboard e.g. :-
  510 INPUT "NUMBER OF DATA POINTS";N
  520 FOR I=0 TO N-1 : INPUT "X-VALUE, U-VALUE";X(I),U(I) : NEXT I
  530 RETURN
  540 :
 4000 REM construct the table
 4010 FOR J=1 TO N-1
 4020         FOR I=J TO N-1
 4030                 DD(I,J)=( DD(I,J-1) - DD(I-1,J-1) )/( X(I) - X(I-J) )
 4040         NEXT I
 4050 NEXT J
 4060 RETURN
 4070 :
10000 REM main segment
10010 GOSUB 500 : REM read data
10020 FOR I=0 TO N-1 : DD(I,0)=U(I) : NEXT I : REM store U-values in  DD( , )
10030 :
10040 GOSUB 4000
10050 :
10060 REM table is stored in lower triangle of DD( , )
10070 REM can print out results now if required
10080 END
```

```
    0 REM P2.2
   10 :
   20 REM  Interpolation using Newton's Forward Formula
   30 :
   40 DIM X(50),U(50),DD(50,50)
   60 GOTO 10000
   70 :
  500 REM read data e.g.
  510 INPUT "NUMBER OF DATA POINTS";N
  520 FOR I=0 TO N-1 : INPUT "X-VALUE, U-VALUE";X(I),U(I) : NEXT I
  530 RETURN
  540 :
 4000 REM compute divided difference table
 4010 FOR J=1 TO N
 4020         FOR I=J TO N-1
 4030                 D=X(I) - X(I-J)
 4040                 IF ABS(D)<1E-30 THEN PRINT "X-VALUES EQUAL" : END
 4050                 DD(I,J)=DD(I,J-1) - DD(I-1,J-1)/D
 4060         NEXT I
 4070 NEXT J
 4080 RETURN
 4090 :
 4100 REM Newton's Forward Formula
 4110 Z=1 : YV=DD(0,0)
 4120 FOR K=1 TO N-1
 4130         Z=Z*( X - X(K-1) ) : REM form product of X differences
 4140         UV=UV + Z*DD(K,K) : REM use DD's down diagonal of table
 4150 NEXT K
 4160 RETURN
 4170 :
10000 REM main segment
10010 GOSUB 500 : REM read data
10020 HX=-1E38 : LX=1E38 : REM set to "impossible" values
10030 FOR I= 0 TO N-1
10040         IF HX<X(I) THEN HX=X(I) : REM find maximum value
10050         IF LX>X(I) THEN LX=X(I) : REM find minimum value
10060 NEXT I
10070 :
10080 FOR I=0 TO N-1
```

```
10090        DD(I,0)=U(I) : REM store U-values in table (as zero order DD's)
10100 NEXT I
10110 :
10120 GOSUB 4000 : REM compute divided differences
10130 :
10140 INPUT "X-VALUE = ";X : REM get X-value for interpolated U
10150 IF X>HX OR X<LX THEN PRINT "EXTRAPOLATED VALUE OF"; : GOTO 10160
10160 PRINT "INTERPOLATED VALUE OF";
10170 GOSUB 4100 : REM use interpolation formula
10180 PRINT " U IS";UV
10190 PRINT "ANOTHER VALUE (Y/N)"; : INPUT X$
10200 IF X$="Y" THEN 10130
10210 END
```

```
   0 REM  P2.3
  10 :
  20 REM  Interpolation using Aitken's Method
  30 :
  40 DIM X(50),U(50),AI(50)
  60 GOTO 10000
  70:
 500 REM read data e.g.
 510 INPUT "NUMBER OF DATA POINTS";N
 520 FOR I=0 TO N-1 : INPUT "X-VALUE, U-VALUE";XX(I),U(I) : NEXT I
 530 RETURN
 540 :
4200 REM Aitken's procedure
4210 FOR J=1 TO N-1
4215        PRINT J-1;"TH APPROXIMATION IS";AI(J-1)
4220        FOR I=J TO N-1
4230               D=X(I) - X(J-1)
4240               IF ABS(D)<1E-30 THEN PRINT "X-VALUES EQUAL" : END
4250               AI(I)=( AI(J-1)*( X(I) - X ) - AI(I)*( X(J-1) -X ) )/D
4260        NEXT I
4280 NEXT J
```

```
 4290 RETURN
 4300 :
10000 REM main segment
10010 GOSUB 500 : REM read data
10020 :
10030 PRINT : INPUT "X-VALUE = ";X
10040 :
10050 FOR I=0 TO N-1 : AI(I)=U(I) : X(I)=XX(I) : NEXT I
10060 :
10070 REM order points on distance of X( )-value from X
10080 FOR I=0 TO N-2 : REM pick element at top
10090         FOR J=I+1 TO N-1 : REM pick succesive elements below
10100                 IF ABS( X - X(I) )<ABS( X - X(J) ) THEN 10110
10110                 D=AI(J) : AI(J)=AI(I) : AI(I)=D : REM swap if X(J)
10120                 D=X(J) : X(J)=X(I) : X(I)=D : REM further from X
10130         NEXT J
10140 NEXT I
10150 :
10160 HX=-1E38 : LX=1E38
10170 FOR I=0 TO N-1
10180         IF HX<X(I) THEN HX=X(I) : REM find max value of X(I)
10190         IF LX>X(I) THEN LX=X(I) : REM find min value of X(I)
10200 NEXT I
10210 :
10220 GOSUB 4200 : REM Aitken's procedure
10230 :
10240 IF X>HX OR X<LX THEN PRINT "EXTRAPOLATED FINAL APPROXIMATION IS"; : GOTO 10270
10250 PRINT "INTERPOLATED FINAL APPROXIMATION IS";
10270 PRINT " ";AI(N-1)
10280 :
10290 INPUT "ANOTHER X-VALUE (Y/N)";X$
10300 IF X$="Y" THEN 10030
10310 END
```

```
   0 REM P2.4
  10 :
  20 REM  Spline interpolation
  30 :
  40 DIM A(50),B(50),C(50),R(50),U(50),U2(50),X(50),H(50)
  50 GOTO 10000
  60 :
 500 REM read data e.g.
 510 INPUT "NUMBER OF POINTS":N : N=N - 1 : REM index goes from 0 to N
 520 PRINT "YOU MUST INPUT THE DATA IN ORDER OF INCREASING X-VALUE"
 530 FOR I=0 TO N : INPUT "X-VALUE, U-VALUE";X(I),U(I) : NEXT I
 540 RETURN
 550 :
 600 REM find out if X is out of range, a node or in range
 610 FL=1
 620 IF X<X(0) OR X>X(N) THEN FL=0 : RETURN
 630 FOR I=0 TO N
 640         IF X=X(I) THEN IN=I : FL=-1 : I=N : RETURN
 650 NEXT I
 660 FOR I=0 TO N-1
 670         IF X > X(I) AND X < X(I+1) THEN IN=I : I=N : RETURN
 680 NEXT I
 690 :
4000 REM  set up 3 diagonals in tridiagonal matrix in arrays A(),B() and C()
4010 FOR I=1 TO N-2 : A(I)=H(I) : C(I)=H(I) : NEXT I
4020 FOR I=1 TO N-1 : B(I)=2*( H(I) + H(I-1) ) : NEXT I
4030 :
4040 REM  set up array R()
4050 FOR I=1 TO N-1
4060         R(I)=6*( ( U(I+1) - U(I) )/H(I) + ( U(I-1) - U(I) )/H(I-1) )
4070 NEXT I
4080 :
4090 REM  calculate second derivative array U2()
4100 U2(0)=0 : U2(N)=0 : REM these are the "natural spline" end conditions
4110 :
4120 REM  lower diagonal in A(), main diagonal in B(), upper diagonal in C()
4130 REM  Gaussian elimination modified for tridiagonal matrix equation
4140 REM  no pivoting - X-values not coincident
```

```
 4150 FOR I=N-1 TO 2 STEP -1 : REM working up from bottom right hand corner
 4160        Z=C(I-1)/B(I)
 4170        B(I-1)=B(I-1) - Z*A(I-1)
 4180        R(I-1)=R(I-1) - Z*R(I) : REM do same to RHS of equation
 4190 NEXT I
 4200 :
 4210 REM  back-substitution
 4220 B(I)=R(1)/B(I)
 4230 FOR I=2 TO N-1
 4240        B(I)=( R(I) - A(I-1)*B(I-1) )/B(I)
 4250 NEXT I
 4260 FOR I=1 TO N-1 : U2(I)=B(I) : NEXT I
 4270 RETURN
 4280 :
 4300 REM  solve for coefficients A,B,C and D and interpolated U
 4310 D=U(IN) : B=U2(IN)/2 : REM  IN denotes the interval of interpolation
 4320 A=( U2(IN+1) - U2(IN) )/( 6*H(IN) )
 4330 C= ( U(IN+1) - D - A*H(IN)*H(IN)*H(IN) - B*H(IN)*H(IN) )/H(IN)
 4340 REM expressions could be bracketted more efficiently but match text
 4350 :
 4360 REM  calculate U for given X
 4370 H=X - X(IN) : U=D + H*( C + H*( B + H*A ) ) : REM Horner's scheme
 4380 RETURN
 4390 :
10000 REM  main segment
10010 GOSUB 500 : REM read data
10020 FOR I=0 TO N-1
10030        DX=X(I+1) - X(I)
10040        IF ABS(DX)<1E-30 THEN PRINT "COINCIDENT X-VALUES" : END
10050        H(I)=DX
10060 NEXT I
10070 :
10080 GOSUB 4000 : REM set up and solve spline equations
10090 :
10100 INPUT "VALUE OF X FOR INTERPOLATION = ":X
10110 :
10120 GOSUB 600 : REM check X for position in or out of interval
10130 :
```

```
10140 IF FL=0 THEN PRINT "OUT OF RANGE" : GOTO 10190
10150 IF FL=-1 THEN PRINT "INTERPOLATED U-VALUE = ";U(IN) : GOTO 10190
10160 :
10170 GOSUB 4300
10180 PRINT "INTERPOLATED U-VALUE = ":U
10190 INPUT "DO YOU WANT TO INTERPOLATE ANOTHER VALUE (Y/N)";V$
10200 IF V$="Y" OR V$="y" THEN 10100
10210 END
```

3

Numerical integration and differentiation

3.1 INTEGRATION

In the previous chapter we saw how interpolation provides a link between the cosy world of smooth, well-behaved functions such as polynomials and the harsher reality of functions which are known at only a finite, and perhaps sparse, number of points. Using the results of Chapter 2, we can calculate approximate values for the definite integrals and derivatives of functions and hence, for example, integrate peak areas in spectroscopic traces, evaluate multiple integrals and solve differential equations of various types (see Chapter 4).

3.1.1 Simpson's Rule

Let us start with the approximation of a definite integral, or quadrature as it is often called, taking the example of $I = \int_0^{0.5} \exp(-x^2)\,\mathrm{d}x$. This, like many other integrals, cannot be evaluated analytically; however, a number of tricks exist which we can try if we are willing to settle for an approximate answer. One approach is to express $\exp(-x^2)$ as a power series in x; provided it obeys certain rules of 'reasonable behaviour' we can then integrate it term by term to obtain another series. Thus in our example

$$
\begin{aligned}
I &= \int_0^{0.5} (1 - x^2 + x^4/2! - x^6/3! + \dots)\,\mathrm{d}x \\
&= [x - x^3/3 + x^5/5.2! - x^7/7.3! + \dots]_0^{0.5} \\
&= (1/2 - 1/24 + 1/320 - 1/5376 + \dots) \\
&= 0.4613 \text{ to four decimal places}
\end{aligned}
$$

In this case the resultant series has converged to the correct answer.

We have performed the common trick of substituting a sequence of easy problems for a hard one. The 'difficult' function $\exp(-x^2)$ is approximated by a polynomial in x which can easily be integrated and then limits put in to evaluate the definite integral. This technique has the advantage that once obtained, we can use the integrated power series to evaluate the integral for other limits. However, power series are not always so easily derived, and if we are prepared to relinquish this advantage then we can get the same result by other methods.

The trick is to use an interpolation polynomial as an approximation to the function to be integrated, integrate that and put in the limits as before. For convenience and clarity of exposition we shall assume the x-values are equispaced; greater accuracy may be achieved if the x-values are dictated by the method, but these techniques such as Gaussian quadrature (see for example, Burden *et al.*, 1981) lie outside the scope of this text. We will derive a formula to approximate $I=\int_{x_{-1}}^{x_1} u\,dx$ where $u=u(x)$ is some function of x. The values $x_{-1}=-h$, $x_0=0$ and $x_1=h$ (Fig. 3.1) are chosen without loss of generality to make the parabola which

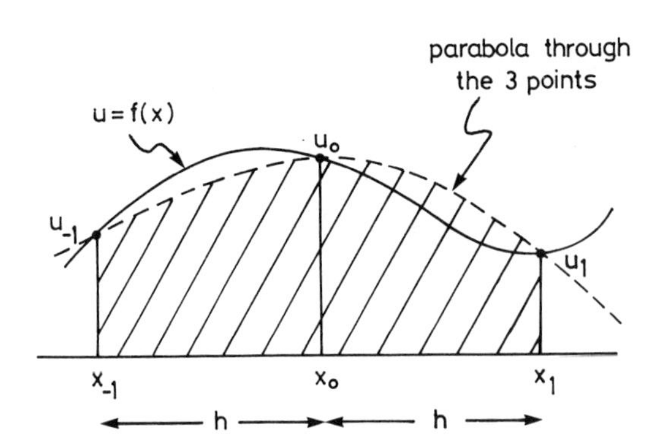

Fig. 3.1 — Illustration of the procedure used to derive Simpson's Rule for integration of a function $f(x)$. The area under the curve $f(x)$ between x_{-1} and x_1 is approximated by the shaded area defined by the interpolation parabola (quadratic) passing through (x_{-1},u_{-1}), (x_0,u_0) and (x_1,u_1).

passes through the three points easy to describe symbolically—the shaded area (Fig. 3.1) will not change in magnitude if we shift the function and ordinates together along the x-axis. The Lagrangian polynomial through the three points is (Chapter 2)

$$P(x)=u_{-1}\frac{(x-x_0)(x-x_1)}{(x_{-1}-x_0)(x_{-1}-x_1)}+u_0\frac{(x-x_{-1})(x-x_1)}{(x_0-x_{-1})(x_0-x_1)}+u_1\frac{(x-x_{-1})(x-x_0)}{(x_1-x_{-1})(x_1-x_0)}$$

Substituting for x_{-1}, x_0, and x_1 and rearranging gives us

$$P(x) = (1/2h^2)\ ((x^2 - hx)u_{-1} - 2(x^2 - h^2)u_0 + (x^2 + hx)u_1) \tag{3.1}$$

All that remains now is to integrate $P(x)$ with respect to x and put in the limits. The approximation to I is then

$$I = (1/2h^2)[(x^3/3 - hx^2/2)u_{-1} - (2x^3/3 - 2h^2x)u_0 + (x^3/3 + hx^2/2)u_1]^h_{-h} \tag{3.2}$$

This reduces to

$$I = (h/3)\ (u_{-1} + 4u_0 + u_1) \tag{3.3}$$

which is the well-known Simpson's Rule and an example of a Cotes (or Newton–Cotes) closed integration formula — so-called closed because the limits of integration include the end points. If we wished to integrate between x_{-1} and x_0, then returning to (3.2) and putting in limits of 0 and $-h$ gives us

$$I = (h/12)\ (5u_{-1} + 8u_0 - u_1) \tag{3.4}$$

which is a partial range formula, so called because the interval of integration is smaller than the region over which the function is approximated.

3.1.2 Error in Simpson's Rule

Following Chapter 2 we may rewrite the quadratic through the three points of Fig. 3.1 in the Newton forward form

$$u = u_0 + \Delta_{0,1}u\ x + \Delta_{0,1,-1}u\ x(x-h) + f^{(3)}(\varepsilon)/3!\ x(x-h)(x+h) \tag{3.5}$$

where the last term on the right-hand side is an error term, $f(x)$ is the function which we are trying to integrate and ε lies within the interpolation interval (section 2.6). If $f(x)$ is a cubic function then $f^{(3)}(\varepsilon)/3!$ is a constant, K, independent of ε. For this case, integrating the error term we find $K(x^4/4 - h^2x^2/2)$ which when we put in the limits $-h$ and h gives a net contribution of zero to the integration. We thus conclude that Simpson's Rule resulting from the parabola defined by the first three terms of (3.5) gives exact results for any cubic curve — a most surprising result. We urge the reader to verify this result by integrating an arbitrary cubic function both analytically and using Simpson's Rule.

If we assume that in the polynomial expansion of $f(x)$ the next most important term contains powers of x up to x^4, then after integration and substitution of limits we find that neglect of this term results in an error proportional to h^5.

3.1.3 Composite rules

Given an interval of integration we may divide it up using equispaced ordinates (Fig. 3.2) and estimate the area (integral) by two simple ways

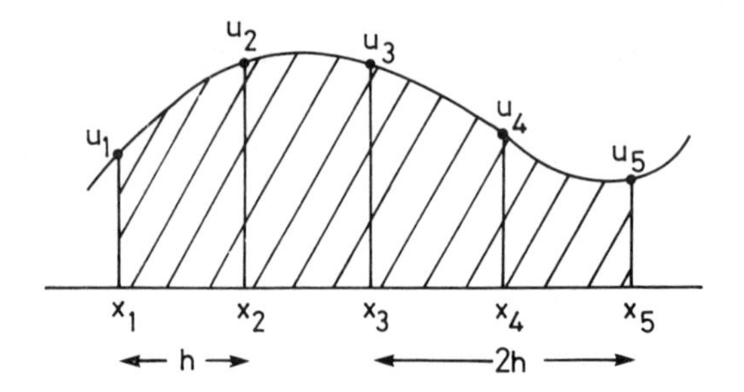

Fig. 3.2 — A composite Simpson's Rule for integration. Dividing the region of integration up into four panels, and applying Simpson's Rule twice with an interval of h, leads to a 16-fold improvement in accuracy compared with a single application using an interval of $2h$.

$$I_0 = (2h/3)\ (u_1 + 4u_3 + u_5) \tag{3.6}$$

$$I_1 = (h/3)(u_1 + 4u_2 + u_3) + (h/3)(u_3 + 4u_4 + u_5)$$

$$= (h/3)(u_1 + 4u_2 + 2u_3 + 4u_4 + u_5) \tag{3.7}$$

Equation (3.7) is known as a composite rule, and is the basis for a number of quadrature methods. The error involved in (3.6) is of the order of $(2h)^5 = 32h^5$ (see previous section) and that involved in each of the two 'two-panel' formula applications of (3.7) is of the order of h^5, that is a total error of the order of $2h^5$. The error in (3.7) is thus about 16 times less than in (3.6). Program P3.1 evaluates an integral by a composite or repeated Simpson's Rule, in which we divide the interval of integration up into an even number of panels and apply the rule over successive pairs of panels. The number of panels is then doubled and the integral re-computed. If the error estimated from the values obtained before and after the doubling (see below) is less than the desired accuracy then the process is terminated, otherwise the doubling is repeated until this condition is met, or the panel width has been reduced below a user-chosen size.

If the 'real' integral is I then, as we have seen in 3.1.2 the error $I - I_0 \simeq Kh^5$ where K is a constant which depends on the fifth derivative of the integrand, and which we will assume does not change appreciably over the interval of integration. Hence $I - I_1 = Kh^5/16$ and $I - I_0 = 16(I - I_1)$ which leads to

$$I_0 - I_1 = 15(I_1 - I) \tag{3.8}$$

Thus, the estimated error is 1/15 of the difference between the two results before and after interval halving. Program P3.2 produces an estimate of the total error assuming smooth functions and this may not apply to other less amenable functions.

Table 3.1 shows a selection of basic quadrature formulae from which other composite rules may be derived. The procedure is to divide the interval of integration

Table 3.1 — Selection of quadrature (integration) formula for evaluating the integral $I_k = \int_{x_o}^{x_k} u(x)\,\mathrm{d}x$, given equispaced data values (x_0,u_0), (x_1,u_1),. . ., etc. The x-interval is h

Integral	Formula	Name	Error order
Closed formulae:			
I_1	$(h/2)(u_0 + u_1)$	trapezoidal	h^3
I_2	$(h/3)(u_0 + 4u_1 + u_2)$	Simpson's Rule	h^5
I_4	$(2h/45)(7u_0 + 32u_1 + 12u_2 + 32u_3 + 7u_4)$	Boole's Rule	h^7
Open formulae:			
I_2	$2hu_1$	mid-ordinate rule	h^3
I_4	$(4h/3)(2u_1 - u_2 + 2u_3)$	Milne's formula	h^5
I_6	$(3h/10)(11u_1 - 14u_2 + 26u_3 - 14u_4 + 11u_5)$	a Steffenson formula	h^7
Partial range formula:			
I_1	$(h/12)\ (5u_0 + 8u_1 - u_2)$		h^4

up into panels and to cover successive groups of one or more panels by one of these formulae. For example the basic trapezoidal rule

$$I = (h/2)(u_i + u_{i+1}) \tag{3.9}$$

may be extended to give the composite trapezoidal rule for $n + 1$ ordinates

$$I = (h/2)(u_1 + u_2) + (h/2)(u_2 + u_3) + (h/2)(u_3 + u_4) + \\ + \ldots + (h/2)(u_{n-1} + u_n)$$

giving after rearrangement

$$I = (h/2)(u_1 + 2u_2 + 2u_3 + \ldots + 2u_{n-1} + u_n) \tag{3.10}$$

This method is easily implemented as a computer program, and this is left as an exercise for the reader. Although a very unsophisticated method, it works satisfactorily where there are a large number of ordinates and the function does not vary too rapidly over an ordinate interval. This is often the case for digitized spectroscopic traces, and the composite trapezoidal method is commonly employed in this situation to evaluate peak areas.

Real-world integration problems often do not involve functions as smooth as polynomials and an inspection of (3.6) and (3.7) suggests the idea of adaptive quadrature. In this approach the ordinate spacing is not fixed, but is continually optimized by the algorithm which subdivides where necessary to obtain the required local accuracy and increases the spacing where this is acceptable. In other words, starting with the five equispaced ordinates in Fig. 3.2 we compare the results of applying (3.6) and (3.7). If the estimated error (3.8) is less than a pre-chosen error limit then we accept the result; if it is not then we add two extra ordinates between either the left or right double panels and re-evaluate the formulae with this reduced step size. We then either accept or reject the result moving on to a new sub-interval (at a different level of subdivision) or further subdividing as required. Thus the step size remains quite large in regions where the function is slowly varying but is progressively reduced where the function exhibits more wildly varying behaviour. This technique is programmed in listing P3.2, and the reader is advised to study it thoroughly in order fully to understand both the method, which is conceptually simple, and its implementation, which is more tricky. The usefulness of this approach is demonstrated in section 3.1.5.

The main considerations to be borne in mind when applying a quadrature method are:

(1) Accuracy, which depends, for a given type of function, on the ordinate spacing and the order of the polynomial which is used to approximate the function. Accuracy can be adversely affected by round-off error if the nature of the function requires that a very large number of panels be used.
(2) Programming simplicity which is determined by the simplicity of the formula, and whether it is implemented in its simple or adaptive form.
(3) Execution time which is affected by several factors including whether we are provided with the function values or have to calculate them. Furthermore, the higher rules need extra multiplications, whereas the simpler ones can be arranged so that sums of ordinates are completed before multiplying by the weights.

Since Simpson's Rule combines cubic accuracy with a very simple formula, it offers a good compromise, especially in its adaptive formulation.

3.1.4 Spline quadrature

Spline interpolation overcomes the problem of 'mismatch' or discontinuities at panel junctions, and the results of Chapter 2 may be extended to the evaluation of definite integrals. If we integrate equation (2.24) then we obtain

$$\begin{aligned} I &= [A_i/4)(x-x_i)^4 + (B_i/3)(x-x_i)^3 + (C_i/2)(x-x_i)^2 + D_i x]_{x_i}^{x_{i+1}} \\ &= A_i h_i^4/4 + B_i h_i^3/3 + C_i h_i^2/2 + D_i h_i \end{aligned} \tag{3.11}$$

as the area of one panel. Substituting the expressions for A_i, B_i, C_i, and D_i in terms of u_i, u_i'', u_{i+1}, and u_{i+1}'' (section 3.8), we find

$$I = (h/2)(u_i + u_{i+1}) - (h_i^3/24)(u_i'' + u_{i+1}'') \tag{3.12}$$

which may be interpreted as the trapezoidal rule with a correction factor. It is programmed in listing P3.3. Its main drawback is that it uses many arrays which consume large amounts of memory, especially when a high degree of panel doubling is required. The problem can be overcome to some extent by dividing the range of integration into sections which are then evaluated separately.

3.1.5 Comparison of methods

The general usefulness of an integration method is best assessed by testing it on functions which we expect will be difficult to integrate, although less robust methods might be perfectly acceptable in particular well-defined situations (such as the example of peak area computation mentioned in section 3.1.3). A 'difficult' function is $u = 1/(1 + 0.25(x - 10)^2)$, an example of what is commonly encountered as the lorenzian lineshape function (Fig. 3.3), which is similar to the Runge function

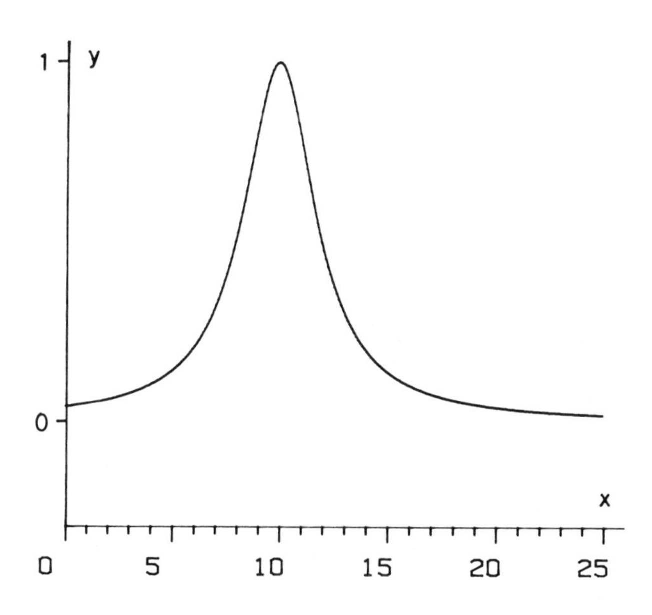

Fig. 3.3 — An example of a lorenzian lineshape function, a 'difficult' function to integrate. This particular plot is of $u = 1/(1 + 0.25(x - 10)^2)$.

that causes so many problems in polynomial interpolation (Chapter 2). Fig. 3.4 shows a plot of the integral estimate, I_n, against number of panels used, n, for $I = \int_0^{30} dx/(1 + 0.25(x - 10)^2)$ using the composite Simpson's Rule. Note that the value of I_n settles down to the 'correct' value when the panels have become sufficiently narrow.

The power of the adaptive approach is emphasized by a simple example. The

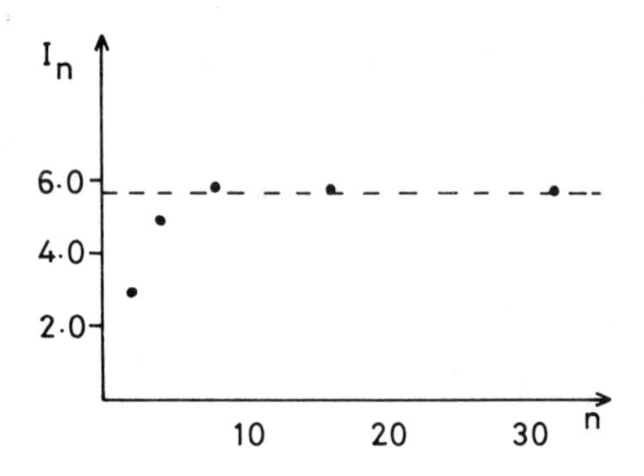

Fig. 3.4 — The effect of increasing the number of panels n in the composite Simpson's Rule approximation I_n of $\int_0^{30} \mathrm{d}x/(1+0.25(x-10)^2)$. The dotted line indicates the correct value of the definite integral.

function $f(x) = \sqrt{(4-x^2)}$ defines a circle of radius two units centred at the origin, and the definite integral

$$I = \int_0^2 \sqrt{(4-x^2)}\, \mathrm{d}x$$

represents the area of a quadrant of this circle (Fig. 3.5). It is easy to show analytically that $I = \pi = 3.14159$ to five decimal places. Using the constant-spacing composite Simpson's Rule (program P3.1) this result, to the stated precision, requires 6138 function calls; the adaptive Simpson's Rule (program P3.2) achieves the same result with only 140 function evaluations. The problem is that the tangent to the circle (slope) rapidly becomes large and negative as $x \rightarrow 2$ (Fig. 3.5) and is in fact infinite at $x = 2$, dictating a very small step size in this particular region. The simple method is forced to use this small spacing throughout the interval of integration so that most of the repeated evaluations are performed to a far greater precision than is required. The adaptive method, in contrast, only subdivides where necessary, as is illustrated in Fig. 3.5 where we see that the number of levels of subdivision increases dramatically close to $x = 2$.

The limitations of the trapezoidal method applied to this demanding example are highlighted by the observation that 16382 function calls are needed to calculate π to five decimal places using the non-adaptive composite method; the adaptive version performs much better but still requires 1035 calls.

3.1.6 Practical applications

Peak area computation in spectroscopy and chromatography is widely used to estimate component concentrations. All the techniques discussed so far assume noise-free data. If we take the simple model of the spectral peak as consisting of noise

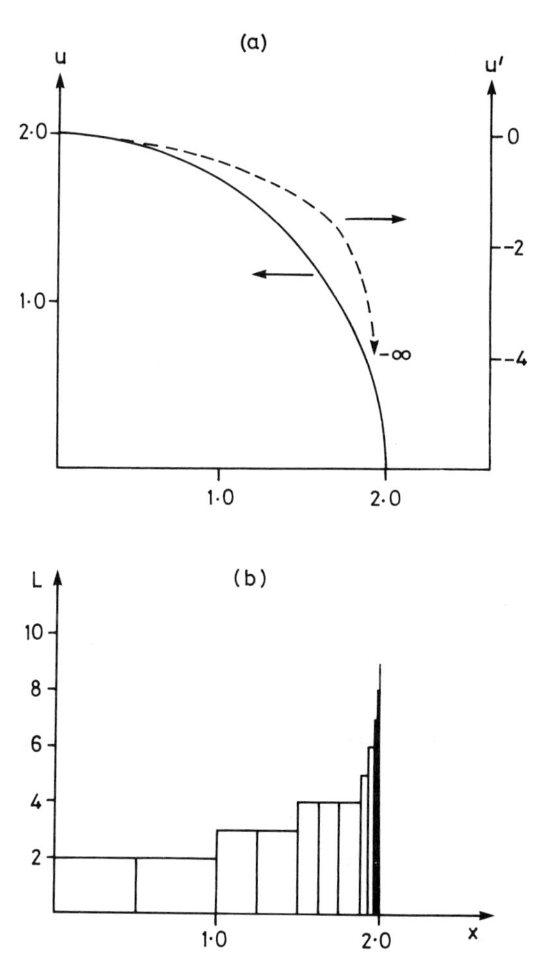

Fig. 3.5 — (a) Plot of the quarter circle defined by $u = \sqrt{(4 - x^2)}$, together with its first derivative u' which becomes infinite at $x = 2$. The area of this quadrant is $\pi = 3.14159$ to five decimal places, and provides a demanding test of numerical integration procedures. (b) How the adaptive Simpson's rule method tackles the integration of the function in (a), to an accuracy of five decimal places. The level of subdivision (L) increases dramatically as we approach $x = 2$, and we need to use smaller and smaller intervals to maintain accuracy.

added to signal, then it might be expected that the noise would integrate to zero. Fig. 3.6 shows the results of integrating a gaussian peak-shape both with and without added noise (see Chapter 5) having an approximately normal or gaussian distribution (not the two occurrences in different contexts of the term 'gaussian'). The integration of the data $Y(J)$ ($J = 1,\ldots,NP$; NP even) is performed using Simpson's method, with a simple program of the form

```
10 H = 0.1 : REM abscissa interval
20 A = 0 : REM area accumulator
```

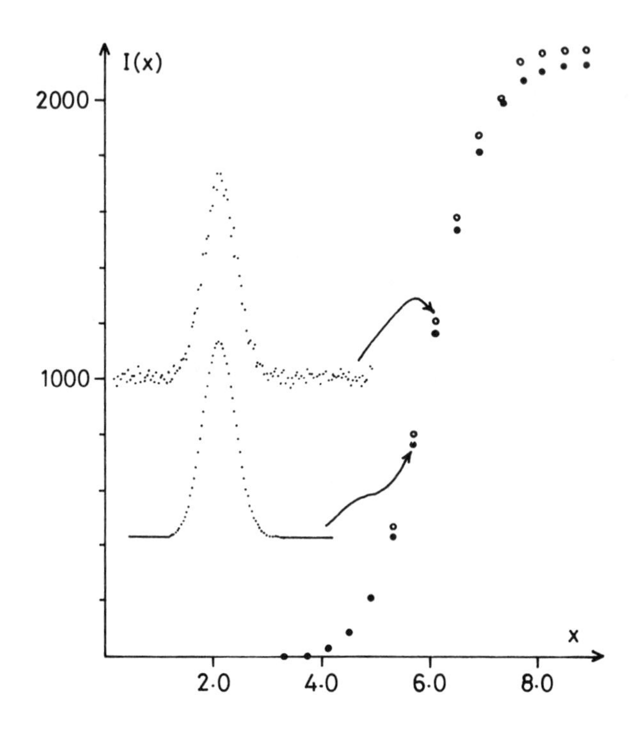

Fig. 3.6 — The effect of experimental noise on the evaluation of the area under a spectral lineshape. A gaussian shape $g(x)$ with added noise (solid circles) and without added noise (open circles) is used and $I(x) = \int_0^x g(t)\,dt$ computed for a range of values of x, using a repeated Simpson's Rule routine (see text). As $x \to \infty$, the plots level off at the approximation to the peak area. Note that these final values are not identical for the two cases.

```
30 FOR J = TO NP STEP 2
40 AR = (Y(J) + 4*Y(J+1) + Y(J+2))*H/3
50 A = A + AR
60 PRINT "X = ";(J+2)*H;"     INTEGRAL = "';A
70 NEXT J
```

Clearly the effect of the noise does not 'cancel out'; the reason is that the property of averaging to zero is one of infinite sample populations, which is never quite the case in real life. A better integration technique for noisy data is to fit a model peak-shape by a least squares minimization method and to integrate analytically the fitted function (see Chapters 5 and 6).

It often happens that we wish to simulate spectra which result from molecules whose axes are distributed over a range of angles $\theta_1 < \theta < \theta_2$. In this application we have an expression for the spectrum as a function of angle, θ, and we evaluate the integral

$$I(x) = \int_{\theta_1}^{\theta_2} f(\theta,x)\ d\theta$$

at successive values of x, which might be the frequency or some other spectral parameter—each point of the spectrum is thus separately computed from an integral expression.

The electron paramagnetic spectrum originating from an unpaired electron in an axial environment may be calculated (Daul *et al.*, 1981) from the integral absorbance expression

$$I(H) \propto \int_0^{\pi/2} \exp(-2((H-(h\upsilon/g\beta))/w)^2)(g_\perp^2((g_\parallel^2/g^2)+1))\ \sin\theta\ d\theta$$

where H is the field (abscissa in the spectral plot), β is the Bohr magneton, h is Planck's constant, υ is the frequency of the applied field and w is a linewidth parameter. The value of g depends on θ and is given by

$$g^2 = g_\parallel^2\ \cos^2\theta + g_\perp^2\ \sin^2\theta$$

The result of computing this spectrum (with $\upsilon = 9$ GHz, $w = 40$ gauss, $g_\perp = 2.0$ and $g_\parallel = 2.1$) using the adaptive Simpson's Rule program P3.2 is shown in Fig. 3.7(b) The dangers of performing the computation with too poor a specified accuracy are clearly demonstrated in Fig. 3.7(a), where physically meaningless structure has now appeared. This example provides a pertinent reminder that one should only attempt such a complicated numerical computation in an area in which one has some expertise—at least then the invalidity of such a result as that shown in Fig. 3.7(a) will be clear.

This example suggests that as a matter of routine one should repeat any such computation with a decreased specified accuracy, and check to see whether the result is affected significantly.

3.1.7 Multiple integrals

Imagine that at some site there is a local concentration of chemical species and that you have measured its concentration at selected points over a sufficiently large region (Fig. 3.8). How can you estimate the total quantity of the material present within the region of measurement? The answer to this question requires the evaluation of the double integral

$$I = \iint u(x,y)\ dx\ dy$$

Following on from the previous discussion of integrals involving one variable, we can

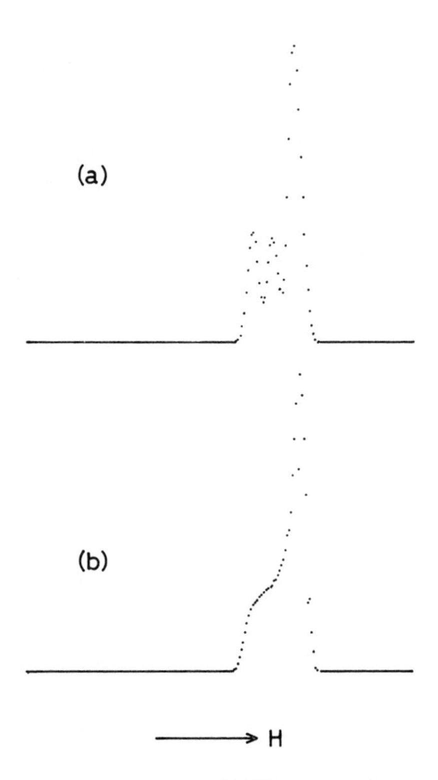

Fig. 3.7 — Electron paramagnetic resonance (EPR) spectrum from an unpaired electron in an axial environment, computed from the integral absorbance equation (section 3.1.6): (a) using too poor an accuracy; (b) using an adequate accuracy.

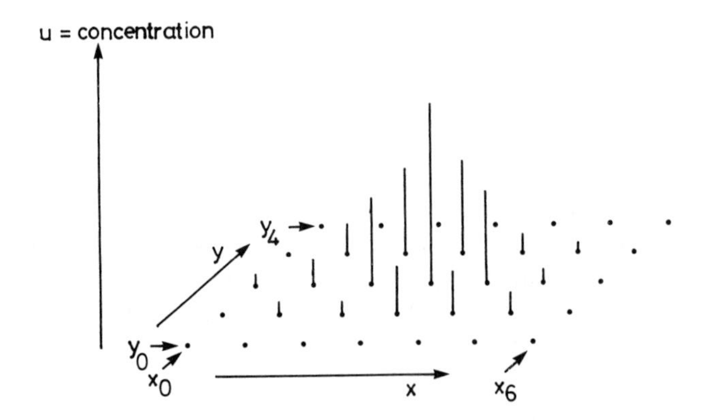

Fig. 3.8 — Example of a situation where double integration becomes necessary. The variation of concentration u of a chemical species over a two-dimensional grid is plotted, and we require to determine the amount of material present in a specified region. Integration must now be performed with respect to x and y.

extend the idea of constructing a polynomial through a set of points (x_0,u_0), (x_1,u_1), etc., to that of constructing a curved surface which passes through a set of points (x_0,y_0, u_0), (x_1,y_1,u_1), etc. The double integral

$$\iint u(x,y)\ \mathrm{d}x\ \mathrm{d}y$$

may thus be interpreted geometrically as the volume between the curved surface and the (x,y) plane. In order to simplify the discussion, we confine ourselves to a rectangular array of points in the (x,y) plane which are regularly spaced in the x- and y-directions, though not necessarily having the same spacing in the two directions.

Returning to Chapter 2, we derived the bilinear interpolation formula (2.28) between four points situated above the corners of a rectangle with sides parallel to the axes of the (x,y)-plane (Fig. 3.9); u_{0y} and u_{1y} are points on straight lines joining

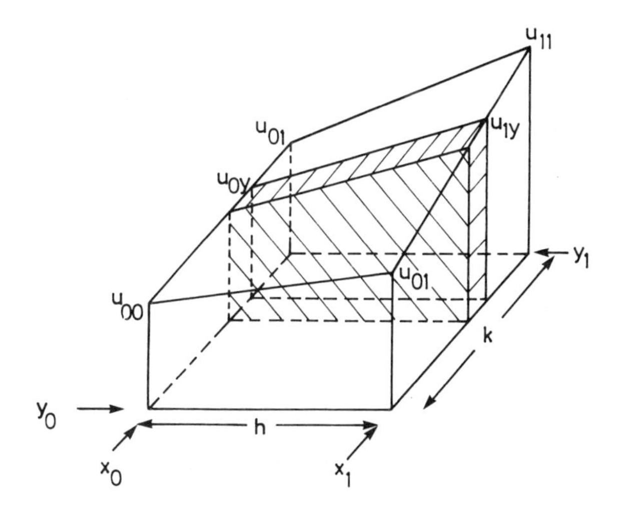

Fig. 3.9 — The use of bilinear interpolation (Chapter 2) to evaluate a double integral. We approximate the integration by summing up the volumes of the shaded slabs.

the tops of ordinates u_{00}, u_{01} and u_{10}, u_{11} respectively. The actual surface passing through u_{00}, u_{01}, u_{10}, and u_{11} is being approximated by this bilinear interpolation and we are going to calculate the volume between this interpolating surface and the (x,y)-plane (should this surface lie beneath the (x,y)-plane it will yield a negative volume). The double integral which we are approximating is

$$I = \int_{y_0}^{y_1} \int_{x_0}^{x_1} u\ \mathrm{d}x\ \mathrm{d}y \tag{3.13}$$

and we first evaluate $\int_{x_0}^{x_1} u \, \mathrm{d}x$ as approximately the area of the large face of the thin plate, shaded in Fig. 3.9; using the trapezoidal rule this is $(h/2)(u_{0y}+u_{1y})$ and (3.13) becomes

$$I=\int_{y_0}^{y_1} (h/2)(u_{0y}+u_{1y}) \, \mathrm{d}y$$

The term to the right of the integral sign is merely the volume of the trapezoidal plate at y. The second integration thus corresponds to adding up the volumes of all the plates or slices to yield the total volume. Applying the trapezoidal rule again

$$\begin{aligned} I &= (k/2)((h/2)(u_{00}+u_{01})+(h/2)(u_{10}+u_{11})) \\ &= (hk/4)(u_{00}+u_{01}+u_{10}+u_{11}) \end{aligned} \tag{3.14}$$

This is the two-variable counterpart of the trapezoidal rule (3.9) and may be represented symbolically by

$$\boxed{\begin{matrix} 1 & 1 \\ \\ 1 & 1 \end{matrix}} \cdot (hk/4)$$

where the numbers inside the square give the weights to multiply the ordinates by, corresponding geometrically to these points in the array. The square denotes a double sum. To solve the problem posed at the beginning of this section, we must extend this method to a composite rule in exactly the same spirit as in the one-variable case. For a 5 by 7 array of data points (Fig. 3.8) the double sum is, symbolically

$$\boxed{\begin{matrix} 1 & 2 & 2 & 2 & 2 & 2 & 1 \\ 2 & 4 & 4 & 4 & 4 & 4 & 2 \\ 2 & 4 & 4 & 4 & 4 & 4 & 2 \\ 2 & 4 & 4 & 4 & 4 & 4 & 2 \\ 1 & 2 & 2 & 2 & 2 & 2 & 1 \end{matrix}} \cdot (hk/4) \tag{3.15}$$

Although this may not look anything like a composite trapezoidal rule, the integers 1,2,4 reflect the fact that when all the volumes of the inter-connected little cells (such as that in Fig. 3.9) are added up, it turns out that the interior coordinates of the ensemble are counted four times, the edge ones twice and those in the corner only once.

We could have gone directly to (2.28) and integrated that term by term, keeping first y constant and integrating with respect to x putting in limits, and then repeating the procedure with respect to y. However, this would not have shown either the relationship between the trapezoidal rule and (3.14) or the relationship between the two-step double integration and calculation of the volume as a stacking together of small plates, each of width dy.

We may now apply the same reasoning to Simpson's Rule. This time we have a rectangular 3 by 3 point array and

$$I = (k/3)((h/3)(u_{00} + 4u_{01} + u_{02}) + (4h/3)(u_{10} + 4u_{11} + u_{12}) + (h/3)(u_{20} + \\ + 4u_{21} + u_{22}))$$

or more conveniently

$$I = \begin{array}{|ccc|} 1 & 4 & 1 \\ 4 & 16 & 4 \\ 1 & 4 & 1 \end{array} \cdot (hk/9) \tag{3.16}$$

Any number of formulae of this type may be made up simply by combining a one-variable sum-of-ordinates rule with itself, or for that matter with another different rule, subject to the same provisos as discussed previously.

The method may be extended to the three-variable case where we integrate a function over the volume of a cuboid. If we seek a geometrical interpretation of this, then we have to imagine something like a 'fog' of variable density and the problem is to determine the total mass of this foggy material lying inside the cuboid.

It is a relatively straightforward matter to extend programs P3.1 and P3.2, the simple and adaptive Simpson's Rule methods respectively, to the two variable case, although the execution times of the resulting programs are much larger due to the much greater number of function evaluations required. We shall consider the modification to the simple Simpson's procedure in some detail; the modified program is giving in listing P3.4. The basic technique is to treat the problem

$$\int_{y_1}^{y_2} \left(\int_{x_1}^{x_2} f(x,y)\, \mathrm{d}x \right) \mathrm{d}y$$

as an integration over x followed by an integration over y (or vice versa). We proceed as follows (see Fig. 3.10):

(1) Determine the area of the profile along the section AB using Simpson's Rule, subdividing the intervals and re-computing the area until the chosen accuracy has been achieved. Repeat along sections FG and HI.

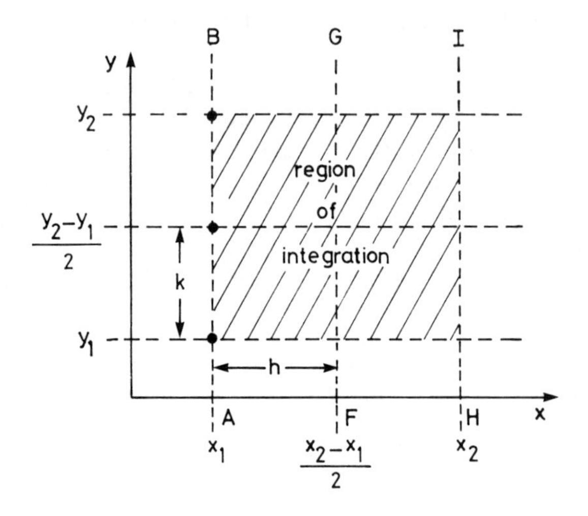

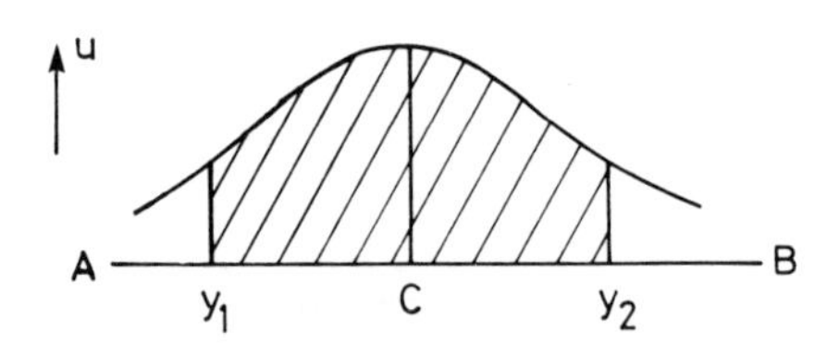

Fig. 3.10 — How to apply Simpson's Rule (simple or adaptive) to the bivariate integral $\int\int u(x,y)\,dx\,dy$, which is the volume under the surface $u(x,y)$. Integrations with respect to y are first performed along AB, FG and HI (an illustrative section along AB is shown), followed by integration with respect to x (see text for details).

(2) Apply the one-variable Simpson's Rule to the set of areas resulting from (1) to estimate the volume between the plane ABIH and the surface defined by $u=f(x,y)$.
(3) Subdivide the interval in the x-direction and so determine the areas of additional profiles equidistant between AB and FG, and between FG and HI, using the procedure described in (1).
(4) Estimate the volume again as in (2). Repeat the subdivision in the x-direction as many times as is necessary.

The extension to the adaptive quadrature method is analogous, if somewhat more complex, and is discussed in section 3.1.8.

We will consider briefly the problem of integrating over a region other than a rectangle with sides parallel to the x- and y-axes. Fig. 3.11 illustrates an example of this where the region of integration is a loop of some exotic function. We firstly integrate to find the areas of profiles such as FG exactly as before, subdividing to obtain progressively greater accuracy. The only new problem to be solved is that of

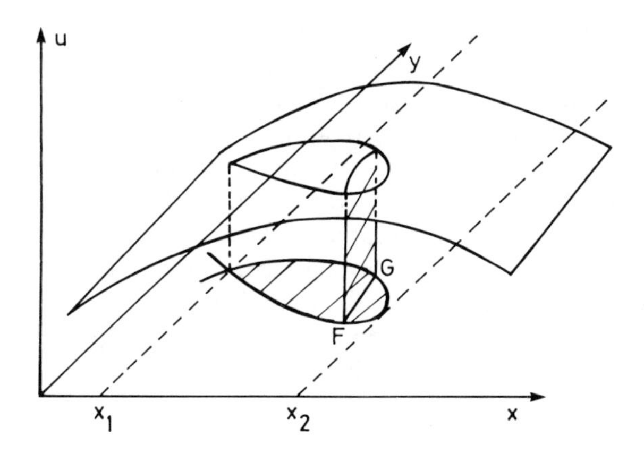

Fig. 3.11 — Bivariate (double) integration $\int\int u(x,y)\,dx\,dy$ over a non-rectangular region.

determining the limits of integration in the y- and x-directions. To do this the 'loopy function' defining the region on the plane must be expressed as two functions — upper and lower limit functions. As an example, if the function is $x^2+y^2=1$, that is we want to integrate over a circular region, then we use an upper limit function $y=+\sqrt{(1-x^2)}$ and a lower limit function $y=-\sqrt{(1-x^2)}$. The x-limits x_1 and x_2 have to be determined from the properties of the curve as well; for the circle $x^2+y^2=1$, they would be $+1$ and -1. For the particular case of a circular region of integration, it might be better to use polar coordinates with limits $r=0,1$ and $\theta=0,2\pi$. For a more detailed discussion of integration over non-rectangular regions, and the general problem of multiple integration where the limits may be functions of the variables, the reader is referred to Burden *et al.* (1981).

3.1.8 A practical application

Our previous EPR example may now be extended to the determination of the lineshape for a polycrystalline powder sample with three different g values g_1, g_2, and g_3, arising from a molecular structure with rhombic symmetry.

Because the symmetry is rhombic rather than axial, we need to integrate explicitly over the polar and azimuthal angles, θ and ϕ. The expression for $I(H)$ is now (Daul *et al.*, 1981).

$$I(H) \propto \int_0^{\pi/2}\int_0^{\pi/2} \exp(-2((H-(h\upsilon/g\beta))/w)^2) \times$$
$$\times\,(g_1^2+g_2^2+g_3^2-(1/g^2)(g_1^4\sin^2\theta\cos^2\phi+g_2^4\sin^2\theta\sin^2\phi+g_3^4\cos^2\theta))\sin\theta\,d\theta\,d\phi$$

The double integral version of the simple composite Simpson's Rule (program P3.4) can be used to evaluate this expression as one steps through the field H, but proves to be impracticable owing to the excessive running times required. The adaptive Simpson method, applied to each of the integrations over θ and ϕ, comes to our aid. As in the one-dimensional case (section 3.1.6), if we specify too low a computational accuracy the 'spectrum' exhibits a wealth of impressive yet physically meaningless structure. This arises because we miss fine structure in the function by stepping through the calculation too coarsely (although we do so with a specified precision). Rather than specifying a lower accuracy, we can overcome this problem by forcing a particular level of subdivision of the step size, whether the computation demands it or not. In this way, sharp features are not missed, and at the same time redundant calculations are avoided. A program incorporating this approach is shown in listing P3.5, and the results displayed in Fig. 3.12.

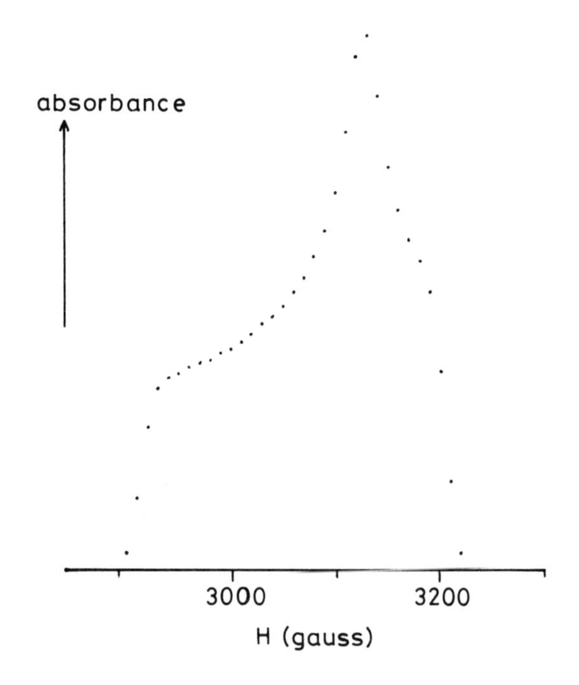

Fig. 3.12—EPR lineshape for a polycrystalline powder sample, calculated using program P3.5. The parameters used were: $g_1 = 2$, $g_2 = 2.05$, $g_3 = 2.2$, linewidth $w = 20$ gauss, microwave frequency $= 9 \times 10^9$ Hz, absolute accuracy $= 0.001$, maximum level of subdivision $= 15$.

3.2 NUMERICAL APPROACHES TO DERIVATIVES AND PARTIAL DERIVATIVES

3.2.1 Use of interpolation polynomials

The results in this section follow directly from the foregoing sections in that numerical differentiation may be accomplished by differentiating, rather than integrating, the Lagrangian or divided difference polynomials. For example, equation (3.2) may be differentiated once with respect to x to give $\mathrm{d}P/\mathrm{d}x$ as an approximation to $\mathrm{d}u/\mathrm{d}x$

$$\mathrm{d}u/\mathrm{d}x = u' = (1/2h^2)((2x - h)u_{-1} - 4xu_0 + (2x + hu_1)$$

and thus

$$(\mathrm{d}u/\mathrm{d}x)_{x_0=0} = (1/2h)(u_1 - u_{-1}) \tag{3.17}$$

$$(\mathrm{d}u/\mathrm{d}x)_{x_{-1}=-h} = (1/2h)(-3u_{-1} + 4u_0 - u_1) \tag{3.18}$$

Differentiating again with respect to x gives

$$\mathrm{d}^2u/\mathrm{d}x^2 = (1/h^2)(u_{-1} - 2u_0 + u_1) \tag{3.19}$$

which applies at any point since we are differentiating a quadratic twice and the result is a constant. Other differentiation formulae of the above type are summarized in Table 3.2.

It is interesting to note that (3.17) uses a quadratic interpolation and yet only needs two function values, a behaviour reminiscent of Simpson's Rule. The geometrical interpretation of (3.17) is that we are approximating the tangent at x_0 by the chord joining two points close and equidistant from x_0 — an 'intuitive' approach which we can in fact adopt without any knowledge of interpolation polynomials (see Chapter 1).

Since differentiation is based on differences and integration is based on sums, the errors in numerical differentiation will be far worse than in numerical integration (section 1.3.4). However, we emphasize that if the only information we have is that the original function, u, passes through the interpolation points, then we cannot say anything definite about the error.

3.2.2 Applications of numerical differentiation

One of the main uses for these methods lies in the approximation of derivative values of complicated functions. For example, Stewart (1967) modified the very efficient Davidson–Fletcher–Powell algorithm (Chapter 6) for minimizing a function of two or more variables so that it used approximations like those above rather than requiring the user to provide a suitable routine to calculate analytically the appropriate derivatives. This enabled the method to be used for problems such as potential energy minimization of molecular structures where the energy is a very complicated function of the atomic coordinates, and its derivatives with respect to these coordinates would be more complicated still.

Also required in this type of calculation is a method for estimating the minimum of a function with respect to one variable. This may be accomplished very effectively by a quadratic interpolation procedure (Chapter 2). The heart of the method is the divided difference polynomial

Table 3.2 — Selection of formulae used for evaluating first and second derivatives at a particular point, given data values at several neighbouring points (equispaced with interval h)

Formula	Error order	
First derivative:		
$u_0' = (1/2h)(-3u_0 + 4u_1 - u_2)$	h^3	
$u_1' = (1/2h)(-u_0 + u_2)$	h^3	central difference (cf. (1.2(b))
$u_0' = (1/6h)(-11u_0 + 18u_1 - 9u_2 + 2u_3)$	h^4	based on cubic through four points
$u_1' = (1/6h)(-2u_0 - 3u_1 + 6u_2 - u_3)$	h^4	
$u_0' = (1/12h)(-25u_0 + 48u_1 - 36u_2 + 16u_3 - 3u_4)$	h^5	
$u_1' = (1/12h)(-3u_0 - 10u_1 + 18u_2 - 6u_3 + u_4)$	h^5	
$u_2' = (1/12h)(u_0 - 8u_1 + 8u_3 - u_4)$	h^5	higher order central difference
Second derivative:		
$u_{0,1,2}'' = (1/h^2)(u_0 - 2u_1 + u_2)$	h^2	based on quadratic, the second derivative of which is constant
$u_0'' = (1/h^2)(2u_0 - 5u_1 + 4u_2 - u_3)$	h^3	
$u_2'' = (1/12h^2)(-u_0 + 16u_1 - 30u_2 + 16u_3 - u_4)$	h^5	

$$u = u_0 + \Delta u_0(x - x_0) + \Delta u_{01}(x - x_0)(x - x_1)$$

which may be differentiated to give

$$u' = \Delta u_0 + \Delta u_{01}(2x - (x_0 + x_1))$$

Thus x_m, the value of x at the turning point where $u' = 0$ is given by

$$x_m = (1/2)((x_0 + x_1) - \Delta u_0/\Delta u_{01}) \tag{3.20}$$

A check on the sign of Δu_{01}, which is proportional to the second derivative, tells us whether it is a minimum or maximum turning point. The method is discussed further in Chapter 6.

In such problems as that of estimating the minimum energy distance between diatomic centres from MO calculations, however (3.20) may be used directly given the energy values at different intermolecular spacings — preferably distributed about the minimum position. Table 3.3 shows the energy of the CO molecule for

Table 3.3—An application of numerical differentiation. Some calculated values of total energy are shown for different interatomic distances in the CO molecule (D. W. Clack, personal communication). The equilibrium distance corresponds to the minimum in the total energy, where the derivative of the energy with respect to distance equals zero

Interatomic distance (Å)	Total energy (AU)
1.150	−23.94450
1.200	−23.95515
1.250	−23.94297

Equilibrium distance = 1.198 Å.
Minimum energy = −23.95516 AU.

several different values of interatomic distance together with the equilibrium distance r_m, computed as above. Given the value of the second derivative of the potential energy function with respect to r at r_m, and the atomic masses, we could also compute the expected vibrational frequency for comparison with the experimental (IR) results.

Numerical differentiation may also be employed to detect hidden fine structure and resolve overlap in composite spectral profiles. However, for such real (i.e. noisy) signals, as opposed to the idealized world of mathematical functions, we are best advised to employ the least squares technique of Savitzky and Golay (1964). The implementation of this approach proves identical with that of least squares smoothing (Savitzky and Golay, 1964) which is deferred until the relevant chapter (Chapter 5).

3.2.3 Multivariable differentiation

Partial differential equations abound in physical chemistry, chemical engineering and physics (for example, Schroedinger's equation), and partial derivatives are also increasingly employed in image processing. We thus need a technique for approximating such quantities as $\partial^2 u/\partial x^2$, $\partial^2 u/\partial x \partial y$ and $\partial^2 u/\partial x^2 + \partial^2 u/\partial y^2 + \partial^2 u/\partial z^2$. Following section 3.2.1 all we need to do is differentiate the interpolation polynomial of Chapter 2.

We assume a rectangular array of function values (Fig. 3.13), with grid spacings h and k, and seek the partial derivatives at (x,y). We firstly interpolate using a quadratic along x_{-1}, x_0, and x_1, giving the three points marked X (Fig. 3.13); we can then interpolate at (x,y) using these points. The resulting biquadratic interpolation polynomial is

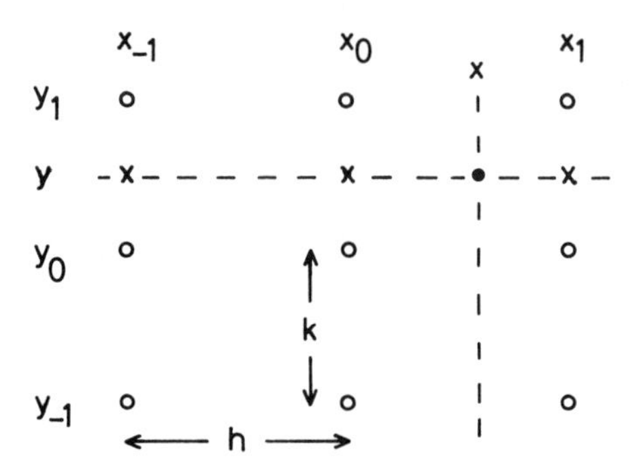

Fig. 3.13 — Grid arrangement used to derive formulae for multivariate differentiation, that is the evaluation of partial derivatives such as $\partial^2 u(x,y)/\partial x\, \partial y$.

$$u_{xy} = u_{x_{-1}} + (1/h)(u_{x_0} - u_{x_{-1}})(y - y_{-1}) + (1/2h^2)(u_{x_1} - 2u_{x_0} - u_{x_{-1}})(y - y_{-1})(y - y_0)$$

where $u_{x_{-1}}$, u_{x_0}, and u_{x_1} are the interpolated function values at the points marked X (Fig. 3.13). Differentiating with respect to y

$$\partial u/\partial y = (1/h)(u_{x_0} - u_{x_{-1}}) + (1/2h^2)(u_{x_1} - 2u_{x_0} - u_{x_{-1}})(2y - y_{-1} - y_0)$$

and again

$$\partial^2 u/\partial y^2 = (1/h^2)(u_x - 2u_{x_0} + u_{x_{-1}})$$

Note that $u_{x_{-1}}$, u_{x_0}, and u_{x_1} depend only on x and are therefore constant in this differentiation. At $x = x_0$ and $y = y_o$

$$(\partial u/\partial y)_{x_0,y_0} = (1/2h)(u_{1,0} - u_{-1,0})$$

Similarly,

$$(\partial u/\partial x)_{x_0,y_0} = (1/2k)(u_{o,1} - u_{0,-1})$$
$$(\partial^2 u/\partial y^2)_{x_0,y_0} = (1/h^2)(u_{1,0} - 2u_{0,0} + u_{-1,0})$$
$$(\partial^2 u/\partial x^2)_{x_0,y_0} = (1/k^2)(u_{0,1} - 2u_{0,0} + u_{0,1})$$
$$(\partial^2 u/\partial x \partial y)_{x_0,y_0} = (1/4hk)(u_{1,1} + u_{-1,-1} - u_{1,-1} - u_{-1,1})$$

These expressions may be better symbolized as (cf. section 3.1.7)

$$(\partial u/\partial x)_{x_0,y_0} = (1/2h) \begin{bmatrix} 0 & 0 & 0 \\ -1 & 0 & -1 \\ 0 & 0 & 0 \end{bmatrix}$$

$$(\partial^2 u/\partial y^2)_{x_0,y_0} = (1/k^2) \begin{bmatrix} 0 & 1 & 0 \\ 0 & -2 & 0 \\ 0 & 1 & 0 \end{bmatrix}$$

$$(\partial^2 u/\partial x \partial y)_{x_0,y_0} = (1/4hk) \begin{bmatrix} -1 & 0 & 1 \\ 0 & 0 & 0 \\ 1 & 0 & -1 \end{bmatrix}$$

and so on. Thus by using weighted sums of grid points we replace partial differential equations by sets of linear equations, where the unknowns u make up the required solution.

It is worth discussing an application of this technique which we mentioned in passing at the beginning of this section — digital image analysis. One problem which often needs to be solved in this rapidly expanding field is the definition of edges and boundaries in images of fibres, particles and chromatography plates, etc. One approach which has been used is to compute and plot some function of the gradients along the x- and y-directions, such as $(\partial u/\partial x)^2 + (\partial u/\partial y)^2$, and look for large values of this function (Pratt, 1978).

For noisy data, one may fit a polynomial function of x and y to the image data (using the methods of Chapter 5) and use this to calculate the derivatives.

3.3 SPECIAL CASES AND SUMMARY

Integrals with one or both limits infinite are frequently encountered, for example

$$\int_0^\infty f(x)\,\mathrm{d}x = \lim_{b\to\infty} \left(\int_0^b f(x)\,\mathrm{d}x \right)$$

It is possible to substitute a new variable in order to compress the infinite interval into a finite one, such as $0 \to 1$. However, this often makes the function itself infinite at some point, leaving us just as badly off. Special methods such as Gauss–Legendre or Gauss–Hermite integration can be used for

$$\int_0^\infty \exp(-x)f(x)\,\mathrm{d}x$$

and we refer the reader to Krylov (1962). Otherwise we use a conventional method to repeatedly evaluate

$$\int_0^b f(x)\,\mathrm{d}x$$

whilst increasing b in stages, so as to extrapolate to a limiting value. Care must be taken to ensure that round-off error does not truncate the calculation prematurely (Chapter 1). Singularities in the integrand $f(x)$ must be checked for, and special methods exist for dealing with these but are beyond the scope of this chapter.

We discussed the Monte Carlo method for evaluation of definite integrals in Chapter 1, since, on the surface, it is a very different approach. For multiple integrals over awkwardly shaped regions, this may be the only workable approach.

In summary, numerical differentiation is a more hazardous procedure than numerical integration, especially when the data is noisy. The least squares approach (Savitzky and Golay, 1964) is better adapted to this case. Numerical integration can often be circumvented by the use of tables, or series expansion such as the power series for

$$\int_{-\infty}^{\infty} (\exp\ -x^2)\,\mathrm{d}x.$$

The vast majority of integrals are not expressible easily, if at all, in a neat analytical form and Simpson's Rule and its relatives offer a good solution to the problem. Numerical differentiation and integration provide a means of approximating the solutions of differential equations, and quickly repay the most modest attention.

```
   0 REM  P3.1
  10 :
  20 REM  Composite (repeated) Simpson's Rule for approximating an integral
  30 GOTO 10000
  40 :
 100 REM function subroutine
 110 U=4 - X*X
 120 IF U<0 THEN U=0 : REM avoid square roots of negative numbers
 130 U=SQR(U)
 140 RETURN
 150 :
5000 REM  Simpson's Rule
5010 AR= U1 + 4*U2 + U3 : REM save operations - multiply by H/3 later
5020 RETURN
5030 :
5100 REM start integration
5110 AO=1E38 : REM set "impossible" value so accuracy check fails first time
5120 AN=0 : H=(XR - XL)/NP : REM initialize area accumulator, reset step size
5130 X=XL : GOSUB 100 : U3=U : REM will be value of first ordinate
5140 :
5200 FOR J=0 TO NP-2 STEP 2 : REM split up interval into pairs of panels
5210         U1=U3 : REM remember old value to save a function calculation
5220         X=XL + (J + 1)*H : GOSUB 100 : U2=U
5230         X=XL + (J + 2)*H : GOSUB 100 : U3=U
5240         GOSUB 5000 : REM apply Simpson's Rule
5250         AN=AN + AR : REM accumulate area
5260 NEXT J
5270 :
5280 AN=AN*H/3 : REM bring H/3 "outside the brackets" of whole calculation
5290 :
5300 ER=ABS(AN - AO)/15 : REM estimate error by comparing successive results
5310 :
5320 IF ER<AC THEN RETURN : REM accuracy check - fails first time
5330 IF NP>2^(MH-1) THEN RETURN : REM too many halvings?
5340 :
5350 NP=2*NP : H=(XR - XL)/NP : REM double number of panels and halve step
5360 :
5370 AO=AN : GOTO 5120
5380 :
```

```
10000 REM  main segment
10010 INPUT "LOWER LIMIT OF INTEGRATION";XL
10020 INPUT "UPPER LIMIT OF INTEGRATION";XR
10030 INPUT "ABSOLUTE ACCURACY REQUIRED";AC
10040 :
10050 NP=2 : MH=10 : REM start with two panels and set max number of halvings
10060 :
10070 GOSUB 5100 : REM do integration
10080 :
10090 IF NP>2^(MH-1) THEN PRINT "TOO MANY HALVINGS - ACCURACY NOT ACHIEVED"
10100 PRINT "VALUE OF INTEGRAL IS";AN;"+/-";ER
10110 END
```

```
    0 REM P3.2
   10 :
   20 REM  Adaptive Simpson's Rule
   30 DIM AL(35),U(35),X(35)
   40 GOTO 10000
  100 REM subroutine to calculate function to be integrated
  110 U=4 - X*X
  120 IF U<0 THEN U=0 : REM avoid square roots of negative numbers
  130 U=SQR(U)
  140 RETURN
  150 :
 5100 REM  Simpson's Rule with interval halving
 5110 ES=0 : IL=0 : S=0 : REM initialize total error estimate, integral and S
 5120 X(1)=XL : X(5)=XR : REM set end points
 5130 X=X(1) : GOSUB 100 : U(1)=U
 5140 X=X(5) : GOSUB 100 : U(5)=U
 5150 :
 5210 X=( X(5) + X(1) )/2 : GOSUB 100 : U(3)=U : X(3)=X
 5220 A=( X(5) - X(1) )*( U(1) + 4*U(3) + U(5) )/6 : REM H is 2*( X(5) - X(1) )
 5230 :
 5240 H=( X(S+5) - X(S+1) )/4 : REM return to here unless finished
```

```
5250 X=X(S+1) + H : GOSUB 100 : U(S+2)=U : X(S+2)=X : REM calc. new
5260 X=X(S+5) - H : GOSUB 100 : U(S+4)=U : X(S+4)=X : REM intermediate values
5270 AL(S/2)=H*( U(S+1) + 4*U(S+2) + U(S+3) )/3 : REM store left hand area
5280 AR=H*( U(S+3) + 4*U(S+4) + U(S+5) )/3 : REM calculate right hand area
5290 AA=AL(S/2) + AR : REM  calculate better area estimate with half step size
5300 ER=(A - AA)/15 : REM error est. if 4th deriv. U is const. in subinterval
5310 IF ABS(ER)<TE*( X(5) - X(1) )/(XR - XL) THEN 5340 : REM accept interval
5320 S=S+2 : IF S>2*LM THEN PRINT "LEVEL LIMIT REACHED" : END
5330 A=AR : GOTO 5380
5340 IL=IL + AA : ES=ES + ER : REM update integral value and error estimate
5350 IF S=0 THEN RETURN : REM calc. completed or H too small
5360 S=S-2 : A=AL(S/2) : REM try larger subinterval
5370 :
5380 U(S+5)=U(S+3) : U(S+3)=U(S+2) : REM renumber values
5390 X(S+5)=X(S+3) : X(S+3)=X(S+2)
5400 GOTO 5240
5410 :
10000 REM main segment
10010 INPUT "LOWER LIMIT":XL
10020 INPUT "UPPER LIMIT":XR
10030 INPUT "ABSOLUTE ACCURACY REQUIRED":TE
10040 INPUT "MAXIMUM NUMBER OF LEVELS":LM : REM don't allow subdivision
10050 REM below round-off error i.e. H must have meaningful value
10060 :
10070 GOSUB 5100 : REM do integration
10080 :
10090 PRINT "VALUE OF INTEGRAL = ";IL;", ESTIMATED ERROR = ";ES
10100 END
```

```
   0 REM P3.3
  10 :
  20 REM  Spline integration
  30 :
  40 DIM A(512),B(512),C(512),R(512),U(512),U2(512)
  50 GOTO 10000
  60 :
 100 REM subroutine to evaluate function to be integrated
 120 U=4 - X*X
 130 IF U<0 THEN U=0 : REM avoid square roots of negative numbers
 140 U=SQR(U)
 150 RETURN
 160 :
4000 REM  set up 3 diagonals in tridiagonal matrix in arrays A(),B() and C()
4010 FOR I=1 TO N-2 : A(I)=H : C(I)=H : NEXT I : REM equidistant points
4020 FOR I=1 TO N-1 : B(I)=4*H : NEXT I
4030 :
4040 REM  set up array R()
4050 FOR I=1 TO N-1
4060          R(I)=6*( U(I+1) - 2*U(I) + U(I-1) )/H
4070 NEXT I
4080 :
4090 REM  calculate second derivative array U2()
4100 U2(0)=0 : U2(N)=0 : REM these are the "natural spline" end conditions
4110 :
4120 REM  lower diagonal in A(), main diagonal in B(), upper diagonal in C()
4130 REM  Gaussian elimination modified for tridiagonal matrix equation
4140 REM  no pivoting - X-values not coincident
4150 FOR I=N-1 TO 2 STEP -1 : REM working up from bottom right hand corner
4160          Z=C(I-1)/B(I)
4170          B(I-1)=B(I-1) - Z*A(I-1)
4180          R(I-1)=R(I-1) - Z*R(I) : REM do same to RHS of equation
4190 NEXT I
4200 :
4210 REM  back-substitution
4220 B(I)=R(1)/B(I)
4230 FOR I=2 TO N-1
4240          B(I)=( R(I) - A(I-1)*B(I-1) )/B(I)
4250 NEXT I
```

```
 4260 FOR I=1 TO N-1 : U2(I)=B(I) : NEXT I
 4270 RETURN
 4280 :
 5000 REM calculate integral from function and second derivative values
 5010 AR=0 : H3=H*H*H
 5020 FOR I=0 TO N-1
 5030         AR=AR + H*( U(I) + U(I+1) )/2 - H3*( U2(I) + U2(I+1) )/24
 5040 NEXT I
 5050 RETURN
 5060 :
10000 REM  main segment
10010 INPUT "LOWER INTEGRATION LIMIT":XL
10020 INPUT "UPPER INTEGRATION LIMIT":XR
10030 INPUT "ABSOLUTE ACCURACY REQUIRED":AC
10040 N=4 : ST=0 : AO=1E38 : REM start on 4 panels, acc. test fails first time
10050 H=(XR - XL)/N
10060 FOR I=ST TO N-ST STEP ST+1 : X=XL + I*H : GOSUB 100 : U(I)=U : NEXT I
10070 REM get U values
10080 :
10090 GOSUB 4000 : REM set up and solve spline equations
10100 :
10110 GOSUB 5000
10120 IF ABS(AO - AR)<AC THEN 10170
10130 IF N=512 THEN PRINT "TOO MANY PANELS" : GOTO 10170
10140 FOR I=N TO 1 STEP -1 : U(2*I)=U(I) : NEXT I : REM save U() for even I
10150 N=2*N : ST=1 : AO=AR: GOTO 10050
10160 :
10170 PRINT "VALUE OF INTEGRAL IS":AR
10180 PRINT "INTEGRATION WAS BASED ON",N,"PANELS"
10190 END
```

```
   0 REM  P3.4
  10 :
  20 REM 2-dimensional Simpson integration over rectangle
  30 REM can be made into an N-dimensional routine by inputting N instead
  40 REM of setting N=2 in the program
  50 REM In this program, what was a subroutine to do the 1-dimensional
  60 REM composite rule, now has to "call itself" because the 2-D composite
  70 REM rule can be thought of as the application of a 1-D rule in the first
  80 REM variable each ordinate of which results from a 1-D rule applied to
  90 REM the second variable. Most BASIC's do not allow a subroutine to call
 100 REM itself (recursion) and this program gets round this by saving the
 110 REM results when the code is used for the first variable before using
 120 REM the same code for the second variable.
 130 :
 140 GOTO 10000
 150 :
 600 REM function to be integrated
 610 REM if it is to be integrated over a boundary within the rectangle
 620 REM you must set U=0 outside the boundary and allow routine to cope
 630 REM tbe resulting discontinuity
 640 IF D<N+1 THEN 5100 : REM get result from another composite rule
 650 REM unless X(1) and X(2) have both been specified
 660 U=LOG( X(1) + 2*X(2) ) : REM example from Burden et al. (1981) p 75
 670 RETURN
 680 :
5000 REM  Simpson's rule in X variable
5010 U=U1(D) + 4*U2(D) + U3(D) : REM factor H()/3 left till ordinates summed
5020 RETURN
5030 :
5100 REM composite Simpson's rule in X(D) variable
5120 UN(D)=0 : H(D)=( XR(D) - XL(D) )/NP : REM set step size
5130 X(D)=XL(D) : D=D+1 : GOSUB 600 : D=D-1 : REM go to next variable
5140 U3(D)=U : REM first ordinate
5150 :
5200 FOR J=0 TO NP-2 STEP 2 : REM split up interval into pairs of panels
5210         J(D)=J
5220         U1(D)=U3(D) : REM remember old value - save a function calc.
5230         X(D)=XL(D) + (J(D) + 1)*H(D) : D=D+1 : GOSUB 600 : D=D-1
5240         U2(D)=U : REM each variable, X(D) has its own set of stores
```

```
 5250        X(D)=XL(D) + (J(D) + 2)*H(D) : D=D+1 : GOSUB 600 : D=D-1
 5260        U3(D)=U : REM otherwise values for D=2 overwrite those for D=1
 5270        GOSUB 5000 : REM apply Simpson's Rule
 5280        UN(D)=UN(D)+U
 5290        J=J(D)
 5300 NEXT J
 5310 :
 5320 U=UN(D) : IF D>1 THEN 670
 5330 REM if D=1 we have a composite rule of composite rules
 5340 GOTO 10110
 5350 :
10000 REM  main segment
10010 N=2
10020 FOR I=1 TO N
10030        PRINT "LOWER AND UPPER INTEGRATION LIMITS IN VARIABLE";I:
10040        INPUT "";XL(I),XR(I)
10050 NEXT I
10060 INPUT "ABSOLUTE ACCURACY REQUIRED":AC
10070 :
10080 NP=2 : MH=5 : REM start with 2x2 panels, set max number of halvings
10090 UO=1E38 : REM set "impossible" value so accuracy test fails first time
10100 D=1 : GOTO 5100 : REM set first variable and do integration
10110 FOR I=1 TO N : U=U*H(I)/3 : NEXT I :PRINT "U";U
10120 ER=(UO - U)/15 : REM estimate error by comparing successive results
10130 AE=ABS(ER)
10140 IF AE<AC THEN 10210 : REM accuracy check - fails first time
10150 IF NP>2^(MH-1) THEN PRINT "STEP SIZE TOO SMALL" : END : REM too many halvings?
10160 :
10170 NP=2*NP : REM double number of panels
10180 :
10190 UO=U : GOTO 10100
10200 :
10210 PRINT "VALUE OF INTEGRAL IS";U;"WITH ESTIMATED ERROR";ER
10220 END
```

```
  0 REM P3.5
 10 :
 20 REM  Adaptive Simpson's Rule applied to calculation of EPR spectrum
 30 DIM AL(35,1),U(35,1),X(35,1),Z(250)
 40 PI=4*ATN(1) : BM=9.2731E-21 : HP=6.6256E-27
 50 REM constants for resonant field calculation
 60 GOTO 10000
 70 :
400 REM plot the results
410 MX=0 : MI=0 : REM set max. and min. values, ensures T-axis on screen
420 FOR I=0 TO NS
430          IF Z(I)>MX THEN MX=Z(I) : REM find max. and min. values
440          IF Z(I)<MI THEN MI=Z(I) : REM this is why an array is needed
450 NEXT I
460 IF ABS(MX - MI)<1E-30 THEN SZ=0 : GOTO 490
470 :
480 SZ=150/(MX-MI) : REM scaling factors for plotting
490 ST=INT(279/NS)
500 HGR : HCOLOR=3
510 FOR I=0 TO NS
520          Y=( Z(I) - MI )*SZ
530          HPLOT I*ST,159 - Y : REM origin at top LHS of screen
540 NEXT I
550 HPLOT 0,159 + MI*SZ TO NS*ST,159 + MI*SZ : REM plot field axis
560 RETURN
570 :
600 REM subroutine to calculate function to be integrated
610 IF D<2 THEN 5100
620 REM XX(0),XX(1) are the polar angles theta, phi
630 SX=SIN( XX(0) ) : SY=SIN( XX(1) ) : CX=COS( XX(0) ) : CY=COS( XX(1) )
640 GA=G1*G1 : GB=G2*G2 : GC=G3*G3 : S1=SX*SX : SC=CX*CX : S2=SY*SY : C2=CY*CY
650 SA=S1*C2 : SB=S1*S2 : GG=GA*SA + GB*SB + GC*SC : G=SQR(GG)
660 T1=2*( (HH - HP*F/(G*BM) )/W )^2 : REM HP*F/(G*BM) is resonant field
670 IF T1>20 THEN U=0 : GOTO 740 : REM avoid unnecessary calculation
680 T1=EXP( -T1 ) : REM apply Gaussian lineshape
685 REM the pre-exponential factor is constant for all angles in  this case
686 REM unless W varies with angle, we could apply the constant after calc.
690 T2=GA + GB + GC
700 T3=GA*GA*SA : T4=GB*GB*SB : T5=GC*GC*SC
```

```
 710 U=T1*( T2 - (T3 + T4 + T5)/GG )*SX : REM mult. by transition probability
 720 REM  SX is weighting factor - the area element of a sphere in polar
 730 REM  coordinates varies with SIN(theta) and tends to 0 (at pole)
 740 RETURN
 750 :
5100 REM  Adaptive integration by Simpson's Rule with interval halving
5102 REM  compare this code with P3.2 - by using arrays with subscript D we
5104 REM  can use the same code for both variables - this amounts to recursion
5106 REM  we remember the current variable values before "re-using" the code
5110 IL(D)=0 : S(D)=0 : REM initialize integral and S(D)
5120 X(1,D)=XL(D) : X(5,D)=XR(D) : REM set end points
5130 XX(D)=X(1,D) : D=D+1 : GOSUB 600 : D=D-1 : U(1,D)=U
5140 XX(D)=X(5,D) : D=D+1 : GOSUB 600 : D=D-1 : U(5,D)=U
5150 :
5210 XX(D)=( X(5,D) + X(1,D) )/2 : D=D+1 : GOSUB 600 : D=D-1
5220 U(3,D)=U : X(3,D)=XX(D)
5230 A(D)=( X(5,D) - X(1,D) )*( U(1,D) + 4*U(3,D) + U(5,D) )/6
5240 :
5250 H(D)=( X(S(D)+5,D) - X(S(D)+1,D) )/4 : REM return to here unless finished
5260 XX(D)=X(S(D)+1,D) + H(D) : D=D+1 : GOSUB 600 : D=D-1
5270 U(S(D)+2,D)=U : X(S(D)+2,D)=XX(D)
5280 XX(D)=X(S(D)+5,D) - H(D) : D=D+1 : GOSUB 600 : D=D-1
5290 U(S(D)+4,D)=U : X(S(D)+4,D)=XX(D)
5300 AL(S(D)/2,D)=H(D)*( U(S(D)+1,D) + 4*U(S(D)+2,D) + U(S(D)+3,D) )/3
5310 AR(D)=H(D)*( U(S(D)+3,D) + 4*U(S(D)+4,D) + U(S(D)+5,D) )/3
5320 AA(D)=AL(S(D)/2,D) + AR(D)
5330 ER(D)=( A(D) - AA(D) )/15 : IF H(D)>( XR(D) - XL(D) )/8 THEN ER(D)=1E10
5335 REM set a lower limit on H(D) to avoid missing peaks in the function
5340 IF ABS( ER(D) )<4*TE*H(D)/( XR(D) - XL(D) ) THEN 5370
5350 S(D)=S(D)+2 : IF S(D)>2*LM THEN PRINT "TOO MANY SUBDIVISIONS" : END
5360 A(D)=AR(D) : GOTO 5410
5370 IL(D)=IL(D) + AA(D) : ES(D)=ES(D) + ER(D)
5380 IF S(D)=0 THEN 5450
5390 S(D)=S(D)-2 : A(D)=AL(S(D)/2,D)
5400 :
5410 U(S(D)+5,D)=U(S(D)+3,D) : U(S(D)+3,D)=U(S(D)+2,D)
5420 X(S(D)+5,D)=X(S(D)+3,D) : X(S(D)+3,D)=X(S(D)+2,D)
5430 GOTO 5250
5440 :
```

```
 5450 U=IL(D) : IF D>0 THEN 740
 5460 IL=U : GOTO 10150
 5470 :
10000 REM main segment
10010 XL(0)=0 : XR(0)=PI/2 : REM limits on first angle variable
10020 XL(1)=0 : XR(1)=PI/2 : REM limits on second angle variable
10030 INPUT "ABSOLUTE ACCURACY REQUIRED":TE
10040 INPUT "MAX. NUMBER OF SUBDIVISIONS = ";LM
10050 REM below round-off error i.e. H must have meaningful value
10060 INPUT "MICROWAVE FREQUENCY IN HZ";F
10070 INPUT "G-VALUES - G1,G2 AND G3";G1,G2,G3
10080 INPUT "FIELD STEP IN GAUSS = ";ST
10090 INPUT "LINEWIDTH = ";W
10100 INPUT "STARTING AND FINISHING FIELDS IN GAUSS";H1,H2
10110 :
10120 NS=0
10130 FOR HH=H1 TO H2 STEP ST
10140         D=0 : FL=0 : GOTO 5100
10150         Z(NS)=IL
10160         PRINT "FIELD = ";HH:" INTENSITY = ";IL
10170         NS=NS + 1
10180 NEXT HH
10190 :
10200 GOSUB 400 : REM plot
10210 END
```

4

Solving differential equations

4.1 INTRODUCTION

In science we often encounter the problem of needing to predict what will happen when a physical, chemical or biological system is 'set up' in some well-defined state and then 'released' to follow some course where variables describing that system rise and fall with time. In chemistry such problems occur very frequently in the study of chemical kinetics where we mix reactants and monitor their decay and the growth of the products with time. Let us start with a simple example from first order kinetics

$$\mathrm{d}u/\mathrm{d}t = -ku \tag{4.1}$$

which represents how the rate of change with time, t, of the molarity, u, of a component varies with the concentration of that component. If (4.1) were an algebraic equation such as $u^2 = -ku$ then we could solve it exactly and directly by rearrangement to give us, in this case, two values of u as the solutions. We could also solve the equation graphically as the points of intersection of two curves (in this case $y = u^2$ and $y = -ku$), and this kind of representation often gives an insight into the nature of the problem. Similarly, a graphical treatment of (4.1) may help us to gain a direct understanding of the nature of a whole class of equations such as this.

4.2 TANGENT FIELD DIAGRAMS

Imagine a board covered with fur. If we could comb the fur very precisely, we could make the direction of each hair represent the value of the slope $\mathrm{d}u/\mathrm{d}t$ (where u and t now refer to the vertical and horizontal variables on the board) at the point to which it is attached to the board. An illustration of this is shown in Fig. 4.1 for equation (4.1) with $k = 1$. If we start at a point such as $t = 0$, $u = 1$ and increase t by a very small amount, then the tangent line segment at (0,1), which has a slope $\mathrm{d}u/\mathrm{d}t = -1$, provides the direction in which we are to move on the plane. Since we have changed our position, i.e. values of u and t, we are now at a new point where the

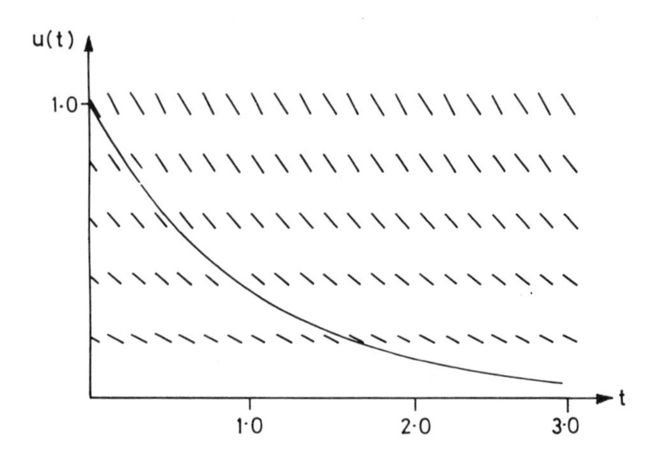

Fig. 4.1 — Tangent field diagram for the differential equation $\mathrm{d}u/\mathrm{d}t = -u$. Each small line segment represents the local slope (tangent). The solid line shows the path traced if we start at the point $(0,1)$ and always move in the direction of the local tangent. In other words it is the solution $u(t)$ of $\mathrm{d}u/\mathrm{d}t = -u$, with the initial condition $u(0) = 1$.

corresponding tangent segment has a slightly different direction. Repeating our moves, we proceed from point to point in the plane changing direction according to the tangent at each point. The process is rather like walking on a constantly changing compass bearing. The line in Fig. 4.1 indicates the path traced out from the point (0,1). Note that it is a curve, not a sequence of jointed line segments, because we have assumed that we are moving, in the spirit of calculus, from point to point by infinitesimal steps. This curve is a solution of $\mathrm{d}u/\mathrm{d}t = -u$. Note the use of the words 'a solution' in the previous sentence — a differential equation generally has an infinity of solutions or maybe no solution at all. We could have chosen a different starting point and obtained a different curve on the u,t plane. In this type of problem, the starting point is known as the initial condition and it is an 'initial value problem'. We shall see that more initial conditions need to be specified if higher derivatives of u are involved.

Since the tangent field diagram is such a useful visual aid, program P4.1 is included for computing and displaying them for arbitrary equations of the form

$$\mathrm{d}u/\mathrm{d}t = f(t,u) \tag{4.2}$$

where $f(t,u)$ is some function of t and u. (Note that the graphics display routine in P4.1 will have to be modified to work on microcomputers other than the Apple II series.) This is an example of an ordinary differential equation of the first order — ordinary as opposed to partial or other type of differential equation, and first order because the highest order of derivative involved is one. Fig. 4.2 shows tangent field plots for two other differential equations of the same form as (4.2). In Fig. 4.2(a) is shown the plot for

$$\mathrm{d}u/\mathrm{d}t = \dot{u} = u' = -ut$$

($\dot{u}$, $\ddot{u}$ etc. represent derivatives of u with respect to time) and the solid line

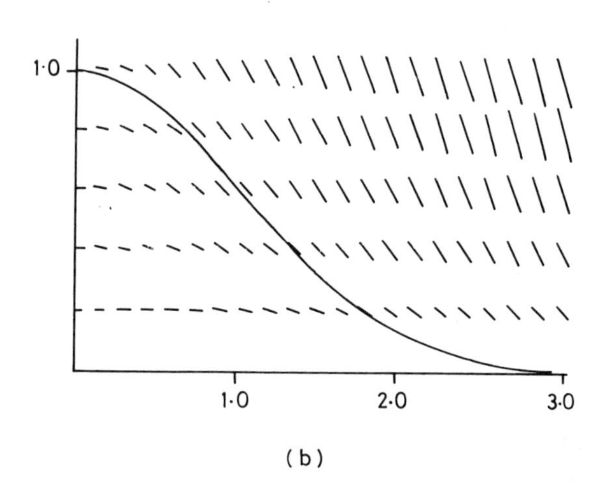

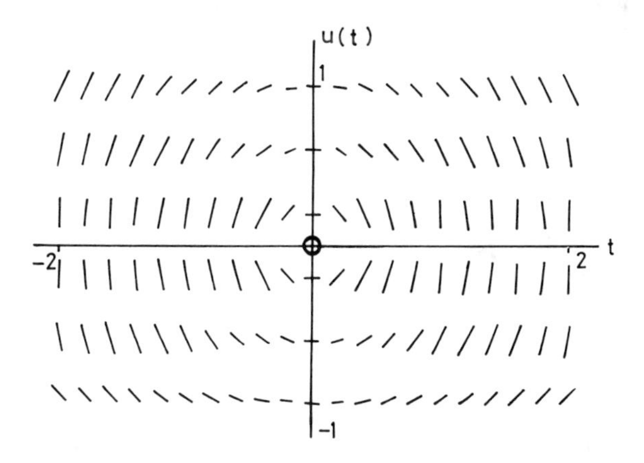

Fig. 4.2 — Tangent field plots for two examples of $\mathrm{d}u/\mathrm{d}t = f(t,u)$: (a) $\mathrm{d}u/\mathrm{d}t = -ut$. The solid line is a typical solution, which is a gaussian lineshape. (b) $\mathrm{d}u/\mathrm{d}t = -t/u$. Solutions are circles centred about the origin (0.0), which is a singular point (see text).

represents a typical solution which is of the gaussian form $u = A\exp(-t^2/2)$ where A is some constant which depends on the initial conditions, e.g. if the initial condition is that $u = 1$ at $t = 0$ then $A = 1$ and the solution is $u = \exp(-t^2/2)$. That this is a solution of $u' = -ut$ should be verified by the reader by differentiation and substitution. Another example is given in Fig. 4.2(b) for the equation $u' = -t/u$ the solutions of which are circles about the origin $t = 0$, $u = 0$. There is no tangent at this point, which is said to be a singular point. The solutions are also singular at all points along the line $u = 0$ (the t-axis) since u' is infinite — a vertical tangent — along this line, except of course at $t = 0$.

4.3 CALCULATING A SOLUTION

4.3.1 Analytical approach

Given an equation of the form

$$u' = \mathrm{d}u/\mathrm{d}t = f(t)$$

where the right-hand side depends only on t, we can rewrite it as

$$\mathrm{d}u = f(t)\,\mathrm{d}t$$

and integrate each side directly. For example, if $f(t) = t^2$ and given the initial condition $u = 1$ at $t = 0$ then

$$\int_1^u \mathrm{d}u = \int_0^t t^2\,\mathrm{d}t$$

and hence $u - 1 = t^3/3$ or $u = t^3/3 + 1$. In these differential equations the problem is really one of 'simple' integration. Similar approaches lead to the solution of equations of the type $\mathrm{d}u/\mathrm{d}t = f(u)$ which can be arranged as $\mathrm{d}u/f(u) = \mathrm{d}t$ and integrated accordingly. The simple kinetic example with which we started is one of these, $\mathrm{d}u/\mathrm{d}t = -u$. Thus $\mathrm{d}t = -\mathrm{d}u/u$ and

$$\int_{t_0}^t \mathrm{d}t = \int_{u_0}^u -\mathrm{d}u/u$$

where the initial condition is $u = u_0$ at $t = t_0$. Hence

$$\begin{aligned} t - t_0 &= [-\ln(u)]_{u_0}^{u} \\ &= \ln(u_0/u) \end{aligned}$$

Thus

$$\exp(t - t_0) = u_0/u$$

giving finally $u = u_0 \exp(-(t - t_0))$. This indeed shows the same behaviour as our dotted solution curve in Fig. 4.1, which is in fact $u = \exp(-t)$.

It is worth noting that many physical systems obey quite simple rate of change laws such as $u' = -u$, and ubiquitous functions such as $u = \exp(-t)$ have arisen naturally out of studies of such classical systems as those of chemical reactions, damped motion and radioactive decay.

Any differential equation which can be re-stated in the form $f(u)\mathrm{d}u = g(t)\mathrm{d}t$ can be similarly treated. We know from bitter experience that analytical integration is rarely an easy process, if it can be performed at all. When the equation takes the form $u' = f(t,u)$ straightforward methods like this are doomed to failure except in special cases. In response to this a whole range of special approximate methods, both analytical and numerical, have been devised.

4.3.2 Taylor series solution

Let us start with a conceptually simple technique which works for most ordinary differential equations, but may be very difficult to implement in practice. Nevertheless, it will serve as a standard method to which others may be referred.

We recall (Chapters 1 and 2) that many functions may be well represented by a polynomial and that in the limit that the interval between the points at which the function is defined approaches zero, the coefficients of the polynomial become derivatives rather than divided differences (Chapter 2). In the nomenclature of the previous examples, we may write the resulting Taylor series

$$u = u_0 + u_0^{(1)}(t-t_0) + u_0^{(2)}(t-t_0)^2/2! + u_0^{(3)}(t-t_0)^3/3! + \ldots + u^{(n)}(t-t_0)^n/n! \quad (4.3)$$

where $u_0^{(n)}$ is the nth derivative of u with respect to t, evaluated at (t_0, u_0).

We will demonstrate the usefulness of this series by obtaining an approximate solution for the equation $u' = -u$.

We already know u' and clearly $u'' = -u'$. Similarly, u''' and higher derivatives are easily found. Taking the initial condition to be $u = 1$ at $t = 0$ and substituting the expressions for u', u'', etc., in (4.3) we find

$$u = 1 - t + t^2/2! - t^3/3! + \ldots + (-t)^n/n! + \ldots$$

which is well-known as the infinite power series representation of $\exp(-t)$. Thus u may be evaluated for any t as accurately as we like (within the rounding error of the computer), by taking enough terms. This example illustrates the simplicity of the concept of the Taylor series approach, but its execution is generally not so facile.

For example, if we wish to solve

$$u' = -u^2 \quad (4.4)$$

subject to the initial condition $u = 1$ at $t = 0$, then differentiating, and recalling the product rule, we obtain expressions for the higher derivatives

$$u^{(2)} = -2uu^{(1)} = 2u^3$$

$$u^{(3)} = -2(u^{(1)})^2 - 2uu^{(2)} = -6u^4$$

$$u^{(4)} = -6u^{(1)}u^{(2)} - 2uu^{(3)} = 24u^5, \qquad \text{etc.}$$

Successive differentiations become increasingly difficult to perform, but in this case we are fortunate that the derivatives again reduce to simple expressions when we substitute for $u^{(1)}$, $u^{(2)}$, $u^{(3)}$, etc. A simple Taylor series results

$$u = 1 - t + t^2 - t^3 + \ldots \quad (4.5)$$

which is the infinite power series for $1/(1+t)$. Equation (4.4) represents the rate equation for a second order reaction with unit rate constant and negligible back-reaction, for example, $2A \xrightarrow{k=1}$ products, where u stands for the concentration of A at time t. We could have solved this equation much more easily by separation of variables so that $-\mathrm{d}u/u^2 = \mathrm{d}t$ followed by the integration of each side. However, this analytical procedure would have failed for the related equation $u' = -u^2 + t$ whereas the successive differentiations in the Taylor series method are little harder than for equation (4.4).

To overcome the difficulty of successive differentiation the Taylor series approach may be implemented using a symbolic manipulation language which has the ability to automate analytic differentiation (Pavelle *et al.*, 1981), but this is beyond the scope of this book.

The problem of having to sum many terms of the infinite series (4.3) may be avoided by using an iterative approach. We obtain the first four terms of the Taylor series as a function of t, the independent variable. For example, given $u' = -ut$ and $u = 1$ at $t = 0$ then

$$\begin{aligned} u^{(1)} &= -ut \\ u^{(2)} &= -u - u^{(1)}t = -u + ut^2 \\ u^{(3)} &= -2u^{(1)} - u^{(2)}t = 3ut - ut^3 \end{aligned}$$

and hence

$$u \simeq u_0 + (-u_0t_0)(t-t_0) + (-u_0 + u_0t_0^2)(t-t_0)^2/2 + \\ + (3u_0t_0 - u_0t_0^3)\ (t-t_0)^3/6$$

This now approximate expression (since we have discarded terms after the fourth) is most accurate close to $t = t_0$, so we consider the value of the solution u_1 at $t = t_0 + h$ where h is the (sufficiently small) step size

$$u_1 = u_0 + (u_0 - u_0t_0)h + (-u_0 + u_0t_0^2)h^2/2 + (3u_0t_0 - u_0t_0^3)h^3/6$$

Thus we have a method for calculating the solution at successive points, using the value at the preceding point, i.e. having calculated u_1 as above, then u_0 on the right-hand side is replaced by u_1 and u_2 calculated, and so on. Fig. 4.3 illustrates this

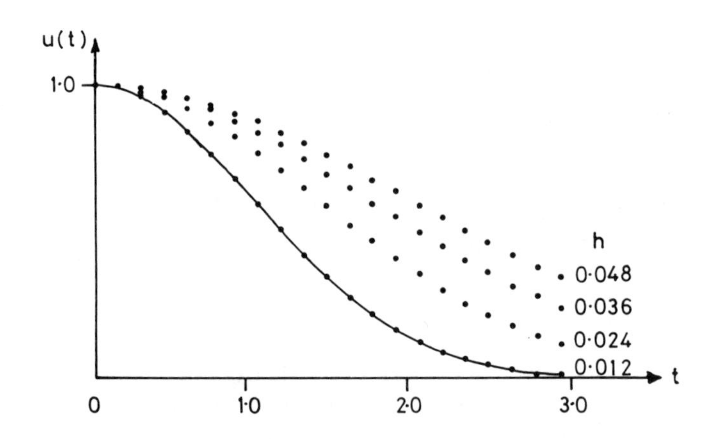

Fig. 4.3 — Taylor series solution of $du/dt = -ut$, computed using a range of step sizes h. For sufficiently small h the calculated solution is a very good approximation to the analytical solution $u = \exp(-t^2/2)$ (solid line).

procedure applied to the example above ($du/dt = -ut$) and compared with the exact solution $u = \exp(-t^2/2)$. This version of the Taylor series technique would be excellent if we did not have to perform the differentiations to obtain the Taylor series coefficients. The methods which are described hereafter can be regarded as techniques for getting round this difficulty.

4.4 EULER'S METHOD

4.4.1 Basic procedure

If we return to equation (4.3) and keep the first two terms of the series, rejecting higher terms, then we obtain $u = u_0 + u_0' (t - t_0)$ or

$$u_1 = u_0 + u_0' h \tag{4.6}$$

where u_0' is the value of $f(t_0, u_0)$ in the general equation $u' = f(t,u)$. Everything on the right-hand side of (4.6) is known and we have Euler's method for predicting the next value, u_1, of the solution at $t_0 + h$ given the previous value u_0 at t_0. Fig. 4.4 shows a

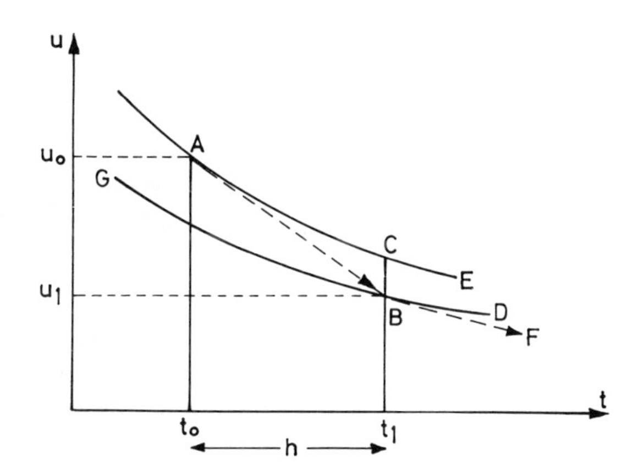

Fig. 4.4 — Geometrical interpretation of Euler's method for the solution of differential equations, showing how it jumps from one solution curve to another as the procedure steps forward in t.

geometrical interpretation of this method. Starting at (t_0, u_0) we approximate the actual solution curve ACE at A by a straight-line tangent AB, which has the equation $u = u_0 + u_0'(t - t_0)$. At $t_1 = t_0 + h$ this tangent approximation has the value $u = u_0 + u_0'h$, and is in error by an amount measured by the distance BC. Note that the point (t_1, u_1), labelled B, lies on a different solution curve, GBD, corresponding to different initial conditions. Our approximate solution may be extended along BF by reapplying the method to obtain u_2, and so on. The approximate solution 'jumps the rails' onto a new track at each step, accumulating errors with each application.

In order to reduce the error involved in truncating the Taylor series solution after the two terms of equation (4.3) it is logical to consider reducing the step size, h, since we know that the error in extrapolation usually increases drastically as we move away from the points at which the function is evaluated (Chapter 2). Fig. 4.5 shows the results of applying program P4.2 which implements this method to $u' = -ut$, starting with $u = 1$ at $t = 0$, using a wide range of step sizes; the true solution $u = \exp(-t^2/2)$ is also plotted for comparison. At large step size the behaviour is erratic, the approximate solution 'curve' oscillating wildly — a so-called unstable solution. As

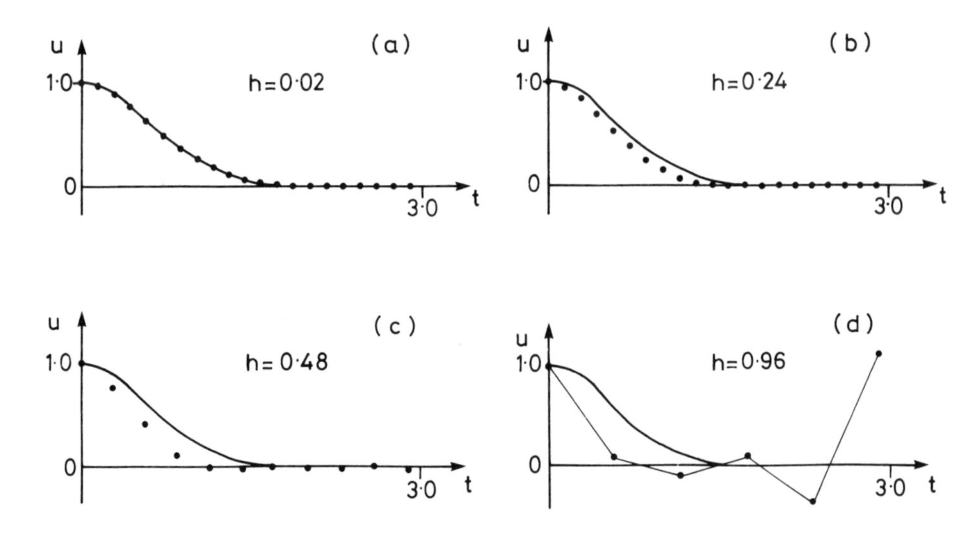

Fig. 4.5 — Results of applying Euler's method to the solution of $du/dt = -ut$, and the effect of using various step sizes h. In each case the correct solution is also shown (solid line). As h is increased, the solution becomes more and more inaccurate, and eventually behaves very erratically (curve (d)).

the step size is reduced the solution damps down to a reasonable approximation to the true solution, the accuracy of the approximation improving with decreasing step size until the approximation is almost identical with the actual curve. However, for very small step sizes we need to make so many calculations to cover the range of interest that the program running time may become unacceptable; more importantly, a different kind of error, rounding error (Chapter 1), becomes apparent because we are not keeping a sufficient number of decimal places in the calculations and are performing a vast number of such inaccurate calculations. We must thus strike a balance between conflicting interests — the step size must be small, but not too small.

Examination of a tangent field diagram for a 'solution' of $u' = -u$ with a large step size (Fig. 4.6) shows clearly why instability occurs, and also demonstrates how below a critical step size the solutions 'funnel in' towards the line $u = 0$, damping the error. If we were solving the equation $u' = u$ instead, this error would have been magnified and the resulting so-called inherent instability would have been due more to the nature of the equation than to the step size.

In general, if the solutions of an equation diverge as t increases then the equation will be unstable in that region. If the solutions converge then any error will tend to be damped out, though this is sometimes a mixed blessing, as we shall see later.

4.4.2 Improving the accuracy

We may extend Euler's method to include the next term in the Taylor series solution (4.3) giving

$$u = u_0 + u_0'h + u_0''h/2 \tag{4.7}$$

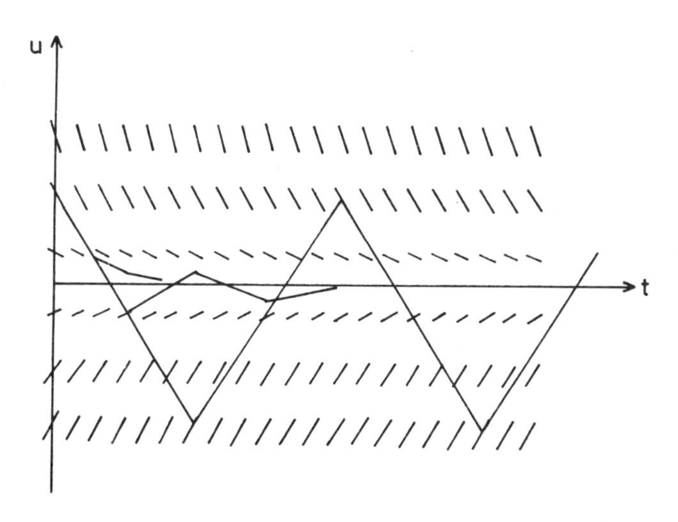

Fig. 4.6 — Tangent field diagram for $du/dt = -u$, and solution paths for different step sizes. The instability arising from using too large a step size is clearly demonstrated.

where u_0 and u'_0 are values of u and u' at $t = t_0$; the former is given by the initial conditions, and the latter calculated from $u' = f(t,u)$ by substituting for u_0 and t_0. The evaluation of u''_0, however, requires either an analytical or numerical differentiation of $f(t,u)$. We have already used the former technique and seen that for a complicated $f(t,u)$ this may be very tedious. Numerical differentiation may be performed in terms of a weighted sum of ordinate values (Chapter 3), but in the case of $u' = f(t,u)$ the required ordinate values are values of the solution we are trying to calculate, since u enters the equation implicitly.

An alternative approach is to rewrite our original differential equation as $du = f(t,u)\,dt$ and integrate between limits, yielding

$$\int_{u_0}^{u_1} du = \int_{t_0}^{t_1} f(t,u)\,dt$$

or

$$u_1 = u_0 + \int_{t_0}^{t_1} f(t,u)\,dt$$

The re-stated problem may then be attacked by the methods of Chapter 3, expressing the integral as a weighted sum of ordinates; but again our equation begs the question: for example, applying the trapezoidal rule gives $u_1 = u_0 + (h/2)(u'_0 + u'_1)$, that is

$$u_1 = u_0 + (h/2)(f(t_0,u_0) + f(t_1,u_1)) \tag{4.8a}$$

which expresses the fact that we need the value of u_1 before we have calculated it in order to evaluate $f(t_1,u_1)$ so that we can calculate u_1. We can interpret (4.8) as an improved version of Euler's method, where the slope at (t_0,u_0) is replaced by the average of the slopes at (t_0,u_0) and (t_1,u_1). Further, $f(t_1,u_1)$ may be estimated using the simple Euler method (4.6) and substituted in (4.8) — this is the basis of the first stage of Heun's method (section 4.5). Instead of averaging the slopes at the end-points we can use the slope at the average point, that is at the mid-point of the interval, to give

$$u_1 = u_0 + hf(t_{1/2},u_{1/2}) \tag{4.8b}$$

where $t_{1/2} = t_0 + (h/2)$ and $u_{1/2}$ is estimated using (4.6) as $u_{1/2} = u_0 + (h/2)f(t_0,u_0)$. This is sometimes referred to as the modified Euler method.

At the heart of the improved methods discussed in the remainder of this chapter lies the assumption, frequently made in numerical analysis, that the solution (u) can be approximated by a quadratic curve. The extension to higher degree approximations is possible but non-trivial, and such results will simply be quoted when required.

4.5 PREDICTOR–CORRECTOR APPROACH

The trapezoidal rule (4.8) is a non-linear equation in u_1, which prevents us from solving it directly for u_1. However, we found in Chapter 1 that a properly formulated method of successive approximations often provides a solution to such an equation, implemented as an iterative formula $x_{k+1} = g(x_k)$.

In the immediate neighbourhood of a solution there is convergence if $|g'(x)| < 1$ (Chapter 1) which in the case of (4.8a) means $|(h/2)\partial f(u_1,t_1)/\partial u_1| < 1$. This condition will always be satisfied if we choose h to be sufficiently small, but we must also recall that as a general rule such methods only work if we start with a sufficiently good estimate for x_0, or in our case u_1. Euler's method may be used to provide a starting value, and the procedure becomes Heun's method

$$\begin{aligned} u_1^{(p)} &= u_0 + hf(u_0,t_0) \\ u_1^{(c)} &= u_0 + (h/2)(f(u_0,t_0) + f(u_1^{(p)},t_1)) \end{aligned} \tag{4.9}$$

where $u_1^{(p)}$ is the predicted value and $u_1^{(c)}$ the corrected value of u_1. Repeating the latter procedure with $u_1^{(p)}$ replaced by the estimated $u_1^{(c)}$ gives us a better corrected value for $u_1^{(c)}$, and so on. When we are satisfied with the accuracy of $u_1^{(c)}$ (it changes less than a chosen amount between corrections) we can replace u_0 and t_0 in the predictor stage by $u_1^{(c)}$ and t_1, repeat the whole procedure to give $u_2^{(c)}$ and so on (program P4.3). If we combine the predictor and single-stage corrector equations we obtain a formula

$$u_1^{(c)} = u_0 + (h/2)(f(u_0,t_0) + f(u_0 + hf(u_0,t_0),t_1))$$

which is identical with a second order Runge–Kutta method (section 4.6).

We can generalize this technique to any pair of formulae derived from Newton–Cotes type integration formulae with a predictor and corrector of the form

$$\begin{aligned} u_{k+1}^{(p)} &= u_k + h(a_0 u'_k + a_1 u'_{k-1} + a_2 u'_{k-2} + \ldots) \\ u_{k+1}^{(c)} &= u_k + h(b_0 u'_{k+1} + b_1 u'_k + b_2 u'_{k-1} + \ldots) \end{aligned} \tag{4.10}$$

respectively. The corrector is an implicit formula for u_{k+1} whereas $u_{k+1}^{(p)}$ is derived explicitly. Notice also that more than one previous value of the solution is used in deriving the next value, and hence such techniques are known as multistep methods. Formulae of even more general form exist for the solution (Lapidus and Seinfeld, 1971), but a discussion of these is beyond the scope of this book.

An example of (4.10) is Adam's fourth-order method

$$u_{k+1}^{(p)} = u_k + (h/24)(55u'_k - 59u'_{k-1} + 37u'_{k-2} - 9u'_{k-3}) \qquad : \text{single step} \tag{4.11}$$

$$u_{k+1}^{(c)} = u_k + (h/24)(9u'_{k+1} + \\ + 19u'_k - 5u'_{k-1} + u'_{k-2}) \qquad : \text{iterated step (cf. (4.9))}$$

The truncation error for the predictor and corrector steps may be derived (cf. section 4.6.3)

$$\begin{aligned} u_{k+1}^{\text{true}} - u_{k+1}^{(p)} &= 251h^5 K/720 \\ u_{k+1}^{\text{true}} - u_{k+1}^{(c)} &= -19h^5 K/720 \end{aligned}$$

where K represents the contribution of fifth-order and higher terms, and is assumed to vary little over the interval of integration. Eliminating the term $h^5 K$ gives us

$$u_{k+1}^{\text{true}} - u_{k+1}^{(c)} = (19/270)(u_{k+1}^{(p)} - u_{k+1}^{(c)})$$

Thus the truncation error in approximating the integral by a sum of ordinates rule turns out to be directly proportional to the difference between the predicted and corrected values. Our estimate for u_{k+1}^{true} may thus be improved by adding $(19/270)(u_{k+1}^{(p)} - u_{k+1}^{(c)})$ to the corrected value $u_{k+1}^{(c)}$. We are only able to do this because the predictor and corrector terms are of equal order. This is a special case of the error cancelling method known as Richardson's extrapolation technique.

The reader will no doubt have spotted a hitch in such multistep predictor–corrector schemes: we need the results of more than one previous step to calculate u_{k+1}. In other words, how do we start the method off, and having done so how do we change the step size when $(u_{k+1}^{(p)} - u_{k+1}^{(c)})$ becomes too large or too small? The easy, if clumsy, answer is to use a Runge–Kutta formula of the same order (see section 4.6) to start and restart the calculation. Another possibility is to switch to a higher order formula when greater accuracy is required, but this does not solve the starting problem.

4.6 RUNGE–KUTTA APPROACH

Runge–Kutta methods seek to attain some of the accuracy of a Taylor series solution without the need to calculate derivatives; the price paid is that they require the evaluation of extra function values at points between (t_0, u_0) and (t_1, u_1). As opposed to the approaches of the previous section, they are single-step methods in that u_1 is computed directly without the need for iteration. An n-order Runge–Kutta method is one which agrees exactly with the Taylor series solution up to and including terms in h^n (where $h = t_1 - t_0$ as before); Euler's basic procedure (4.6) is thus a first order Runge–Kutta method.

The advantage of Runge–Kutta methods over predictor–corrector methods is that they are self-starting, needing no values previous to (t_0, u_0), and one can easily change the step size h as necessary during the course of a computation; one disadvantage is that it is not as easy to estimate the accuracy of the procedure.

4.6.1 Mathematical derivation

If we examine (4.6), (4.8) and (4.9), we find that each is of the basic form

$$u_1 = u_0 + hF(f(t_0,u_0),f(t^*,u^*)) \tag{4.12}$$

where t^* and u^* are values of t and u at some point between (t_0,u_0) and (t_1,u_1), and $F(f_0,f^*)$ is some function of the slopes at (t_0,u_0) and (t^*,u^*).

For example, (4.9) has $t^* = t_0 + (h/2)$ and $F(f_0,f^*) = f(t_{1/2},u_0 + (h/2)f_0)$, whereas the simple Euler method (4.6) has $t^* = t_0 + h$ and $F(f_0,f^*) = f_0$. We are led to try a generalization of these special cases by writing $t^* = t_0 + b_0h$ and $u^* = u_0 + b_1hf_0$ and

$$F(t^*,u^*) = a_0f_0 + a_1f^* = a_0f(t_0,u_0) + a_1f(t_0 + b_0h, u_0 + b_1hf_0)$$

Hence

$$u_1 = u_0 + h(a_0f_0 + a_1f(t_0 + b_0h, u_0 + b_1hf_0) \tag{4.13}$$

where the parameters a_0, a_1, b_0 and b_1 are to be determined by expressing (4.13) as a Taylor series-type approximation and then comparing it term by term with the truncated series (4.7). Recalling the definition of an nth-order Runge–Kutta method, we are thus seeking a second-order procedure.

The total differential of $f(t,u)$ is given by (Chapter 1)

$$df(t,u) = (\partial f/\partial t)\,\mathrm{d}t + (\partial f/\partial u)\,\mathrm{d}u \tag{4.14}$$

and may be interpreted as the change in f produced by small changes in t and u.

Applying it in the neighbourhood of (t_0, u_0) and writing $\mathrm{d}t = t - t_0$ and $\mathrm{d}u = u - u_0$, where u is close to u_0, then $\mathrm{d}f = f(t,u) - f(t_0,u_0)$ giving

$$f(t,u) = f(t_0,u_0) + (\partial f/\partial t)_0 (t - t_0) + (\partial f/\partial u)_0 (u - u_0)$$

and hence

$$f(t_0 + b_0 h, u_0 + b_1 h f_0) = f_0 + (\partial f/\partial t)_0 b_0 h + (\partial f/\partial u)_0 f_0 b_1 h$$

Finally, (4.13) becomes

$$u_1 = u_0 + (a_0 + a_1) h f_0 + a_1 b_0 (\partial f/\partial t)_0 h^2 + a_1 b_1 f_0 (\partial f/\partial u)_0 h^2 \tag{4.15}$$

Now the truncated Taylor series (4.7) contains $u'' = f'$ which we find by dividing (4.14) throughout by $\mathrm{d}t$ to give

$$\begin{aligned} u_0'' &= (\mathrm{d}f/\mathrm{d}t)_0 = (\partial f/\partial t)_0 + (\partial f/\partial u)_0 (\mathrm{d}u/\mathrm{d}t)_0 \\ &= (\partial f/\partial t)_0 + (\partial f/\partial u)_0 f_0 \end{aligned}$$

and thus (4.7) becomes

$$u_1 = u_0 + h f_0 + (h^2/2)(\partial f/\partial t)_0 + (h^2/2)(\partial f/\partial u)_0 f_0 \tag{4.16}$$

Comparing (4.15) and (4.16) and equating coefficients of f_0, $(\partial f/\partial t)_0$, and $(\partial f/\partial u)_0$ yields the following set of equations

$$\begin{aligned} a_0 + a_1 &= 1 \\ a_1 b_0 &= 1/2 \\ a_1 b_1 &= 1/2 \end{aligned} \tag{4.17}$$

4.6.2 Examples of Runge–Kutta formulae

Since we have three equations in four unknowns, we must choose a value for one of the unknowns. Fixing $b_0 = 1/2$ gives us $a_1 = 1, a_0 = 0$ *and* $b_1 = 1/2$. Substituting these values into (4.13) leads to

$$u_1 = u_0 + h(f(t_0 + h/2, u_0 + (h/2) f(t_0, u_0))) \tag{4.18}$$

which is the modified Euler method mentioned earlier (4.8b). The evaluation of u_1 naturally divides into two steps

$$u_{1/2} = u_0 + (h/2)(f(t_0, u_0))$$

which uses Euler's method to estimate the solution at a point halfway between t_0 and t_1, followed by

$$u_1 = u_0 + h(f(t_{1/2}, u_{1/2}))$$

which is an example of the mid-point rule for estimating an integral applied to

$$\int f(t,u)\,\mathrm{d}t\,.$$

The method is illustrated in Fig. 4.7, and this pictorial representation provides us

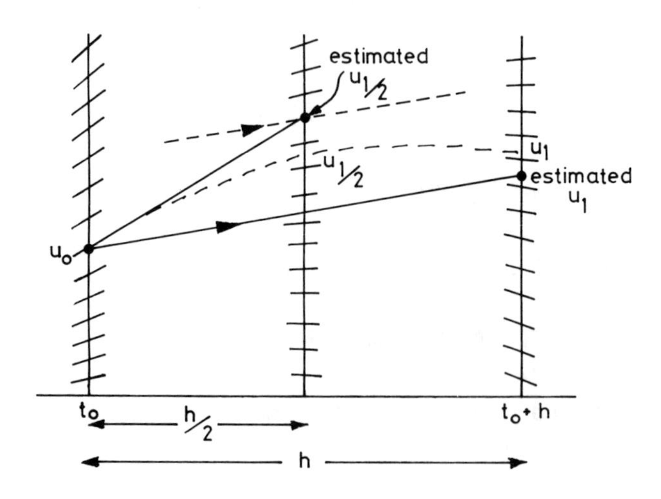

Fig. 4.7 — Graphical illustration of a Runge–Kutta formula (4.13) with $a_1 = 1$, $a_0 = 0$ and $b_1 = 1/2$, also known as the modified Euler method. The slope at $t = t_0 + (1/2)h$ is used to estimate u_1 at $t = t_0 + h$, starting at $t = t_0$.

with another way of thinking about the Runge–Kutta formula (4.18):

(1) given initial conditions calculate u'_0, the slope of the tangent to the solution curve at (t_0, u_0);
(2) choose an intermediate point $t_{1/2} = t_0 + (h/2)$;
(3) use u'_0 to estimate $u_{1/2}$ using Euler's method;
(4) calculate an estimate for $u'_{1/2}$, the slope of the tangent at $t = t_{1/2}$;
(5) draw a line of slope $u'_{1/2}$ passing through $(t_0,. u_0)$ — where it intersects the ordinate defined by $t = t_1$ provides the estimated value of u_1.

In (5) we could have used a formula involving u'_0 as well as $u'_{1/2}$. Indeed we could generalize the procedure further and introduce more than one intermediate value of t and use weighted sum of ordinates integration formulae (Chapter 3) with increasing order to estimate u and hence u' further and further towards u based on previous estimates of u'.

If we had chosen $b_0 = 1$ in (4.17) we would have obtained $a_1 = 1/2$, $b_1 = 1$, and $a_0 = 1/2$ giving from (4.13) Heun's method with a single corrector step (see section 4.5)

$$u_1 = u_0 + (h/2)(f(t_0, u_0) + f(t_0 + h, u_0 + hf_0))$$

or, in two steps

$$u_1^* = u_0 + hf(t_0, u_0)$$

followed by

$$u_1 = u_0 + (h/2)(f(t_0, u_0) + f(t_1, u_1^*)) \tag{4.19}$$

where u_1^* is a first estimate using Euler's formula (4.6) and u_1 is the final estimate obtained from a trapezoidal-type rule.

We could have chosen b_0 to fix the intermediate value of t at any point in the interval so deriving an infinity of possible formulae, all of them giving agreement between the final value of u_1 and the Taylor series solution to terms in h^2. This is typical of Runge–Kutta formulae.

If we were to consider third-order formulae by taking into account terms up to h^2 then we would end up with a set of six equations in eight unknowns (cf. (4.17)). A typical Runge–Kutta method which agrees with the Taylor series expansion up to and including terms in h^4, a fourth-order formula, is implemented as follows

(1) estimate the solution at $t_0 + h/3$

$$u_{1/3} = u_0 + (h/3)u_0'$$

(2) estimate the solution at $t_0 + 2h/3$

$$u_{2/3} = u_0 + h((-1/3)u_0' + u_{1/3}')$$

(3) estimate the solution at $t_0 + h$

$$u_1 = u_0 + h(u_0' - u_{1/3}' + u_{2/3}')$$

(4) finally, obtain a refined estimate of u_1

$$u_1 = u_0 + (h/8)(u_0' + 3u_{1/3}' + 3u_{2/3}' + u_1') \tag{4.20}$$

4.6.3 Estimating the error

In Chapter 3 we saw how the error in numerical integration, by such as Simpson's Rule, could be estimated by taking a smaller step size, integrating across the same interval, and comparing the two results. An analogous procedure may be followed in the case of the numerical solution of differential equations, leading to the expression (for interval halving)

$$E_k = (u_{k+1}^{(h)} - u_{k+1}^{(2h)})/(2^{n+1}-2) \tag{4.21}$$

where E_k is the estimated error at u_k. The calculated values of the solution at t_{k+1} based on a step of $2h$ from t_{k-1} and two steps of h from t_{k-1} to t_{k+1} *via* t_k are represented by $u_{k+1}^{(2h)}$ and $u_{k+1}^{(h)}$ respectively; n is the order of the method.

As an example, consider the solution of the differential equation $u' = ut$ with initial condition $u_0 = 1$ at $t = 0$, using the second-order Runge–Kutta formula (4.19). The error in u_1 at $t = 1$ is estimated by calculating the solution at $t = 2$ in two ways

(1) for a step size $h = 2$, $u_2^{(2h)} = 3.0$
(2) for a step size $h = 1$ then repeated application of (4.19) gives $u_1 = 1.5$ and $u_2^{(h)} = 3.75$.

Hence the error at u_1 is estimated from (4.21) with $n = 2$ to be $0.75/6 = 0.125$. Since the analytic solution of $u' = ut$ is $u = \exp(t^2/2)$, then the true value of u_1 is $\exp(1/2) = 1.649$, and the real error at u_1 is $(1.649 - 1.500) = 0.149$ to three decimal places. Had we chosen the equation $u' = t^2$ with $u_0 = 1$ at $t = 0$, then we would have had an exact estimate for the error since the solution is $u = 1 + t^3/3$ and the error term contains no higher powers of t than cubic (see Chapter 3). It is interesting to note that our trapezoidal-type rule (4.19) actually becomes the trapezoidal rule if u' is a function of t alone.

Another approach to the estimation of the error is to calculate u_{k+1} from u_k and then recalculate u_k from the estimate u_{k+1} of the solution by using $(-h)$ instead of h in the formula for u_{k+1}. The difference between the original value of u_k and the recalculated value is equal to twice the estimated error for the step if the error term is of the form Kh^{n+1} with n even. If n is odd then $Kh^{n+1} = K(-h)^{n+1}$ and the backward procedure exactly cancels the forward error. The quantity K takes into account all the higher terms apart from the $(n + 1)$th, and is assumed to be essentially constant over the step taken.

It is important to realize that what is determined above is a local error estimate rather than a global one. If we were using the solution of a differential equation to predict, for example, the position of a projectile after some lapse of time from its launch, then we would want to know where that projectile was going to be rather than the local error from step to step. It might be that the local errors cancelled over the interval of integration, but it is equally likely that the errors will propagate and increase as the 'solution' slips from one track of the true solution to another and finally 'goes off the rails' (see section 4.5).

Program P4.4 implements a fourth-order Runge–Kutta procedure. It estimates the local error as just described and halves the current step size if it exceeds the local error bound supplied by the user. This halving is repeated until the estimated error is small enough. In a manner akin to the adaptive quadrature approaches discussed in Chapter 3, the procedure P4.4 steps through the solution, attempting a doubling of the step size when this is possible and further halving the step size when it proves necessary. As most kinetic simulations require outputs at fixed equidistant graph points, the working step size is not allowed to exceed this required step size, and the solution is only stored at the required points.

The same Runge–Kutta method is used in program P4.5 to provide the starting values for the fourth-order Adams predictor–corrector procedure (section 4.5) which then advances the solution. At each step the accuracy of the solution is improved by adding the estimate of the error calculated as described in section 4.5. A refinement, which is left to the reader to implement, is to monitor the difference between the predictor and first corrector; if this exceeds the user-provided error bound then the Runge–Kutta routine is used to halve the step size until the error criterion in the restarted Adams solution is met. Step size doubling, when the solution is advancing satisfactorily, may also be incorporated into the program. The advantage of using a predictor–corrector method at each step, if possible, rather than a Runge–Kutta formula, is the reduction in the number of $f(t,u)$ evaluations and the corresponding increase in speed of execution.

4.7 EXTRAPOLATION APPROACH

This final basic approach which we consider is an extension of the adaptive quadrature techniques used in Chapter 3, together with the error-cancelling method of the previous section. One may apply it locally as the solution is being pushed forward, or globally after the solution has been estimated over the entire required range. We shall consider the implementation of the latter option.

We first choose a suitable step size h and integrate the differential equation over the required range, thus obtaining approximate values of u at the defined mesh points. We then halve the step size and obtain another set of values at the same mesh points as before (thus discarding half of the calculated values). This is repeated for a further halving of the step size. Let the three values of the 'solution' so calculated for any mesh point k be designated $u_k^{(h)}(0)$, $u_k^{(h/2)}(0)$, and $u_k^{(h/4)}(0)$, where the index in brackets indicates to which level of the extrapolation process the values correspond (we start at level 0). Following the same type of procedure as for the error-cancelling technique of the previous sections, and calling the exact solution u_k, we have

$$\begin{aligned} u_k - u_k^{(h)}(0) &= Kh^3 + O(h^4) \\ u_k - u_k^{(h/2)}(0) &= Kh^3/4 + O(h^4) \quad &&: 2 \times \text{number intervals, } 8 \times \text{accuracy} \\ u_k - u_k^{(h/4)}(0) &= Kh^3/16 + O(h^4) \quad &&: 4 \times \text{number intervals, } 64 \times \text{accuracy} \end{aligned}$$

where $O(h^4)$ represents terms in h^4 and above. Multiplying through by the right-hand side denominators and combining the equations in pairs we get

$$\begin{aligned} u_k &= (4u_k^{(h/2)}(0) - u_k^{(h)}(0))/3 + O(h^4) \\ u_k &= (4u_k^{(h/4)}(0) - u_k^{(h/2)}(0))/3 + O(h^4) \end{aligned}$$

We have 'cancelled' the first and, we hope, the largest error term. We proceed by rewriting the equations as

$$u_k = u_k^{(h)}(1) + K'h^4 + O(h^5)$$
$$u_k = u_k^{(h/2)}(1) + K'h^4/16 + O(h^5)$$

where the definitions of $u_k^{(h)}(1)$ and $u_k^{(h/2)}(1)$ are clear by comparison with the previous equations. We have taken out $k'h^4$ and $K'h^4/16$ as representing the next highest term to be eliminated by a similar process to give

$$u_k = (16u_k^{(h/2)}(1) - u_k^{(h)}(1))/15 + O(h^5)$$

and ignoring terms in h^5 and above we obtain the final estimate of the solution

$$u_k^{(h)}(2) = (16u_k^{(h/2)}(1) - u_k^{(h)}(1))/15$$

We can conveniently 'plot' the progress of this calculation in a triangular table

$$\begin{array}{llll}
u_k^{(h)}(0) \searrow & & & \\
u_k^{(h/2)}(0) \rightarrow & u_k^{(h)}(1) = \dfrac{4u_k^{(h/2)}(0) - u_k^{(h)}(0)}{3} \searrow & & \\
\quad \searrow & & & \\
u_k^{(h/4)}(0) \rightarrow & u_k^{(h/2)}(1) = \dfrac{4u_k^{(h/4)}(0) - u_k^{(h/2)}(0)}{3} & \rightarrow u_k^{(h)}(2) = \dfrac{16u_k^{(h/2)}(1) - u_k^{(h)}(1)}{15} &
\end{array}$$

In general

$$u_k^{(h)}(i) = (4^i u_k^{(h/2)}(i-1) - u_k^{(h)}(i-1))/(4^i - 1) \tag{4.22}$$

and we can extend the table to include a larger number of values (at the expense of more calculations and the need to store more results), but the bottom right-hand value gives the best estimate of the true value of the solution. The method might be implemented in practice as follows:

(1) compute the solution at a given step size and store the results;
(2) halve the step size and recompute the solution, storing every other result corresponding to the same mesh points as in (1);
(3) use the Richardson technique to combine the two estimates at each point;
(4) if the Richardson estimate for any mesh point differs too much from the equivalent result in (2) re-compute the solution after a further halving of step size, keeping only those values corresponding to the original mesh;
(5) compute another Richardson estimate using the new values and the previous row of the triangular table, thus adding a new row to the table;
(6) repeat from (4) until convergence is obtained.

The global extrapolation method is programmed in P4.6, using the trapezoidal rule as the corrector and Euler's formula as the predictor. The method is quite expensive on computer time and storage, but is very tolerant of 'difficult' equations.

4.8 SYSTEMS OF EQUATIONS

In practical chemical problems, we often deal with more complicated situations than those involving a single chemical species, and thus need to solve systems of simultaneous differential equations. Consider the example

$$\mathrm{A}_1 \underset{k_{21}}{\overset{k_{12}}{\rightleftharpoons}} \mathrm{A}_2 \underset{k_{32}}{\overset{k_{23}}{\rightleftharpoons}} \mathrm{A}_3$$

where A_1, A_2, and A_3 are chemical species, with corresponding concentrations a_1, a_2, and a_3, and k_{12}, k_{21}, k_{23}, and k_{32} are the back and forward rate constants for the linked equilibria. We may write down the rate equations as

$$\begin{aligned}
\mathrm{d}a_1/\mathrm{d}t &= -k_{12}a_1 + k_{21}a_2 \\
\mathrm{d}a_2/\mathrm{d}t &= -k_{23}a_2 + k_{32}a_3 + k_{12}a_1 - k_{21}a_2 \\
\mathrm{d}a_3/\mathrm{d}t &= -k_{32}a_3 + k_{23}a_2
\end{aligned} \tag{4.23}$$

(Such equations where the independent variable t occurs only in the derivatives and not explicitly are known as autonomous equations). Applying Euler's method to (4.23) gives us

$$\begin{aligned}
a_1^{(h)} &= (1 - k_{12}h)a_1 + k_{21}ha_2 \\
a_2^{(h)} &= k_{12}ha_1 + (1 - (k_{21} + k_{23})h)a_2 + k_{32}ha_3 \\
a_3^{(h)} &= k_{23}ha_2 + (1 - k_{32}h)a_3
\end{aligned}$$

where the superscript (h) denotes the next value of the solution after a step h in time. In this case of coupled first-order reactions we may conveniently put the equations in matrix form

$$\begin{bmatrix} a_1^{(h)} \\ a_2^{(h)} \\ a_3^{(h)} \end{bmatrix} = \begin{bmatrix} 1 - k_{12}h & k_{21}h & \\ k_{12}h & 1 - (k_{21} + k_{23})h & k_{32}h \\ & k_{23}h & 1 - k_{32}h \end{bmatrix} \begin{bmatrix} a_1 \\ a_2 \\ a_3 \end{bmatrix} \tag{4.24}$$

Calculating the solution to this set of equations then reduces to multiplying the vector of current concentrations $[a]$ at time $t = t_0$ by the matrix of rate constant terms to give the vector of new concentrations $[a^{(h)}]$ at time $t = t_0 + h$. The 'old' vector is replaced by the 'new' vector and the process repeated as many times as is necessary. Note that the sum of any column of the rate matrix is unity, ensuring that the total concentration of A_1, A_2, and A_3 is conserved — no particles are gained or lost.

Let us now set k_{21} and k_{32} equal to zero and examine the simplified case of consecutive reactions. We now have

$$\dot{a}_1 = -k_{12}a_1$$
$$\dot{a}_2 = +k_{12}a_1 - k_{23}a_2$$
$$\dot{a}_3 = +k_{23}a_2$$

where $\dot{a} = \mathrm{d}a/\mathrm{d}t$. Eliminating a_1 between the first two equations yields

$$\dot{a}_1 + \dot{a}_2 = -k_{23}a_2$$

and differentiating the second with respect to t gives

$$\ddot{a}_2 = k_{12}\dot{a}_1 - k_{23}\dot{a}_2$$

We may now eliminate $\dot{a}_1$ to give finally

$$\ddot{a}_2 + (k_{12} + k_{23})\dot{a}_2 + k_{12}k_{23}a_2 = 0$$

which is a second-order differential equation. Two such coupled first-order equations may often be combined to give a second-order equation in one of the variables. We could take the process even further and derive a third-order differential equation in a_3. We could not, however, write the system as a second or third-order equation in a_1 since there is no feedback to a_1; $\dot{a}_2$ depends on a_1 but not vice versa. Physically, this represents the fact that A_1 will decay as first order in essentially the same way whatever happens to its decay products.

Although in this example we could have gone on to produce an analytical expression for a_2 as a function of time, in general we have to resort to numerical methods.

The reverse process of reducing an nth-order differential equation to a set of n first order equations is always possible and may be numerically very useful. We merely introduce new variables so that, for example, $\ddot{u} + t\dot{u} + u = t^2$ becomes

$$\dot{u} = v \quad \text{and} \quad \dot{v} = -tv - u + t^2$$

It may sometimes be desirable to express a first-order differential equation as two simultaneous equations. For instance, the equation $\mathrm{d}u/\mathrm{d}v = -v/u$ has as its solutions a set of circles about the origin, as may easily be seen by drawing its tangent field diagram. It is also a non-linear equation (u occurs as $1/u$) and if we tried to compute the solutions numerically we would rapidly get into trouble since $\mathrm{d}u/\mathrm{d}v = \infty$ for $u = 0$. We may solve the problem, and the equation, by converting it to a pair of equations and introducing another variable, t, and writing

$$\mathrm{d}u/\mathrm{d}t = -v \quad \text{and} \quad \mathrm{d}v/\mathrm{d}t = u$$

which is analogous to the use of parametric equations in coordinate geometry.

4.9 SOME ILLUSTRATIVE EXAMPLES

4.9.1 Single rate equation

To demonstrate some of the ideas discussed in this chapter, we first consider the example of a chemical species introduced periodically into a system, perhaps photochemically, decaying by a second-order process and with the resulting rate equation

$$du/dt = A\cos^2(2\pi ft) - ku^2$$

where u is the concentration of the species at time t, A is the maximum rate of production of the species (for example at maximum intensity of illumination), f is the frequency of successive maxima and k is the rate constant for the decay. Such a scheme might form the basis of a simple attempt to simulate a reaction occurring in the upper atmosphere where the periodic illumination is provided by sunlight. In this case the term $A\cos^2(2\pi ft)$ represents the diurnal variation of photon flux — better (more realistic) and more complex expressions could be formulated if desired. Fig. 4.8 shows a typical plot of concentration against time computed using Heun's

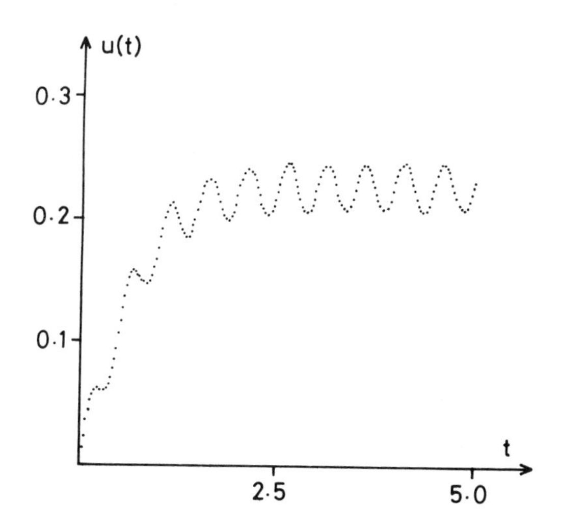

Fig. 4.8 — Solution of $du/dt = 0.5\cos^2(2\pi t) - 5u^2$, within initial condition $u(0)=0$. Heun's method is used with an $h=0.025$ and a specified accuracy of 10^{-4}. The differential equation represents the second order decay of a species introduced periodically into a system (see text).

method (P4.3) with $A=0.5$, $f=1$, $k=5$, $u_0=0$ and a specified accuracy of 10^{-4}; the step size used was 0.025. The general behaviour shows an initial rise which quickly settles down to a mean equilibrium value, modulated by the $\cos^2(2\pi ft)$ term; in this example the peak-to-peak amplitude of the modulation is 0.038. The same result is obtained using the Runge–Kutta method (P4.4), but at the expense of longer execution times and a more complex algorithm. Furthermore, the Runge–Kutta method does not perform so well for the 'stiff' case (section 4.9.3), where k is large.

4.9.2 Systems of equations

The four programs P4.3–P4.6 may be adapted in a straightforward manner to the solution of systems of first-order differential equations, and this is illustrated in program P4.7 for the fourth-order Runge–Kutta method of listing P4.4. The user supplies a set of program subroutines which calculate the various derivatives

$$
\begin{aligned}
u'(1) &= f_1(t, u(1), u(2), \ldots, u(m)) \\
u'(2) &= f_2(t, u(1), u(2), \ldots, u(m)) \\
&\vdots \\
u'(m) &= f_m(t, u(1), u(2), \ldots, u(m))
\end{aligned}
$$

where $u(1), u(2), \ldots, u(m)$ represent the different variables, and not the values of a single variable at different times. For the fourth-order Runge–Kutta method with two simultaneous differential equations we have, following (4.20)

$$
\begin{aligned}
&\text{(a)} & u(1)_{1/3} &= u(1)_0 + (h/3)u'(1)_0 \\
& & u(2)_{1/3} &= u(2)_0 + (h/3)u'(2)_0 \\
&\text{(b)} & u(1)_{2/3} &= u(1)_0 + h(-(1/3)u'(1)_0 + u'(1)_{1/3}) \\
& & u(2)_{2/3} &= u(2)_0 + h(-(1/3)u'(2)_0 + u'(2)_{1/3}) \\
&\text{(c)} & u(1)_1^* &= u(1)_0 + h(u'(1)_0 - u'(1)_{1/3} + u'(1)_{2/3}) \\
& & u(2)_1^* &= u(2)_0 + h(u'(2)_0 - u'(2)_{1/3} + u'(2)_{2/3}) \\
&\text{(d)} & u(1)_1 &= u(1)_0 + (h/8)(u'(1)_0 + 3u'(1)_{1/3} + 3u'(1)_{2/3}) + u'(1)_1^*) \\
& & u(2)_1 &= u(2)_0 + (h/8)(u'(2)_0 + 3u'(2)_{1/3} + 3u'(2)_{2/3}) + u'(2)_1^*)
\end{aligned}
$$

The function evaluations in step (d), for example, are now

$$
\begin{aligned}
u'(1)_0 &= f_1(t_0, u(1)_0, u(2)_0) \\
u'(2)_0 &= f_2(t_0, u(1)_0, u(2)_0) \\
u'(1)_{1/3} &= f_1(t_{1/3}, u(1)_{1/3}, u(2)_{1/3}) \\
u'(2)_{1/3} &= f_2(t_{1/3}, u(1)_{1/3}, u(2)_{1/3}) \\
u'(1)_{2/3} &= f_1(t_{2/3}, u(1)_{2/3}, u(2)_{2/3}), \qquad \text{etc.}
\end{aligned}
$$

Instead of single variables u_0, u'_0, etc., we now need arrays with an element for each suffix 0, 1/3, 2/3, 1, etc.

This Runge–Kutta formulation is explicit, that is the only unknowns in the expressions appear on the left-hand side; the reader is referred elsewhere for a discussion of implicit Runge–Kutta formulae (see, for example, Lapidus and Seinfeld, 1971). In contrast, the predictor–corrector methods have an implicit corrector stage and the Adams method corrector of (4.11) becomes for two equations

$$
\begin{aligned}
u(1)_{k+1}^{(c)} &= u(1)_k + (h/24)(9f_1(t_{k+1}, u(1)_{k+1}, u(2)_{k+1}) + (h/24)(19u'(1)_k - \\
&\qquad 5u'(1)_{k-1} + u'(1)_{k-2}) \\
u(2)_{k+1}^{(c)} &= u(2)_k + (h/24)(9f_2(t_{k+1}, u(1)_{k+1}, u(2)_{k+1}) + (h/24)(19u'(2)_k - \\
&\qquad 5u'(2)_{k-1} + u'(2)_{k-2})
\end{aligned}
$$

where the third terms on the right-hand side collect together all the known derivative values. In general, these equations represent a set of non-linear algebraic equations. We may solve them by successive approximations, and this is best done by running through the equations one-by-one using improved values of $u(1)_{k+1}$ and $u(2)_{k+1}$ as soon as they become available; this is reminiscent of Gauss–Siedel rather than Jacobi iteration (Chapter 1).

We now apply this modified program P4.7 to a kinetics-type problem. Bradley *et al.* (1973) investigated the reaction of OH radicals, generated by titrating against NO_2 the H atoms produced in a microwave discharge, with some hydrocarbons. Fast competitive reactions consume the OH radicals, leading to the system of reactions involving the OH species

$$\mathrm{H + NO_2 \xrightarrow{k_1} OH + NO} \qquad k_1 = 2.9 \times 10^{13}\ \mathrm{cm^3\,mol^{-1}\,s^{-1}}$$

$$\mathrm{OH + OH \xrightarrow{k_2} H_2O + O} \qquad k_2 = 1.55 \times 10^{12}\ \mathrm{cm^3\,mol^{-1}\,s^{-1}}$$

$$\mathrm{O + OH \xrightarrow{k_3} O_2 + H} \qquad k_3 = 1.1 \times 10^{13}\ \mathrm{cm^3\,mol^{-1}\,s^{-1}}$$

The system of differential equations derived from these reactions is

$$\begin{aligned}
\mathrm{d[H]}/\mathrm{d}t &= -k_1[\mathrm{H}][\mathrm{NO_2}] + k_3[\mathrm{O}][\mathrm{OH}] \\
\mathrm{d[NO_2]}/\mathrm{d}t &= -k_1[\mathrm{H}][\mathrm{NO_2}] \\
\mathrm{d[OH]}/\mathrm{d}t &= k_1[\mathrm{H}][\mathrm{NO_2}] - k_2[\mathrm{OH}][\mathrm{OH}] - k_3[\mathrm{O}][\mathrm{OH}] \\
\mathrm{d[NO]}/\mathrm{d}t &= k_1[\mathrm{H}][\mathrm{NO_2}] \\
\mathrm{d[H_2O]}/\mathrm{d}t &= k_2[\mathrm{OH}][\mathrm{OH}] \\
\mathrm{d[O]}/\mathrm{d}t &= k_2[\mathrm{OH}][\mathrm{OH}] - k_3[\mathrm{O}][\mathrm{OH}] \\
\mathrm{d[O_2]}/\mathrm{d}t &= k_3[\mathrm{O}][\mathrm{OH}]
\end{aligned}$$

where the square brackets symbolize the concentration of the species. Although this system of equations looks very complex and intimidating, it is easily solved (that is, the evolution with time of the concentrations of all the species are computed) using P4.7, where all the function calls for this system are explicitly included. The computation is performed with $[\mathrm{H}]_0 = 4.5 \times 10^{-10}\ \mathrm{mol\,cm^{-3}}$ and $[\mathrm{NO_2}]_0 = 5.6 \times 10^{-10}\ \mathrm{mol\,cm^{-3}}$, and for the interval $t = 0$ to $t = 10$ ms; the results are plotted in Fig. 4.9. This is an excellent example of a complex problem which defies analytical solution, yet may be solved numerically using a procedure little more complicated than for the simplest case. Moreover, the changes one would have to make to the program in order to incorporate extra reaction steps or to modify the present ones would be trivial.

4.9.3 'Stiff' systems

An equation of system of equations which has both very rapidly and very slowly varying components in its solution is said to exhibit 'stiffness'. As an example consider the system

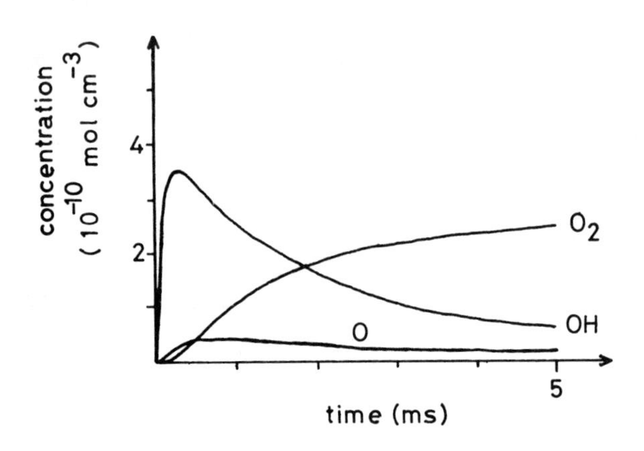

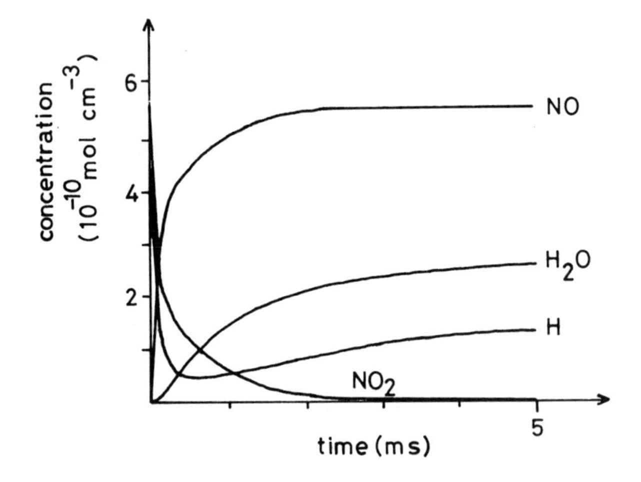

Fig. 4.9 — Solution curves for a system of chemical rate equations taken from Bradley *et al.* (1973). Full details are given in the text.

$$\mathrm{A} \underset{k_2}{\overset{k_1}{\rightleftharpoons}} \mathrm{B} \overset{k_3}{\rightarrow} \mathrm{C}$$

The rate equations for this are

$$\begin{aligned} \mathrm{d}a/\mathrm{d}t &= -k_1 a + k_2 b \\ \mathrm{d}b/\mathrm{d}t &= k_1 a - k_2 b - k_3 b \\ \mathrm{d}c/\mathrm{d}t &= k_3 b \end{aligned}$$

where a, b and c are the concentrations of A, B and C respectively at time t. We can easily set up program P4.7 to solve this system. It is a trivial matter to isolate the simple decay step, by setting $k_1 = k_2 = 0$, and the result (plot of b against t) is shown in Fig. 4.10. We set $b(0) = 100$, $k_3 = 0.1$ and chose an initial step size of $h = 1$ s in the

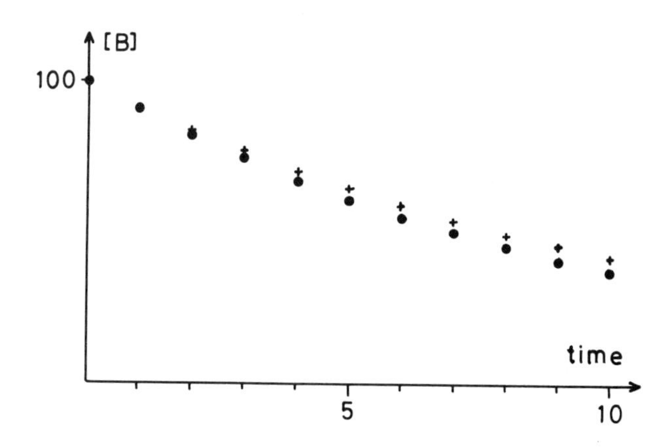

Fig. 4.10 — Variation of [B] (concentration of B) with time for the system A⇌B→C, calculated using the adaptive Runge–Kutta approach (P4.7). Closed circles: with the equilibrium step 'turned off'. Crosses: with the equilibrium step active, the rate constants for the equilibrium stage being chosen so that the system exhibits 'stiffness'.

interval $0 < t < 10$s with an accuracy parameter of 10^{-6}. This adaptive procedure (P4.7) performed the solution without needing to reduce h below the initial value. If we now 'turn on' the equilibrium A⇌B by setting $k_1 = 100$, $k_2 = 10$, $a(0) = 10$ and $b(0) = 100$ we can solve over the same time range with the same accuracy parameter and initial value of h. This time the adaptive procedure immediately chooses a step size of 0.015–0.031 which it maintains throughout the time range. Since the step A⇌B is always close to equilibrium (the initial values of a and b were in fact deliberately chosen to start out in this position) one might have intuitively expected it to have little effect on the solution (which is plotted in Fig. 4.10). However, the ratio between the fastest and slowest rate constants is 1000, and it is the fastest step which controls the choice of step size despite the fact that the reaction step in which it is involved is always almost at equilibrium. This observation reminds us quite dramatically of the dynamic nature of the equilibrium. Heun's method is generally recommended for solving systems of stiff equations, rather than the Runge–Kutta procedure (P4.7) we have used. It is left to the reader to write the Heun's method version of P4.7.

For a more detailed discussion of the problems involved in solving stiff systems of differential equations, the reader is referred to Chandler *et al.* (1982) and Carley and Morgan (1989).

4.9.4 Final remarks

Another type of instability arises when we go to higher order methods of the predictor–corrector kind, when values from previous points are 'fed back' into the process of finding the current value of the solution. The analogy here is with an amplifier system feeding back from the speaker to the microphone a delayed copy of the speaker output. Under certain circumstances the system 'howls', with the slightest noise growing and overwhelming the true signal. In mathematical terms, we have replaced our first order differential equation by a higher order difference equation which has more than the one solution which we seek. The other 'parasitic' solutions may grow and swamp the true solution unless the method is designed to damp them out.

A number of ways around the problem exist, for example:

(1) Use a stable method. Implicit, low order, formulae such as the trapezoidal rule are good in this respect. Either use a global extrapolation method to give good accuracy with a reasonable step size, or use the basic method with a small step size. Euler's method reformatted in backward form

$$u_1 = u_0 + hf(t_1, u_1)$$

where the function value used on the right-hand side is that at the end of the interval rather than the beginning, is the lowest order implicit formula.

(2) Use an accurate higher order method and starting from an estimated value of the solution at the end of the interval (for example, one may know that all solutions tend to zero) work backwards (using a step size $-h$) to the beginning of the interval. The procedure is repeated with new 'starting' values (that is, values at the end of the interval) until the required (known) initial value is reached. This in essence makes the whole solving process an implicit one.

4.10 PARTIAL DIFFERENTIAL EQUATIONS

4.10.1 Basic procedure

Consider a horizontal transparent tube closed at one end and containing an inert dilute gel. If we place the open end in contact with a concentrated dye solution then we know that diffusion will take place and that the colour of the dye will gradually permeate the tube. How can we calculate the dye concentration at various points along the tube and at various times after the start of the diffusion process? In Chapter 1 we examined the Monte Carlo method; in this chapter we tackle the problem from the macroscopic viewpoint.

We shall assume that Fick's law of diffusion holds, that is that the rate of transport of the dye along the direction of the tube at any point is proportional to the concentration gradient (rate of change of concentration with distance) at that point. Imagine the tube as being made up of a set of small interconnected compartments and let us consider three such compartments somewhere near the middle of the tube (Fig. 4.11). The variables m_i and c_i are mass and concentration in the ith compartment, h is the width of each compartment, x is the displacement along the tube and t is time, measured from the start of the diffusion. Then the average concentration gradients between the centres of the compartments $(i-1)$ and i, and between i and $(i+1)$ are

$$(\Delta c/\Delta x)_{i-1,i} = (c_i - c_{i-1})/h$$

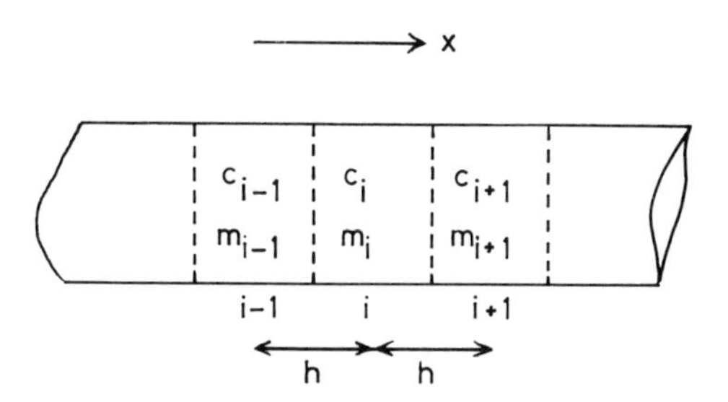

Fig. 4.11 — Definition of the variables used in the numerical derivation of the diffusion equation, for diffusion from a dye bath down a tube of gel.

$$(\Delta c/\Delta x)_{i,i+1} = (c_{i+1}-c_i)/h$$

We can approximate Fick's law by

$$\Delta m/\Delta t = -DA(\Delta c/\Delta x) \tag{4.25}$$

where A is the cross-sectional area of the dashed boundary surface between cells (Fig. 4.11), D is the diffusion coefficient and Δm is the mass transported in time Δt. If we set $k=\Delta t$ then from cell $(i-1)$ to cell (i) we have

$$(1/k)\Delta m_{i-1\to i} = -DA(c_i-c_{i-1})/h$$

and for $i\to i+1$

$$(1/k)\Delta m_{i\to i+1} = -DA(c_{i+1}-c_i)/h$$

Cell i gains mass $(\Delta m)_{i-1\to i}$ and loses $(\Delta m)_{i\to i+1}$; thus the net gain of mass in cell i is given by

$$(\Delta m)_i = DAk(c_{i+1}-2c_i+c_{i-1})/h$$

Dividing each side by Ah, the volume of the cell, we obtain the change in concentration in cell i after time $k=\Delta t$ has elapsed

$$(\Delta c)_i = (Dk/h^2)(c_{i+1}-2c_i+c_{i-1}) \tag{4.26}$$

Our approximation can best be diagrammed as in Fig. 4.12 where our cells are represented as points, each of which has a value of c associated with it. Successive rows $(j-1)$, j, $(j+1)$, $(j+2)$, etc., represent the concentrations at points (that is, centres of cells) all the way along the tube at times $t_{j-1}=(j-1)k$, $t_j=jk$, $t_{j+1}=(j+1)k$, etc., and the columns similarly show how the concentration at a fixed x value (for example, $x_{i+1}=(i+1)h$) changes with time. One can imagine rods of height $c_{i,j}$, $c_{i+1,j}$, etc., erected at each point and the tops of these ordinate rods touching a surface showing how the concentration evolves in time and space. In Fig. 4.12 we are saying that the value of the concentration at time $(j+1)$ and position i can be predicted from

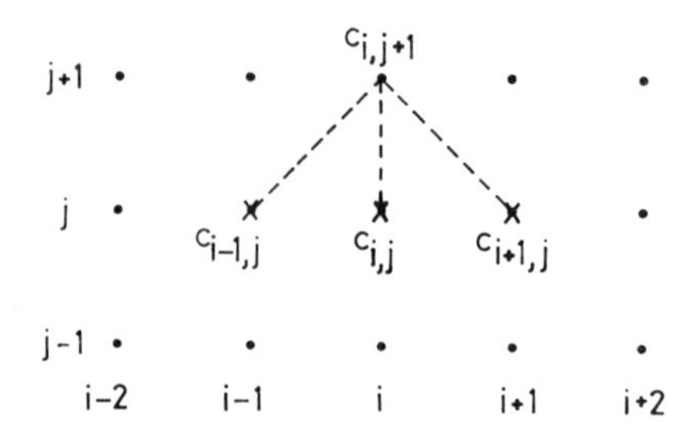

Fig. 4.12 — Grid representation of the evolution with x and t of the solution $c(x,t)$ of the diffusion equation. The indices i and j label steps in x (space) and t (time) respectively.

the values at time j and positions $(i-1)$, i and $(i+1)$; in terms of (4.26) $(\Delta c)_i = c_{i,j+1} - c_{i,j}$ or

$$(1/k)(c_{i,j+1}-c_{i,j}) = (D/h^2)(c_{i+1,j}-2c_{i,j}+c_{i-1,j}) \tag{4.27}$$

A quick backward look at Chapter 1 reminds us that the left-hand side of (4.27) is an approximation for $(\partial c/\partial t)_{x_i}$ and as $k \to 0$ tends to the value of this partial derivative. In like fashion

$$(c_{i+1,j} - 2c_{i,j} + c_{i-1,j})/h^2 \to (\partial^2 c/\partial x^2)_{x_i,t_i}$$

as $h \to 0$. Equation (4.27) is thus an approximation to

$$\partial c/\partial t = D(\partial^2 c/\partial x^2) \tag{4.28}$$

which is known as 'the diffusion equation', and can in fact be derived analytically from Fick's law. We could derive a large number of other formulae all of which would tend to (4.28) as $h, k \to 0$. Equation (4.27) is an explicit expression which can be used to march the solution forward from t_j to t_{j+1} and in which $c_{i,j+1}$ can be expressed explicitly in terms of other previous values; the procedure is analogous to the Euler and Runge–Kutta methods used in initial value ordinary differential equations.

Equation (4.27) may be rearranged to give

$$c_{i,j+1} = \alpha c_{i-1,j} + (1-2\alpha)c_{i,j} + \alpha c_{i+1,j} \tag{4.29}$$

where $\alpha = (kD/h^2)$. Fig. 4.13 illustrates what happens for a row of values where $c_{0,j}$ and $c_{6,j}$ are the values in fictional cells just beyond the ends of the tube. Obviously we will have to specify values $c_{1,j}$ to $c_{5,j}$ and if $j = 0$ (that is, time zero) these are our initial conditions; however, also for time $(j+1)$ we will need values of $c_{0,j+1}$ and $c_{6,j+1}$, etc., that is all values at the ends of the tube for all time after the start. These are boundary conditions which may vary with time if we wish. There is, of course, nothing to prevent us specifying other points between the ends as having fixed or varying values with time — in our example these would represent sources of dye other than that at the end. The values of $c_{i,j+1}$, $c_{i,j+2}$, etc., and thus the solution will

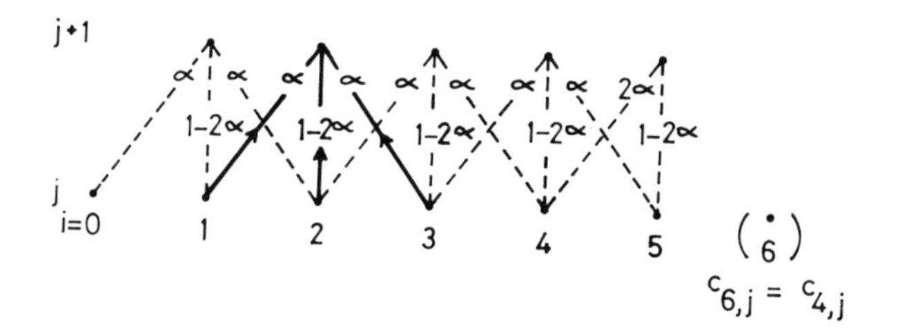

Fig. 4.13 — Representation of how concentrations at time $j+1$ are derived from concentrations at time j, for diffusion down a tube of gel. A fictional cell at $i=6$ is shown with a concentration given by $c_{6,j}=c_{4,j}$ in order to accommodate the closed end in the solution procedure.

depend critically on our initial conditions and our boundary conditions, and any general solution we could obtain would be so arbitrary as to be useless.

In our original problem we had the tube closed at one end with the other end dipped in a dye of known concentration. For the open end we set, say, $c_{0,j}$ always equal to the dye bath concentration, but in the case of the closed end although we know that the initial concentration is zero, we also realize that this cannot be true after some time has elapsed. What is certain, however, is that no dye will diffuse in or out of the closed end of the tube, or in other words the value of $\partial c/\partial x$ at the outer partition wall of the fifth cell (see Fig. 4.13) is zero becaue that wall is impenetrable. We can simulate this by inventing an extra cell and always assuming that its fictional concentration is the same as that in the last but one cell. Since $\partial c/\partial x = 0$ then according to Fick's law $\partial c/\partial t = 0$. It is as if we had put a 'mirror' halfway across the fifth cell and reflected the concentrations inside the tube of gel to give the fictional concentration in cell 6. Thus $(c_{6,j} - c_{4,j})/2h = 0$ is a second order approximation to $(\partial c/\partial x)$ in cell 5.

We can now write down expressions for each $c_{i,j+1}$ and if we look upon Fig. 4.13 as the flow graph of a matrix operation we get

$$\underset{\text{for } j+1}{\begin{bmatrix} c_1 \\ c_2 \\ c_3 \\ c_4 \\ c_5 \end{bmatrix}} = \begin{bmatrix} 1-2\alpha & \alpha & 0 & 0 & 0 \\ \alpha & 1-2\alpha & \alpha & 0 & 0 \\ 0 & \alpha & 1-2\alpha & \alpha & 0 \\ 0 & 0 & \alpha & 1-2\alpha & \alpha \\ 0 & 0 & 0 & \alpha & 1-2\alpha \end{bmatrix} \underset{\text{for } j}{\begin{bmatrix} c_1 \\ c_2 \\ c_3 \\ c_4 \\ c_5 \end{bmatrix}} + \begin{bmatrix} \alpha c_0 \\ 0 \\ 0 \\ 0 \\ \alpha c_6 \end{bmatrix} . \tag{4.30}$$

This matrix form seems not of immediate use to us but it does tell us several fundamental things about our solution. In Chapter 1 we were introduced to the notion of regarding a series of measurements such as concentration made at values of some other variable, such as wavelength, as a vector. In other words we can look at a table of function values, or a set of ordinates, as a vector in a space of as many diminsions as there are values. We can even look upon a function or its graph as a vector, or point, in an infinite dimensional space with one axis for each possible ordinate. To get a perfectly accurate solution we would need an infinitely small step size h and k, and consequently an infinite matrix. However, whether infinite or not, this matrix shares the property that differentiation has of preserving linearity, that is $[M]([\upsilon_1]+[\upsilon_2]) = [M][\upsilon_1]+[M][\upsilon_2]$. Thus, if we have two different initial vectors of concentrations we can add the solutions calculated separately to get the same final solution as we would have got by starting with the sum of the two initial concentration vectors. This property is the well-known 'superposition principle'.

4.10.2 Error and stability analysis

A detailed mathematical treatment of the factors affecting the stability of the procedure described in the previous section is outside the scope of this book, and the reader is referred elsewhere (Froeberg, 1969; Carley and Morgan, 1989). It is found that in order to ensure stability, the quantity $\alpha = Dk/h^2$ must satisfy

$$\alpha = Dk/h^2 < 1/2 \tag{4.31}$$

If $\alpha > 1/2$ then stability cannot be guaranteed, although it may not necessarily be lost. The contrast between stability and accuracy is well illustrated by this diffusion example since we can make h and k very small, and, assuming very accurate arithmetic, give a very small truncation error for the approximations to the derivatives. However, if $(Dk/h^2) > 1/2$ we can get instability despite this accuracy.

We can 'flesh out' the inequality (4.31) with some physical insight by noticing an interesting link with the Monte Carlo simulation of diffusion which we explored in section 1.6. There we encountered the Einstein–Smoluchowski equation which related the diffusion coefficient (a macroscopic quantity) with the jump distance and jump time used in the microscopic modelling. If we replace these two quantities with h and k, then the Einstein–Smoluchowski equation states that the diffusion coefficient D_M which would be equivalent to a Monte Carlo simulation on this particular distance–time mesh is given by

$$D_M = h^2/2k$$

Thus the inequality (4.31) reduces to

$$D < D_M$$

and the quantity D_M can be regarded as a sort of upper limit on the rate of diffusion that can be coped with for given values of h and k. Since the Monte Carlo method is simply following Nature's way, this conclusion has an intuitive 'correctness' to it.

We thus try to keep $\alpha = (Dk/h^2) < 1/2$ for stability, but a limitation of this explicit method is that if we attempt to increase the accuracy by reducing h to, say, $h/10$ then we must reduce k to $k/100$ to keep α constant; as a consequence we need to use 1000 times more points. The program in listing P4.8 implements this explicit method for our gel diffusion problem. Fig. 4.14 shows the effect of instability arising from too large a value for α and also compares the values of the solution as h and k are reduced, demonstrating the convergence to the true solution. In practice, we reduce

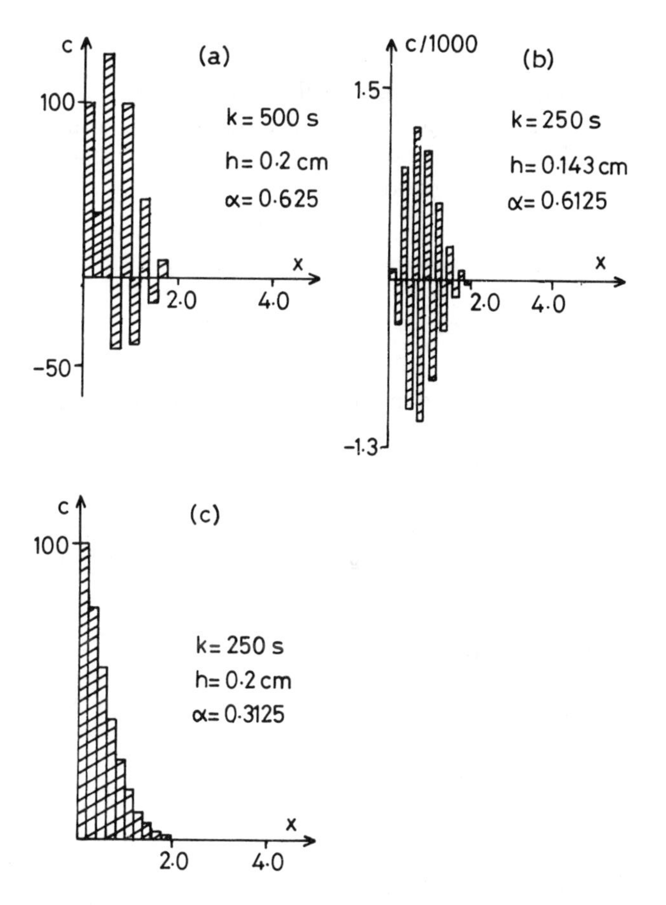

Fig. 4.14 — Solution plots for the gel diffusion problem, for various combinations of h and k, and consequent values of $\alpha = Dk/h^2$ ($D = 5\times10^{-5}$ cm^2 s^{-1}). A reasonable solution in these examples is obtained only if $\alpha = 0.3125$ (<0.5).

h and k (maintaining $Dk/h^2 < 0.5$) until we judge that convergence is occurring. This we achieve in the usual manner (see sections 4.6 and 4.7 and Chapter 3, for example), comparing the two solutions at common mesh points before and after the step reduction.

More complex approaches to the solution of partial differential equations lead to methods which overcome the stability constraint (4.31), the most well-known of which is due to Crank and Nicholson. A discussion of these methods will be found elsewhere (Carley and Morgan, 1989; Jost, 1960; Crank *et al.*, 1981).

4.10.3 Example 1: diffusion and reaction

In chemistry we are more likely to be interested in problems involving diffusion coupled with reaction, where the rate of change of concentration $\partial c/\partial t$ is made up of diffusion and reaction terms. For example, we might have

$$\partial c/\partial t = D(\partial^2 c/\partial x^2) - qc^m \tag{4.32}$$

where q is a reaction rate constant and m is the reaction order.

We shall consider the system consisting of a tube closed at both ends, containing two reactants A and B. A and B start off in opposite halves of the tube and diffuse into each other, reacting according to

$$\mathrm{A+B} \xrightarrow{k_1} \mathrm{C}$$

$$\mathrm{C+A} \xrightarrow{k_2} \text{products}$$

We shall be particularly interested in the concentration profile of C along the tube, and how it evolves with time as A and B interdiffuse and react. We also allow C to diffuse, in order to make the example realistic. The program to do this is given in P4.9, and the results of applying it are in Fig. 4.15, where we plot concentration profiles of A, B and C at various times. Note how the C profile becomes asymmetric with time as the step, C+A→products, gains force. This example uses the following parameters

$$k_1 = 0.1. \qquad k_2 = 0.01 \qquad D_A = D_B = 0.1 \qquad D_C = 0.05$$

and the initial ($t=0$) conditions

[A] = 2 throughout one half of the tube

[B] = 2 throughout other half of tube

[C] = 0 throughout all the tube

Analytical solutions are possible only for the simpler diffusion/reaction systems, such as the first order decay of a single component coupled with diffusion (Jost, 1960).

4.10.4 Example 2: Laplace's equation

The equations (4.28) and (4.32) are examples of what are known as parabolic equations which are very often solved in regions such as illustrated in Fig. 4.16(a). There are hyperbolic equations such as the wave equation

$$\partial^2 c/\partial t^2 = p^2(\partial^2 c/\partial x^2)$$

representing a vibrating wire and other oscillating systems, which in the same region need an extra initial condition since we have $\partial^2 c/\partial t^2$ instead of $\partial c/\partial t$ (Fig. 4.16(b), and elliptic equations such as

$$(\partial^2 c/\partial x^2) + (\partial^2 c/\partial y^2) = 0$$

which we came across in section 1.6 and which account for equilibrium systems like equilibrium diffusion in a flat plate (Fig. 4.16(c)). It is interesting to note that

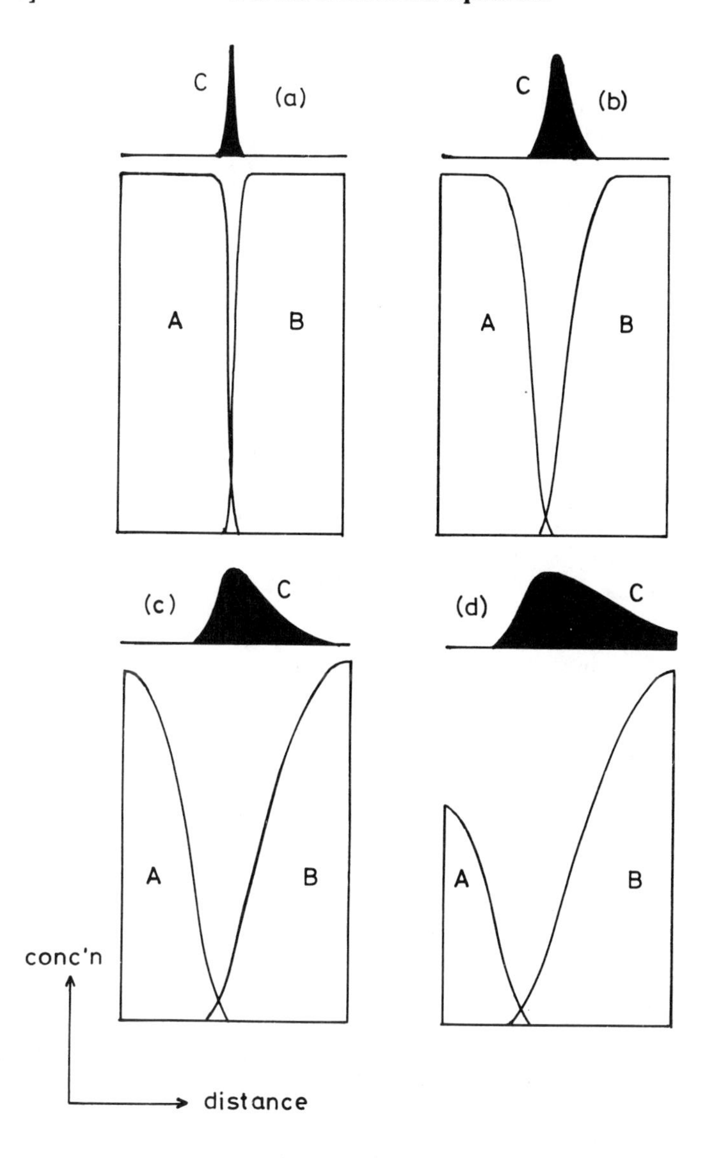

Fig. 4.15 — An example of diffusion with reaction. A and B start off in opposite halves of a closed tube, and diffuse into each other. They react to give C, which also diffuses along the tube and may be consumed by reaction with A. Concentration profiles are shown for A, B, and C at times (a) 30 s (b) 520 s (c) 2740 s (d) 8010 s.

Schroedinger's 'wave' equation is really a diffusion-type equation with an imaginary coefficient, D.

We can easily extend this treatment of the diffusion equation to further space dimensions, for example

$$\partial c/\partial t = D((\partial^2 c/\partial x^2)+(\partial^2 c/\partial y^2)) \tag{4.33}$$

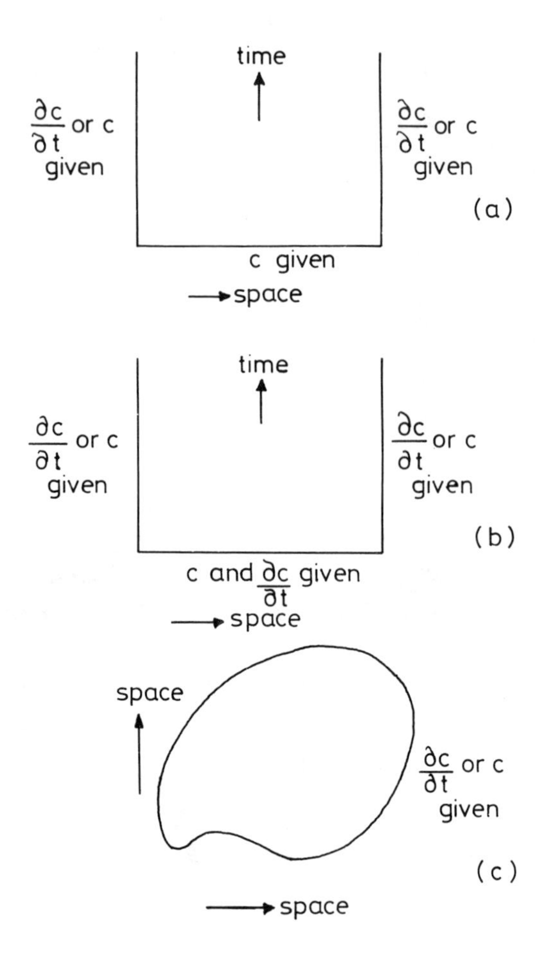

Fig. 4.16 — Regions, initial conditions and boundary conditions appropriate to the solution of different types of partial differential equation: (a) parabolic equations, such as the diffusion equation $\partial c/\partial t = D\partial^2 c/\partial x^2$; (b) hyperbolic equations, such as the wave equation $\partial^2 c/\partial t^2 = p^2\partial^2 c/\partial x^2$; (c) elliptic equations such as $\partial^2 c/\partial x^2 + \partial^2 c/\partial y^2 = 0$.

which can be replaced by

$$c_{j+1,l,m} - c_{j,l,m} = (Dk/h^2)(c_{j,l-1,m} + c_{j,l+1,m} - 4c_{j,l,m} + c_{j,l,m-1} + c_{j,l,m+1}) \quad (4.34)$$

where j is the time subscript and l,m are the x,y subscripts (compare (4.27)). Initial conditions are specified over, and boundary conditions specified around the border of, some closed region for all times subsequent. Equation (4.34) leads to an explicit expression for $c_{j+1,l,m}$ in terms of previous values, each new value being founded on a 'cross' of five values below it, with even more restrictive constraints on the value of $\alpha = Dk/h^2$.

We shall conclude this section by examining a way of approximating the solution of (4.33) when $t \to \infty$, that is at equilibrium when $\partial c/\partial t = 0$. For example, consider a shallow tank containing electrolyte, a rectangular outer electrode and a vertical wire inner electrode. If we connect a low voltage a.c. source to this cell what will be the peak-to-peak variation of voltage (E) through the electrolyte medium? This will be governed by Laplace's equation

$$(\partial^2 E/\partial x^2)+(\partial^2 E/\partial y^2) = 0 \tag{4.35}$$

with boundary conditions stating that the potentials at the outer and inner electrodes are fixed at E_O and E_I. We divide up the region into a mesh of points as shown in Fig. 4.17. The finite difference equation at each point is easily obtained from (4.34)

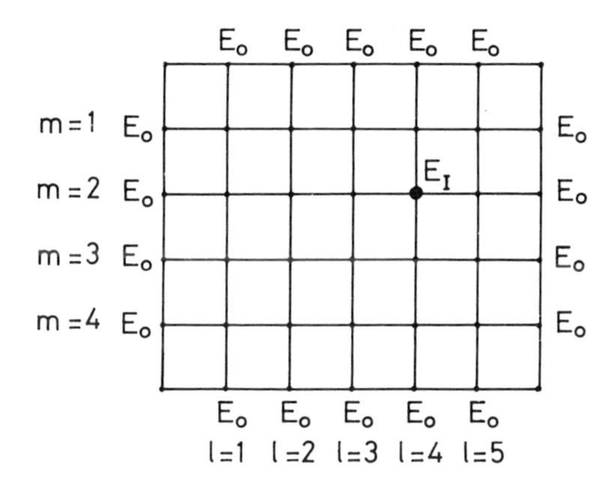

Fig. 4.17 — Mesh representing an electrolytic tank with a rectangular outer electrode at potential E_O, and an inner wire electrode at potential E_I. The variation of potential from mesh point to mesh point is described by Laplace's equation, which thus has to be solved at each point.

by setting the left-hand side equal to zero and substituting $E_{l,m}$ for $c_{j,l,m}$, etc. Note that the expression

$$E_{l-1,m}+E_{l+1,m}-4E_{l,m}+E_{l,m-1}+E_{l,m+1} = 0 \tag{4.36}$$

does not depend on h. If we use different step sizes h and k in the x and y directions then

$$\beta^2 E_{l-1,m}+\beta^2 E_{l+1,m}+E_{l,m-1}+E_{l,m+1}-2(1+\beta^2)E_{l,m} = 0 \tag{4.37}$$

where $\beta = k/h$. Equation (4.36) says that $E_{l,m}$ is the average of the four points around it and if we were to interpret the grid in Fig. 4.17 as a grid of wires of uniform resistance then we could express (4.36) as

$$(E_{l-1,m}-E_{l,m})+(E_{l+1,m}-E_{l,m})+(E_{l,m-1}+E_{l,m})+(E_{l,m+1}+E_{l,m}) = 0$$

which says that the sum of the potential differences between the centre and the surrounding points is zero. In other words the current flowing into the centre will equal that flowing out — this is simply a statement of Kirchhoff's first law.

Each interior point in Fig. 4.17 (including the inner electrode E_{42}) gives an equation which is summarized by

$$
\left[\begin{array}{cccccccccccccccccccc}
4 & -1 & 0 & 0 & -1 & 0 & 0 & 0 & 0 & 0 & 0 & 0 & 0 & 0 & 0 & 0 & 0 & 0 & 0 & 0 \\
-1 & 4 & -1 & 0 & 0 & -1 & 0 & 0 & 0 & 0 & 0 & 0 & 0 & 0 & 0 & 0 & 0 & 0 & 0 & 0 \\
0 & -1 & 4 & -1 & 0 & 0 & -1 & 0 & 0 & 0 & 0 & 0 & 0 & 0 & 0 & 0 & 0 & 0 & 0 & 0 \\
0 & 0 & -1 & 4 & 0 & 0 & 0 & -1 & 0 & 0 & 0 & 0 & 0 & 0 & 0 & 0 & 0 & 0 & 0 & 0 \\
-1 & 0 & 0 & 0 & 4 & -1 & 0 & 0 & -1 & 0 & 0 & 0 & 0 & 0 & 0 & 0 & 0 & 0 & 0 & 0 \\
0 & -1 & 0 & 0 & -1 & 4 & -1 & 0 & 0 & -1 & 0 & 0 & 0 & 0 & 0 & 0 & 0 & 0 & 0 & 0 \\
0 & 0 & -1 & 0 & 0 & -1 & 4 & -1 & 0 & 0 & -1 & 0 & 0 & 0 & 0 & 0 & 0 & 0 & 0 & 0 \\
0 & 0 & 0 & -1 & 0 & 0 & -1 & 4 & 0 & 0 & 0 & -1 & 0 & 0 & 0 & 0 & 0 & 0 & 0 & 0 \\
0 & 0 & 0 & 0 & -1 & 0 & 0 & 0 & 4 & -1 & 0 & 0 & -1 & 0 & 0 & 0 & 0 & 0 & 0 & 0 \\
0 & 0 & 0 & 0 & 0 & -1 & 0 & 0 & -1 & 4 & -1 & 0 & 0 & -1 & 0 & 0 & 0 & 0 & 0 & 0 \\
0 & 0 & 0 & 0 & 0 & 0 & -1 & 0 & 0 & -1 & 4 & -1 & 0 & 0 & -1 & 0 & 0 & 0 & 0 & 0 \\
0 & 0 & 0 & 0 & 0 & 0 & 0 & -1 & 0 & 0 & -1 & 4 & 0 & 0 & 0 & -1 & 0 & 0 & 0 & 0 \\
0 & 0 & 0 & 0 & 0 & 0 & 0 & 0 & -1 & 0 & 0 & 0 & 4 & 0 & 0 & 0 & -1 & 0 & 0 & 0 \\
0 & 0 & 0 & 0 & 0 & 0 & 0 & 0 & 0 & 0 & 0 & 0 & 0 & 4 & 0 & 0 & 0 & 0 & 0 & 0 \\
0 & 0 & 0 & 0 & 0 & 0 & 0 & 0 & 0 & 0 & -1 & 0 & 0 & 0 & 4 & -1 & 0 & 0 & -1 & 0 \\
0 & 0 & 0 & 0 & 0 & 0 & 0 & 0 & 0 & 0 & 0 & -1 & 0 & 0 & -1 & 4 & 0 & 0 & 0 & -1 \\
0 & 0 & 0 & 0 & 0 & 0 & 0 & 0 & 0 & 0 & 0 & 0 & -1 & 0 & 0 & 0 & 4 & -1 & 0 & 0 \\
0 & 0 & 0 & 0 & 0 & 0 & 0 & 0 & 0 & 0 & 0 & 0 & 0 & 0 & 0 & 0 & -1 & 4 & -1 & 0 \\
0 & 0 & 0 & 0 & 0 & 0 & 0 & 0 & 0 & 0 & 0 & 0 & 0 & 0 & -1 & 0 & 0 & -1 & 4 & -1 \\
0 & 0 & 0 & 0 & 0 & 0 & 0 & 0 & 0 & 0 & 0 & 0 & 0 & 0 & 0 & -1 & 0 & 0 & -1 & 4
\end{array}\right]
\begin{bmatrix} E_{11} \\ E_{12} \\ E_{13} \\ E_{14} \\ E_{21} \\ E_{22} \\ E_{23} \\ E_{24} \\ E_{31} \\ E_{32} \\ E_{33} \\ E_{34} \\ E_{41} \\ E_{42} \\ E_{43} \\ E_{44} \\ E_{51} \\ E_{52} \\ E_{53} \\ E_{54} \end{bmatrix}
=
\begin{bmatrix} 2E_O \\ E_O \\ E_O \\ 2E_O \\ E_O \\ 0 \\ 0 \\ E_O \\ E_O \\ E_I \\ 0 \\ E_O \\ E_O+E_I \\ 4E_I \\ E_I \\ E_O \\ 2E_O \\ E_O+E_I \\ E_O \\ 2E_O \end{bmatrix}
\tag{4.38}
$$

There are two main points to note about the matrix of (4.38). Clearly, about 20% of the matrix elements are non-zero and those that are are arranged fairly predictably. The matrix is not completely regular and the degree of regularity is dependent on the ordering of the mesh nodes in Fig. 4.17. We can make use of the sparseness of the matrix to cut down on the work of solving (4.38) only if we avoid an elimination method which would rapidly introduce non-zero elements; there are some sparse matrices which remain so after LU factorization (see Chapter 1), but not in general. We have used a large value of h in Fig. 4.17 so that (4.38) would be manageable to write down. If we had halved the step size we would have had a 80×80 matrix which would certainly pose a sizeable computational problem by elimination which takes time proportional to N^3 for a $N \times N$ matrix.

In cases like this iterative methods often save the day. In Chapter 1 we discussed the Gauss–Seidel and Jacobi iterative methods. The former method is to be preferred since it converges faster than the Jacobi procedure. It is also more natural to apply by computer program, since we just overwrite the kth values of the solution with the $(k+1)$th values as they are computed, whereas in the Jacobi method we need an intermediate store to hold all the $(k+1)$th values before they can replace the kth values in a new cycle.

For an interior point we merely replace $E_{l,m}$ by the average of the four points at the 'N, S, E, and W' positions immediately surrounding it; that is

$$E_{l,m}^{(k+1)} = (1/4)\,(E_{l-1,m}^{(k+1)} + E_{l,m-1}^{(k+1)} + E_{l+1,m}^{(k)} + E_{l,m+1}^{(k)})$$

but which of the values of E on the right-hand side actually carry the subscript $(k+1)$, and have thus been the subject of a previous row operation, will depend on the order in which the mesh points are dealt with. Next to the boundaries some of the E values will, of course, be fixed boundary values or false boundary points, created to simulate derivative boundary conditions. As we remarked earlier, these iterative methods are most useful when applied to large sparse systems with patterns of easily generated coefficients. Thus, a general-purpose Gauss–Seidel program would be rather uneconomical to write. It is interesting to make an analogy between this iterative solution and the repeated application of the original time-dependent equation (4.29), with c_{ij} replaced by E_{ij}, to find E_{ij} as j (time)$\to\infty$. We can think of each iterative cycle as also representing a step forward in time towards ∞, but with the intermediate solutions now being non-realistic. The use of (4.29) would give the same answer at large time, but would need to take more, shorter time steps in order to keep $\alpha < 1/2$. Of course, as a bonus we obtain the real solution as a function of time.

In Chapter 1 we tackled the identical problem using a Monte Carlo method. The outstanding feature of this approach is that it solves for an individual interior mesh point with no regard for the solution values at any surrounding points. The method of this section, in contrast, must solve for the complete set of mesh points, which may or may not be an advantage, depending on one's requirements.

```
    0 REM P4.1
   10 :
   20 REM tangent field plotter
   30 :
   40 GOTO 10000
   50 :
  400 REM compute derivative (slope) at (TZ, DZ)
  410 DZ= -UZ
  420 REM if any divisions by zero are likely to occur, protect the routine
  430 REM by resetting the divisor to a small, but finite, number
  440 RETURN
  450 :
10000 REM main segment
10010 INPUT "T RANGES FROM ";T1 : INPUT "TO";T2
10020 INPUT "U RANGES FORM ";U1 : INPUT "TO";U2
10030 INPUT "NUMBER OF VERTICAL STEPS";NU : NT=2*NU
10040 :
10050 HT=(T2 - T1)/NT : REM horizontal step size
10060 HU=(U2 - U1)/NU : REM vertical step size
10070 ST=250/(T2 - T1 + HT) : SU=150/(U2 - U1 + HU) : REM scaling factors
10080 REM extra HT and HU allows margin for plotting at edge
10090 :
10100 HGR : HCOLOR=3 : REM Apple graphics commands
10110 FC = .1 : TC=ST*FC*HT : UC=SU*FC*HU : REM  FC controls cross size
10120 FOR IT=0 TO NT
10130         TZ=T1 + IT*HT
10140         FOR IU=0 TO NU
10150                 UZ=U1 + IU*HU
10160                 TA=TZ - T1 : UA=UZ - U1
10170                 :
10180                 TS=ST*TA : US=SU*UA
10190                 HPLOT TS - TC,159 - US TO TS + TC,159 - US : REM cross
10200                 HPLOT TS,159 - US + UC TO TS,159 - US - UC : REM grid pt
10210                 :
10220                 GOSUB 400 : REM get derivative value
10230                 D=1/SQR(1 + DZ*DZ) : REM calc. cosine/sine(slope angle)
10240                 ET=.5*HT*D : EU=.5*HU*DZ*D
10250                 TB=(TA + ET)*ST : UB=(UA + EU)*SU
10260                 TA=TA*ST : UA=UA*SU
```

```
10270                 HPLOT TA,159-UA TO TB,159-UB
10280         NEXT IU
10290 NEXT IT
10300 END
```

```
    0 REM  P4.2
   10 :
   20 REM basic Euler method for solving an initial value problem
   30 :
   40 DIM U(100)
   50 GOTO 10000
   60 :
  400 REM differential equation DZ=F(TZ,UZ)
  410 REM example from Chapra and Canale (1985) p 498
  420 DZ=4*EXP(.8*TZ) - .5*UZ
  430 RETURN
  440 :
10000 REM main segment
10010 INPUT "STARTING TIME = ":T0
10020 INPUT "STARTING VALUE OF U =";U(0)
10030 INPUT "FINISHING TIME = ":TF
10040 INPUT "NO. OF STEPS = ":NS : IF NS>100 THEN PRINT "TOO MANY" : GOTO 10040
10050 H0=(TF - T0)/NS : REM calculate step size in time
10060 :
10070 FOR I=1 TO NS
10080         TZ=T0 + (I - 1)*H0 : UZ=U(I-1)
```

```
10090          GOSUB 400 :  REM get value of derivative
10100          U(I)=U(I-1) + H0*DZ : REM Euler step
10110 NEXT I
10120 :
10130 REM plot the results
10140 MX=0 : MI=0 : REM set max. and min. values, ensures T-axis on screen
10150 FOR I=0 TO NS
10160          IF U(I)>MX THEN MX=U(I) : REM find max. and min. values
10170          IF U(I)<MI THEN MI=U(I) : REM this is why an array is needed
10180 NEXT I
10190 IF ABS(MX - MI)<1E-30 THEN SU=0 : GOTO 10220
10200 :
10210 SU=150/(MX-MI) : REM scaling factors for plotting
10220 ST=INT(279/NS)
10230 HGR : HCOLOR=3
10240 FOR I=0 TO NS
10250          Y=( U(I) - MI )*SU
10260          HPLOT I*ST,159 - Y : REM origin at top LHS of screen
10270 NEXT I
10280 HPLOT  0,159 + MI*SU TO NS*ST,159 + MI*SU : REM plot T-axis
10290 END
```

```
    0 REM  P4.3
   10 :
   20 REM  Heun's method for first order initial value problems
   30 :
   40 DIM U(100)
   50 GOTO 10000
   60 :
  400 REM routine to calculate derivative from differential equation
  410 DZ= -UZ : REM simple first order decay
  420 RETURN
  430 :
 6000 REM  Heun's procedure
 6010 IC=0 : UZ=U0 :TZ=T0 + (I - 1)*H0 : GOSUB 400 : D1=DZ :REM reset IC, get deriv.
 6020 UP=U0 + H0*DZ : REM  Euler step as a Predictor
 6030 :
 6040 TZ=T0 + I*H0
 6050 UZ=UP : GOSUB 400 : REM get new derivative value
 6060 UC=U0 + H0*(D1 + DZ)/2 : REM  Corrector step
 6070 IF ABS(UP - UC)<AC THEN RETURN : REM iterate Corrector to convergence
 6080 REM number of iterations needed is a measure of need for smaller H0
 6090 REM it is possible to raise or lower step size depending on whether
 6100 REM fewer or greater than, say, 2 or 5 iterations have been needed
 6110 IC=IC + 1 : IF IC>MC THEN PRINT "CORRECTOR FAILED TO CONVERGE" : END
 6120 UP=UC : GOTO 6050 : REM use old corrector as new predictor
 6130 :
10000 REM main segment
10010 INPUT "STARTING TIME = ";T0
10020 INPUT "STARTING VALUE OF U = ";U(0)
10030 INPUT "FINISHING TIME = ";TF
10040 INPUT "NUMBER OF STEPS = ";NS : IF NS>100 THEN 10040
10050 REM can also dimension array using NS
10060 INPUT "ACCURACY OF CORRECTOR STEP = ";AC
10070 :
10080 H0=(TF - T0)/NS : MC=20 : REM calculate step size, set max. corr. cycles
10090 FOR I=1 TO NS
10100         U0=U(I-1)
10120         GOSUB 6000 : REM  Heun's procedure
10130         U(I)=UC
10140 NEXT I
```

```
10150 :
10160 REM now plot results
10170 MX=0 : MI=0 : REM initialize max. and min.,ensure T-axis on screen
10180 FOR I=0 TO NS
10190         IF U(I)>MX THEN MX=U(I) : REM find max. and min. values
10200         IF U(I)<MI THEN MI=U(I) : REM this is why we need an array U()
10210 NEXT I
10220 IF ABS(MX - MI)<1E-30 THEN SU=0 : GOTO 10250 : REM if U() all 0
10230 :
10240 SU=150/(MX - MI) : REM scale factors
10250 ST=INT(279/NS)
10260 HGR : HCOLOR=3
10270 FOR I=0 TO NS
10280         Y=( U(I) - MI )*SU
10290         HPLOT I*ST,159 - Y : REM origin at top left of screen
10300 NEXT I
10310 HPLOT 0,159 + MI*SU TO NS*ST,159 + MI*SU : REM plot T-axis
10320 END
```

```
    0 REM  P4.4
   10 :
   20 REM  Runge-Kutta method with interval doubling and halving
   30 :
   50 DIM U(100)
   60 NM=4 : REM order of method
   70 GOTO 10000
   80 :
  400 REM first order differential equation to be solved
  410 DZ=4*EXP(.8*TZ) - .5*UZ : REM ref. Chapra and Canale (1985) p 498
  420 RETURN
  430 :
 6200 REM  4th order Runge-Kutta procedure
 6210 TZ=T : UZ=U : GOSUB 400 : D0=DZ
 6220 U1=U + (H/3)*D0
 6230 :
 6240 TZ=T + H/3 : UZ=U1 : GOSUB 400 : D1=DZ
 6250 U2=U + H*( D1 - D0/3 )
 6260 :
 6270 TZ=T + 2*H/3 : UZ=U2 : GOSUB 400 : D2=DZ
 6280 U3=U + H*(D0 - D1 + D2)
 6290 :
 6300 TZ=T + H : UZ=U3 : GOSUB 400 : D3=DZ
 6310 UN=U + (H/8)*(D0 + 3*D1 + 3*D2 + D3)
 6320 RETURN
 6330 :
10000 REM main segment
10010 INPUT "STARTING TIME = ":T0
10020 INPUT "STARTING VALUE FOR U = ":U0
10030 INPUT "FINISHING TIME = ":TF
10040 INPUT "NUMBER OF STEPS = ":NS
10050 INPUT "ABSOLUTE ACCURACY REQUIRED = ":AC
10060 :
10070 H0=(TF - T0)/NS : REM calculate largest step size
10080 U(0)=U0 : CT=0 : P=10 : L0=2^P : L=L0 : REM L0= no. smallest steps in H0
10090 LR=L0
10100 FOR M=0 TO 1 : REM start of simulated REPEAT .... UNTIL loop
10110          :
10120          FOR K=0 TO 1 : REM start of simulated REPEAT .... UNTIL loop
```

```
10130                   H=H0*L/L0 : REM halve interval each time L=L/2
10140                   U=U0 : T=T0 : GOSUB 6200 : UA=UN : REM first R-K step
10150                   :
10160                   H=H/2
10170                   GOSUB 6200 : U=UN : REM T is still T0
10180                   :
10190                   T=T0 + H : GOSUB 6200 : UB=UN
10200                   ER=(UB - UA)/(2^(NM + 1) - 2)
10210                   REM compare est. error with tolerance - H=H/2 if too big
10220                   K=(ABS(ER)<AC) : IF K=0 THEN L=L/2 : K=1 if expression TRUE
10230                   IF L<1 THEN PRINT "STEP SIZE TOO SMALL" : END
10240           NEXT K
10250           :
10260           T0=T0 + 2*H
10270           U0=UB
10280           LR=LR - L : REM LR is no. of smallest possible sub-intervals left
10290           IF LR>0 THEN 10120
10300           CT=CT + 1 : U(CT)=U0 : PRINT "VALUE IS ":U0:" AT T= ":T0
10310           IF L<L0 THEN L=2*L : REM double step size for new interval if <H0
10320           LR=L0
10330           M=(CT>=NS) : REM TF reached? : M=1 if CT>=NS TRUE in Applesoft
10340 NEXT M
10350 :
10360 REM plot results
10370 MX=0 : MI=0 : REM ensure that max. and min. values never both > or <0
10380 FOR I=0 TO NS
10390           IF U(I)>MX THEN MX=U(I)
10400           IF U(I)<MI THEN MI=U(I)
10410 NEXT I
10420 IF ABS(MX - MI)<1E-30 THEN SU=0 : GOTO 10450 : REM stop divide by 0
10430 :
10440 SU=150/(MX - MI)
10450 ST=INT(279/NS) : REM scaling factors to fit plot to plot window
10460 HGR : HCOLOR=3
10470 FOR I= 0 TO NS
10480           Y=( U(I) - MI )*SU
10490           HPLOT I*ST,159 - Y
10500 NEXT I
10510 HPLOT 0,159 + MI*SU TO NS*ST,159 - MI*ST : REM draw T-axis
10520 END
```

```
   0 REM  P4.5
  10 :
  20 REM  Adams 4th order Predictor-Corrector method with Runge-Kutta starter
  30 :
  50 DIM U(100)
  60 NM=4 : REM order of method
  70 GOTO 10000
  80 :
 400 REM first order differential equation to be solved
 410 DZ=8.5 + TZ*(-20 + TZ*(12 -2*TZ) ) : REM Horner's scheme
 420 REM  ref. Chapra and Canale (1985) p 493
 430 RETURN
 440 :
1000 REM plot results
1010 MX=0 : MI=0 : REM ensure that max. and min. values never both > or <0
1020 FOR I=0 TO NS
1030         IF U(I)>MX THEN MX=U(I)
1040         IF U(I)<MI THEN MI=U(I)
1050 NEXT I
1060 IF ABS(MX - MI)<1E-30 THEN SU=0 : GOTO 1090 : REM stop divide by 0
1070 :
1080 SU=150/(MX - MI)
1090 ST=INT(279/NS) : REM scaling factors to fit plot to plot window
1100 HGR : HCOLOR=3
1110 FOR I= 0 TO NS
1120         Y=( U(I) - MI)*SU
1130         HPLOT I*ST,159 - Y
1140 NEXT I
1150 HPLOT 0,159 + MI*SU TO NS*ST,159 + MI*SU : REM draw T-axis
1160 RETURN
1170 :
6200 REM  4th order Runge-Kutta procedure
6210 TZ=T : UZ=U : GOSUB 400 : D0=DZ
6220 U1=U + (H/3)*D0
6230 :
6240 TZ=T + H/3 : UZ=U1 : GOSUB 400 : D1=DZ
6250 U2=U + H*( D1 - D0/3 )
6260 :
6270 TZ=T + 2*H/3 : UZ=U2 : GOSUB 400 : D2=DZ
```

```
6280 U3=U + H*(D0 - D1 + D2)
6290 :
6300 TZ=T + H : UZ=U3 : GOSUB 400 : D3=DZ
6310 UN=U + (H/8)*(D0 + 3*D1 + 3*D2 + D3)
6320 RETURN
6330 :
7100 REM use Runge-Kutta to find the next NS points
7110 FOR M=0 TO 1 : REM start of simulated REPEAT .... UNTIL loop
7120         :
7130         FOR K=0 TO 1 : REM start of simulated REPEAT .... UNTIL loop
7140                 H=H0*L/L0 : REM halve interval each time L=L/2
7150                 U=U0 : T=T0 : GOSUB 6200 : UA=UN : REM first R-K step
7160                 :
7170                 H=H/2
7180                 GOSUB 6200 : U=UN : REM T is still T0
7190                 :
7200                 T=T0 + H : GOSUB 6200 : UB=UN
7210                 ER=(UB - UA)/(2^(NM + 1) - 2)
7220                 REM compare est. error with tolerance - H=H/2 if too big
7230                 K=(ABS(ER)<AC) : IF K=0 THEN L=L/2
7240                 IF L<1 THEN PRINT "STEP SIZE TOO SMALL" : END
7250         NEXT K
7260         :
7270         T0=T0 + 2*H
7280         U0=UB
7290         LR=LR - L : REM LR is no. of smallest possible sub-intervals left
7300         IF LR>0 THEN 7130 : REM still some of interval left
7310         CT=CT + 1 : U(CT)=U0
7320         IF L<L0 THEN L=2*L : REM double step size if < H0
7330         LR=L0
7340         M=(CT>=NS) : REM check if NS points have been found
7350 NEXT M
7360 RETURN
7370 :
10000 REM main segment
10010 INPUT "STARTING TIME = ":TA
10020 INPUT "STARTING VALUE FOR U = ":U0
10030 INPUT "FINISHING TIME = ":TF
10040 INPUT "NUMBER OF STEPS = ":NT
```

```
10050 IF NS>100 THEN PRINT "TOO MANY" : GOTO 10040 : REM can use NT to DIM U()
10060 INPUT "ABSOLUTE ACCURACY REQUIRED IN R-K STARTER = ":AC
10070 :
10080 H0=(TF - TA)/NT : REM calculate step size
10090 U(0)=U0 : CT=0 : P=8 : L0=2^P : L=L0 : REM L0= no. smallest steps in H0
10100 LR=L0
10110 REM  start using Runge-Kutta
10120 NS=3 : U0=U(0) : T0=TA : GOSUB 7100
10140 LC=20 : REM set upper limit on number of correction cycles
11000 REM start Adams' method which needs 4 points to start it
11010 FOR I=3 TO NT-1
11020         FOR K=I-3 TO I
11030                 UZ=U(K) : TZ=TA + K*H0 : GOSUB 400
11040                 D(3-I+K)=DZ
11050         NEXT K
11060         :
11070         P=U(I) + (H0/24)*( 55*D(3) - 59*D(2) + 37*D(1) -9*D(0) )
11080         NC=0 : REM this is the Adams predictor - set correction counter
11090         UZ=P : TZ=TA + (I + 1)*H0 : CO=P
11100         GOSUB 400
11110         :
11120         C=U(I) + (H0/24)*( 9*DZ + 19*D(3) - 5*D(2) + D(1) )
11130         REM this is  the Adams corrector
11140         NC=NC + 1 : IF NC>LC THEN PRINT "TOO MANY CORRECTIONS" : END
11150         IF ABS(CO - C)>AC THEN CO=C : UZ=C : GOTO 11100
11160         U(I+1)=C : REM accept new value
11165         :
11170         REM error cancelling step
11180         PC=(19/270)*(P - C) : REM estimated truncation error
11190         IF ABS(PC)>ABS(C) THEN PRINT "UNSTABLE SOLUTION" : END
11200         U(I+1)=U(I+1) + PC : REM 19/270 left so to cf. text
11210         REM could use this error estimate as measure of need to reduce or
11220         REM increase  Adams step size - doubling would need no new values
11230         REM (unless right at start) but halving would need R-K restart
11240 NEXT I
11250 :
11260 NS=NT : GOSUB 1000 : REM plot results - NS is now no. pts. in plot
11270 END
```

```
   0 REM  P4.6
  10 :
  20 REM  Extrapolation applied to the trapezoidal rule for initial value
  30 REM  problems
  40 LOMEM: 16384 : REM Applesoft command to avoid clash with memory
  50 DIM R1(100,10),R2(100,10),U(100)
  60 L=0 : REM counter giving level of subdivision
  70 GOTO 10000
  80 :
 400 REM differential equation expression
 410 DZ=4*EXP(.8*TZ) - .5*UZ : REM example from Chapra and Canale (1985) p498
 420 RETURN
 430 :
1000 REM plot results
1010 MX=0 : MI=0 : REM ensure that max. and min. values never both > or <0
1020 FOR I=0 TO NS
1030         IF U(I)>MX THEN MX=U(I)
1040         IF U(I)<MI THEN MI=U(I)
1050 NEXT I
1060 IF ABS(MX - MI)<1E-30 THEN SU=0 : GOTO 1090 : REM stop divide by 0
1070 :
1080 SU=150/(MX - MI)
1090 ST=INT(279/NS) : REM scaling factors to fit plot to plot window
1100 HGR : HCOLOR=3
1110 FOR I= 0 TO NS
1120         Y=( U(I) - MI)*SU
1130         HPLOT I*ST,159 - Y
1140 NEXT I
1150 HPLOT 0,159 + MI*SU TO NS*ST,159 + MI*SU : REM draw T-axis
1160 RETURN
1170 :
6400 REM store current row as penultimate row
6410 FOR I=1 TO NS : FOR K=1 TO L+1 : R1(I,K)=R2(I,K) : NEXT K : NEXT I
6420 RETURN
6430 :
6500 REM trapezoidal rule for initial value problems is Heun's method with
6510 REM only one iteration of the corrector
6520 H=H0/(2^L) : REM returns value at next mesh point defined by H0 ( H<H0 )
6530 FOR M=1 TO 2^L
```

```
 6540         UZ=UC : TZ=TC : GOSUB 400 : DC=DZ
 6550         UZ=UC + H*DC : TZ=TC + H : GOSUB 400 : DN=DZ : REM Euler predict
 6560         UN=UC + (H/2)*(DC + DN)
 6570         UC=UN : TC=TC + H
 6580 NEXT M
 6590 RETURN
 6600 :
10000 REM main segment
10010 INPUT "STARTING TIME = ";T0
10020 INPUT "STARTING VALUE FOR U = ":U0
10030 INPUT "FINISHING TIME = ";TF
10040 INPUT "NUMBER OF STEPS = ":NS
10050 IF NS>100 THEN PRINT "TOO MANY" : GOTO 10040 : REM can use NT to DIM U()
10060 INPUT "ABSOLUTE ACCURACY REQUIRED = ";AC
10070 :
10080 H0=(TF - T0)/NS : REM calculate step size
10090 U(0)=U0
10100 :
10110 FOR I=1 TO NS
10120         UC=U(I-1) : TC=T0 + (I - 1)*H0
10130         GOSUB 6500
10140         R1(I,1) = UN : REM penultimate row
10150         U(I)=UN
10160 NEXT I
10170 :
10180 L=L + 1 : IF L>9 THEN PRINT "STEP TOO SMALL" : END : REM try halving step
10190 PRINT "LEVEL",L
10200 FOR I=1 TO NS
10210         UC=U(I-1) : TC=T0 + (I - 1)*H0
10220         GOSUB 6500
10230         R2(I,1)=UN : REM bottom row of triangle
10240         U(I)=UN
10250 NEXT I
10260 :
10270 REM calculate extrapolated values
10280 FOR I=1 TO NS
10290         FOR K=2 TO L+1
10300                 NK=4^(K - 1)
10310                 R2(I,K)=( NK*R2(I,K-1) - R1(I,K-1) )/(NK - 1)
```

```
10320          NEXT K
10330 NEXT I
10340 :
10350 FL=0 : REM if any point is too inaccuarate, FL will be set to 1
10360 FOR I=1 TO NS
10370          IF ABS( R2(I,L+1) - R1(I,L) )>AC THEN FL=1
10380 NEXT I
10390 IF FL=1 THEN GOSUB 6400 : GOTO 10180
10400 :
10410 FOR I=1 TO NS : U(I)=R2(I,L) : NEXT I
10420 GOSUB 1000 : REM plot results
10430 END
```

```
  0 REM  P4.7
 10 :
 20 REM  Runge-Kutta method with interval doubling and halving
 30 REM  adapted for a system of equations
 40 :
 50 LOMEM: 16384 : REM avoid clash with graphics memory
 60 DIM UU(10,100) : REM if over 10 equations were involved - DIM UZ() etc.
 70 NM=4 : REM order of method
 80 GOTO 10000
 90 :
400 REM set of simultaneous first order differential equations to be solved
410 REM  example from Bradley et al. (1973)
420 ON J GOSUB 510,520,530,540,550,560,570 : REM others may be added
430 RETURN
440 :
500 REM expressions for derivatives of UZ() with TZ
510 DZ= -K1*UZ(1)*UZ(2) + K3*UZ(6)*UZ(3) : RETURN
520 DZ= -K1*UZ(1)*UZ(2) : RETURN
530 DZ= K1*UZ(1)*UZ(2) - K2*UZ(3)*UZ(3) - K3*UZ(6)*UZ(3) : RETURN
540 DZ= K1*UZ(1)*UZ(2) : RETURN
550 DZ= K2*UZ(3)*UZ(3) : RETURN
```

```
 560 DZ= K2*UZ(3)*UZ(3) - K3*UZ(6)*UZ(3) : RETURN
 570 DZ= K3*UZ(6)*UZ(3) : RETURN
 580 :
 6200 REM  4th order Runge-Kutta procedure
 6210 TZ=T : FOR J=1 TO NE : UZ(J)=U(J) : NEXT J : REM compare with P4.4
 6220 FOR J=1 TO NE
 6230          GOSUB 400 : D0(J)=DZ
 6240          U1(J)=U(J) + (H/3)*D0(J)
 6250 NEXT J
 6260 :
 6270 TZ=T + H/3 : FOR J=1 TO NE : UZ(J)=U1(J) : NEXT J
 6280 FOR J=1 TO NE
 6290          GOSUB 400 : D1(J)=DZ
 6300          U2(J)=U(J) + H*( D1(J) - D0(J)/3 )
 6310 NEXT J
 6320 :
 6330 TZ=T + 2*H/3 : FOR J=1 TO NE : UZ(J)=U2(J) : NEXT J
 6340 FOR J=1 TO NE
 6350          GOSUB 400 : D2(J)=DZ
 6360          U3(J)=U(J) + H*( D0(J) - D1(J) + D2(J) )
 6370 NEXT J
 6380 :
 6390 TZ=T + H : FOR J=1 TO NE : UZ(J)=U3(J) : NEXT J
 6400 FOR J=1 TO NE
 6410          GOSUB 400 : D3(J)=DZ
 6420          UN(J)=U(J) + (H/8)*( D0(J) + 3*D1(J) + 3*D2(J) + D3(J) )
 6430 NEXT J
 6440 RETURN
 6450 :
10000 REM main segment
10010 INPUT "NUMBER OF COMPONENTS = ":NE
10020 INPUT "STARTING TIME = ";T0
10030 :
10040 FOR J=1 TO NE
10050          PRINT "COMPONENT":J:": STARTING VALUE FOR U = ":: INPUT "";U0(J)
10060 NEXT J
10070 :
10080 INPUT "FINISHING TIME = ";TF
10090 INPUT "NO. STEPS = ":NS : IF NS>100 THEN 10040 : REM or DIM UU(NE,NS)
```

```
10100 INPUT "ABSOLUTE ACCURACY REQUIRED = ":AC
10110 :
10120 K1=2.9E13 : K2=1.55E12 : K3=1.1E13
10130 H0=(TF - T0)/NS : REM calculate largest step size
10140 FOR J=1 TO NE : UU(J,0)=U0(J) : NEXT J
10150 CT=0 : P=8 : L0=2^P : L=L0 : REM L0= no. smallest steps in H0
10160 LR=L0
10170 FOR M=0 TO 1 : REM start of simulated REPEAT .... UNTIL loop
10180       :
10190       FOR K=0 TO 1 : REM start of simulated REPEAT .... UNTIL loop
10200             H=H0*L/L0 : REM halve interval each time L=L/2
10210             T=T0 : FOR J=1 TO NE : U(J)=U0(J) : NEXT J
10220             GOSUB 6200
10230             FOR J=1 TO NE : UA(J)=UN(J) : NEXT J : REM 1st R-K step
10240             :
10250             H=H/2
10260             GOSUB 6200
10270             FOR J=1 TO NE : U(J)=UN(J) : NEXT J : REM T is still T0
10280             :
10290             T=T0 + H
10300             GOSUB 6200
10310             FOR J=1 TO NE : UB(J)=UN(J) : NEXT J
10320             FL=1 : REM set flag - will = 0 if one of NE errors >AC
10330             FOR J=1 TO NE
10340                   ER=( UB(J) - UA(J) )/(2^(NM + 1) - 2)
10350                   IF ABS(ER)>AC THEN FL=0
10360             NEXT J
10370             REM compare est. error with tolerance - H=H/2 if too big
10380             K=FL : IF K=0 THEN L=L/2
10390             IF L<1 THEN PRINT "STEP SIZE TOO SMALL" : END
10400       NEXT K
10410       :
10420       T0=T0 + 2*H
10430       FOR J=1 TO NE : U0(J)=UB(J) : NEXT J
10440       LR=LR - L : REM LR is no. of smallest possible sub-intervals left
10450       IF LR>0 THEN 10190
10460       CT=CT + 1 : FOR J=1 TO NE : UU(J,CT)=U0(J) : NEXT J
10470       IF L<L0 THEN L=2*L : REM double step size for next interval if <H0
10480       LR=L0
```

```
10490         M=(CT>=NS) : REM check if TF reached, CT>=NS is 1 if true
10500 NEXT M
10510 :
10520 REM plot results - modified to replace U() by UU(,)
10530 MX=0 : MI=0 : REM ensure that max. and min. values never both > or <0
10540 FOR I=0 TO NS
10550         FOR J=1 TO NE
10560                 IF UU(J,I)>MX THEN MX=UU(J,I)
10570                 IF UU(J,I)<MI THEN MI=UU(J,I)
10580         NEXT J
10590 NEXT I
10600 IF ABS(MX - MI)<1E-30 THEN SU=0 : GOTO 10450 : REM stop divide by 0
10610 :
10620 SU=150/(MX - MI)
10630 ST=INT(279/NS) : REM scaling factors to fit plot to plot window
10640 HGR : HCOLOR=3
10650 FOR I= 0 TO NS
10660         FOR J=1 TO NE
10670                 Y=( UU(J,I) - MI)*SU
10680                 HPLOT I*ST,159 - Y
10690         NEXT J
10700 NEXT I
10710 HPLOT 0,159 + MI*SU TO NS*ST,159 + MI*SU : REM draw T-axis
10720 END
```

```
0 REM  P4.8
10 :
20 REM basic solution of diffusion equation by simple explicit method
30 :
40 DIM CO(101),CN(101)
50 GOTO 10000
60 :
500 REM set initial conditions
510 FOR II=1 TO NC+1 : CO(II)=0 : NEXT II
520 RETURN
530 :
600 REM plot concentration profile along tube
610 MX=CO(0) : REM diffusion can never lead to a value higher than this
620 HGR : HCOLOR=3
630 IF ABS(MX)<1E-30 THEN SC=0 : GOTO 650
640 SC=150/MX
650 FOR II=0 TO NC
660         Y=SC*CO(II)
670         HPLOT II,159 TO II,159 - Y
680 NEXT II
690 RETURN
700 :
10000 REM main segment
10010 INPUT "LENGTH OF TUBE = ";LE
10020 INPUT "NUMBER OF CELLS = ";NC
10030 INPUT "SOURCE CONCENTRATION = ";C0
10040 INPUT "DIFFUSION COEFFICIENT = ";DC
10050 INPUT "DURATION OF SIMULATION = ";TM
10060 INPUT "NUMBER OF TIME STEPS = ";NT
10070 :
10080 H=LE/NC : K=TM/NT : CY=1 : REM space step and time step
10090 AL=K*DC/(H*H) : REM check that AL will be =< .5 to ensure stability
10100 IF AL>.2 THEN K=K/2 : CY=CY*2 : GOTO 10090 : REM AL<<.5 for more accuracy
10110 GOSUB 500
10120 J=0 : CO(0)=C0 : REM boundary condition at source end
10130 GOSUB 600 : Z=1 - 2*AL : REM plot initial values and set multiplier
10140 FOR L=0 TO 1 : REM start of simulated REPEAT....UNTIL loop
10150         :
10160         FOR I=1 TO NC
```

```
10170                 CN(I)=AL*( CO(I-1) + CO(I+1) ) + Z*CO(I)
10180         NEXT I
10190         :
10200         CN(NC+1)=CN(NC-1) : REM boundary condition at closed end
10210         FOR I=1 TO NC : CO(I)=CN(I) :   NEXT I
10220         J=J+1
10230         IF J=CY*INT(J/CY) THEN GOSUB 600 : PRINT "T =  ";K*J
10240         :
10250         L=(K*J>TM) : REM finished?
10260         :
10270 NEXT L
10280 END
```

```
  0 REM  P4.9
 10 :
 20 REM simulation of diffusion with reaction by simple explicit method
 30 REM reactants A and B are initially at either end of a sealed vessel
 40 REM A + B goes to C with rate constant K1, C reacts with more A to form
 50 REM products with rate constant K2 - time step K must be small enough to
 60 REM give stability with respect to the rate process and the diffusion
 70 :
 80 LOMEM 16384 : REM avoid clash with graphics memory
 90 DIM AO(101),AN(101),BO(101),BN(101),CO(101),CN(101)
100 GOTO 10000
110 :
500 REM set initial conditions
510 FOR I=1 TO NC : AO(I)=0 : BO(I)=0 : CO(I)=0 : NEXT I
520 N2=INT(NC/2)
530 FOR I=1 TO N2 : AO(I)=A0 : NEXT I
540 FOR I=N2+1 TO NC : BO(I)=B0 : NEXT I
550 :
560 AO(0)=A0 : AO(NC+1)=0 : BO(0)=0 : BO(NC+1)=B0 : CO(0)=0 : CO(NC+1)=0
570 RETURN
580 :
```

```
600 REM plot concentration profile along tube
610 MA=0
620 FOR I=1 TO NC
630         IF AO(I)>MA THEN MA=AO(I)
640         IF BO(I)>MA THEN MA=BO(I)
650         IF CO(I)>MA THEN MA=CO(I)
660 NEXT I
670 :
680 HGR : HCOLOR=3
690 ST=100/NC : SY=75/MA
700 FOR II=1 TO NC
710         YA=SY*AO(II)
720         YB=SY*BO(II)
730         YC=SY*CO(II)
740         IS=II*ST
750         HPLOT IS,159 TO IS,159 - YA
760         HPLOT IS + 140,159 TO IS + 140,159 - YB
770         HPLOT IS + 140,80 TO IS + 140,80 - YC
780 NEXT II
790 RETURN
800 :
10000 REM main segment
10010 INPUT "LENGTH OF TUBE = ":LE
10020 INPUT "NUMBER OF CELLS = ":NC
10030 INPUT "SOURCE CONCENTRATION A = ":A0
10040 INPUT "SOURCE CONCENTRATION B = ":B0
10050 INPUT "DIFFUSION COEFFICIENT FOR A = ":DA
10060 INPUT "DIFFUSION COEFFICIENT FOR B = ":DB
10070 INPUT "DIFFUSION COEFFICIENT FOR C = ":DC
10080 INPUT "RATE CONSTANTS: K1,K2":K1,K2
10090 INPUT "DURATION OF SIMULATION = ":TM
10100 INPUT "TIME STEP = ":K
10110 INPUT "PLOTTING CYCLE = ":PC
10120 :
10130 H=LE/NC : REM space step
10140 AA=K*DA/(H*H) : AB=K*DB/(H*H) : AC=K*DC/(H*H)
10150 GOSUB 500
10160 J=0 : GOSUB 600 : PRINT "T = 0"
10170 ZA=(1 - 2*AA) : ZB=(1 - 2*AB) : ZC=(1 - 2*AC)
```

```
10180 FOR L=0 TO 1 : REM start of simulated REPEAT....UNTIL loop
10190         :
10200         FOR I=1 TO NC : REM reaction is calculated by Euler's approx.
10210                 Q1=K1*AO(I)*BO(I)*K : Q2=K2*AO(I)*CO(I)*K
10220                 AN(I)=AA*( AO(I-1) + AO(I+1) ) + ZA*AO(I) - Q1 - Q2
10230                 BN(I)=AB*( BO(I-1) + BO(I+1) ) + ZB*BO(I) - Q1
10240                 CN(I)=AC*( CO(I-1) + CO(I+1) ) + ZC*CO(I) + Q1 - Q2
10250         NEXT I
10260         :
10270         FOR I=1 TO NC : REM new values become basis for next time step
10280                 AO(I)=AN(I) : BO(I)=BN(I) : CO(I)=CN(I)
10290         NEXT I
10300         AO(0)=AO(1) : AO(NC+1)=AO(NC-1) : REM boundary conditions
10310         BO(0)=BO(1) : BO(NC+1)=BO(NC-1) : REM concentration gradient at
10320         CO(0)=CO(1) : CO(NC+1)=CO(NC-1) : REM ends is made to be zero
10330         J=J+1
10340         IF J=PC*INT(J/PC) THEN GOSUB 600 : PRINT "T =  ";K*J
10350         :
10360         L=(K*J>TM) : REM finished?
10370         :
10380 NEXT L
10390 END
10400 REM it is straightforward to alter or expand lines 10210 to 10240 to
10410 REM simulate other reaction schemes, make the diffusion rates vary with
10420 REM other factors etc. but the scheme is limited by the fact that
10430 REM  D*K/(H*H) must be =<.5 to be stable (perhaps less for accuracy) and,
10440 REM if K1, K2 or other rate constants are large, this will place another
10450 REM limit on K - essential to compare results as H,K become smaller
```

5

Fitting straight lines and polynomials to experimental data

5.1 INTRODUCTION

Many experiments consist of the measurement of some physical quantity y as a function of a second quantity x, which we are able either to vary in a controlled manner or to measure accurately. An example of the former is an acid–base titration where we measure pH as a function of volume of liquid added, and of the latter the measurement of the concentration of a reacting species in solution as a function of time. The result of such an experiment is thus a set of pairs of values (x_i,y_i; $i = 1, \ldots, n$) and we often wish to establish what (if any) relationship exists between the dependent variable y and the independent variable x. This process usually reduces to testing the validity of a particular model, that is, determining how well a theoretical or empirical relationship or equation explains or fits the experimental data. If there is a choice of possible models then, all other considerations being equal, we would be disposed towards the one which fits the data 'best'.

The problem with fitting a model to 'normal' experimental data is complicated by the presence of experimental error or noise in the data. Fig. 5.1 shows a typical diagram that might be obtained by plotting the (x_i,y_i) pairs as points on a graph: the points are scattered about a trend line. We can draw a band enclosing the points, and the narrower the band the better the correlation between the y and x values; we leave a mathematical description of correlation until later in this chapter.

The model we wish to test against this data will yield us a relationship between y and x involving a number of parameters p_j

$$y_i^* = F(p_1, p_2, \ldots, p_m; x_i) \qquad (5.1)$$

where y_i^* is the predicted value of y at $x = x_i$. This function defines a curve whose exact shape depends on the parameters p_j.

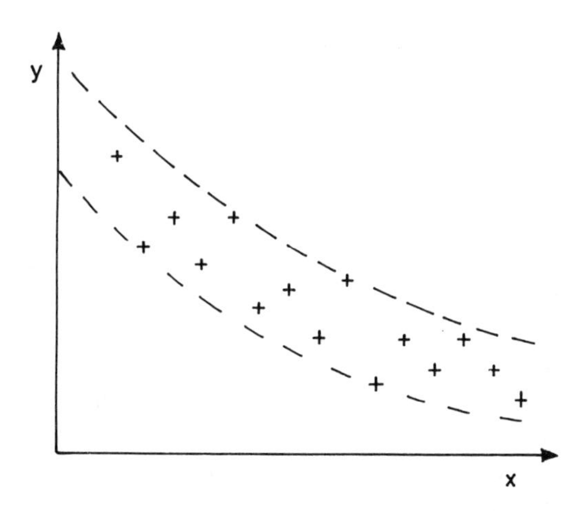

Fig. 5.1 — Plot of some typical noisy experimental data: a quantity y is measured as another quantity x is varied in a controlled manner. The points are observed to lie within a band (dotted lines), indicating some overall trend.

The aim of the exercise is then to choose the values of p_j which yield the curve which 'best' fits the experimental data; the parameters so obtained may then be given some physical interpretation the validity of which depends, naturally, on the validity of the original model (5.1). Our first important task is to give a precise mathematical meaning to the term 'goodness of fit'. We can then proceed to derive procedures for choosing the parameters p_j which optimize this quantity.

Intuitively, the curve which best fits a set of data plotted on a graph is that which leaves the experimental points most 'evenly balanced', in some ill-defined way, about that curve. In the simplest case, where the theoretical curve is a straight line, one can place a ruler on the graph paper and move it around until this criterion appears to be met and draw the line, a procedure most schoolchildren have gone through. In practice the eye and brain combination performs remarkably well in this straightforward situation, providing the scatter of points is not too great. For anything more complex, numerical methods have to be employed. A major drawback with the manual method is that it gives no quantitative idea of the goodness of fit, and no understanding of the processes involved.

Least squares analysis is based on the basic assumption that experimental noise (that is errors) obeys a normal or gaussian distribution, that is the probability $p(\Delta y)$ that a particular measurement is displaced by an amount Δy from its unknown but true (error-free) value is (section 6.6)

$$p(\Delta y) = \exp(-(\Delta y)^2/2\sigma^2)/(2\pi)^{1/2}\sigma \tag{5.2}$$

where σ is the standard deviation in the datum. Under suitable circumstances this property follows as a fundamental prediction of the Central Limit Theorem.

5.2 CENTRAL LIMIT THEOREM

A detailed discussion of this principle is beyond the scope of this book and we shall confine ourselves to the relevant essentials. The theorem states that if n_s samples are drawn at random from a parent population of mean M and standard deviation σ then the means M_s of the sample sets form a population of mean M and the standard deviation tends to $\sigma/(n_s)^{1/2}$ as n_s tends to infinity. Whether or not the original population is normal, the sample means tend to a normal distribution as n_s tends to infinity. Furthermore, the approximation is generally acceptable for $n_s \geqslant 15$ (see Lee and Lee, 1982). This theorem may appear quite esoteric to many readers, and of no obvious application, so we will give an example of the theorem in action before proceeding to show its application to the particular problem of noise distribution in experimental data.

Most microcomputers have a pseudo-random number generator as a feature of the BASIC interpreter (see Chapter 1). It is usually implemented by a statement of the form

```
X = RND(1)
```

which on the Apple II micro returns in X a random number between 0 and 1. A perfect generator would have a flat distribution, the probability $p(x)$ of generating a random number x being constant and independent of x. The observed distribution for the random number generator on the Apple II, derived by generating a large number of random numbers and partitioning them into small intervals $(x, x + \Delta x)$, is shown in the histogram plot in Fig. 5.2(a) and indicates satisfactory uniformity. If we consider the large number of values of x generated in the derivation of Fig. 5.2(a) to be the parent population referred to in the Central Limit Theorem (CLT), then clearly it has an extremely non-gaussian form. Choosing a value of $n_s = 12$, we derive a large set of sample means by generating 12 random numbers, taking the arithmetic average and repeating the process many times in order to produce the distribution shown in Fig. 5.2(b). This plot has the well-known bell-shaped form of the normal (gaussian) distribution (and with unit standard deviation for a value of $n_s = 12$), and in fact using the techniques of Chapter 6 a gaussian curve fits the derived data of Fig. 5.2(b) extremely closely. This demanding test provides a dramatic demonstration of one of the main predictions of the CLT. This observation also allows the construction of a gaussian noise generator (listing P5.1) which is useful when synthesizing 'real' data for investigating and demonstrating aspects of least squares methods. We shall make extensive use of such artificial data sets later in this chapter and also in Chapter 6.

The relevance of the CLT to the analysis of experimental measurements requires a little more explanation. The parent population we identify with the set of all possible measurements of the quantity y for a particular value of x. Hence if we repeat a particular measurement n_s times and take the average M_s of these measurements, then the population of all such means will form a normal distribution. As a consequence the probability that a given value M_s deviates from the mean of the sample means M (which equals the 'true' value of y for this specific value of x) by an amount dy is represented by the gaussian function (5.2) as required.

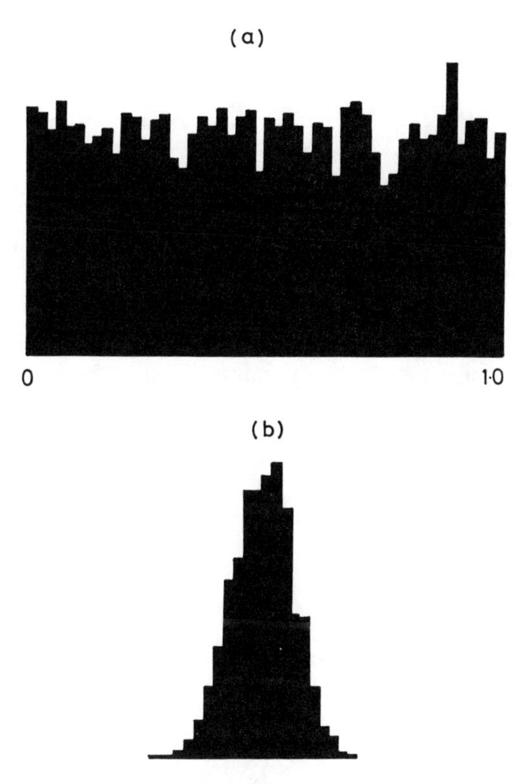

Fig. 5.2 — (a) Histogram plot showing the distribution of random numbers generated by a large number of calls (*c.* 5500) to the Applesoft (Apple BASIC) random number generator. (b) Distribution of the arithmetic means of sets of random numbers, exhibiting a characteristic gaussian peak-shape.

Furthermore, the CLT allows us to make some comments on experimental procedure:

(1) Values of the dependent variable y should be the result of averaging preferably more than 15 measurements made at each value of x if we are to ensure a normal distribution of experimental error and thus have confidence in the subsequent least squares analysis. The closer the 'natural' distribution of the errors is to being gaussian, the less stringent is this prescription.
(2) As well as improving the reliability of the least squares analysis, the use of large values of n_s leads to a reduction in data scatter since the standard deviation of the distribution of the sample means M_s is proportional to $1/(n_s)^{1/2}$. In spectral analysis this relationship is often expressed in terms of the quantity signal-to-noise (S/N) which is the ratio of the peak signal to the standard deviation of the noise (the noise often conveniently being measured in the baseline region), i.e. S/N is proportional to $(n_s)^{1/2}$ and hence repeated measurements (signal averag-

ing) lead to an increase in S/N and a resultant enhancement of the quality of the spectrum.

5.3 GOODNESS OF FIT

Assume we have made some measurements (x_i,y_i ; $i = 1, \ldots, n$) and have a physical model which yields a theoretical relationship (5.1) which we wish to fit to the experimental data, i.e. to choose the 'best' values of p_j ($j = 1, \ldots, m$). We will further assume that we may neglect any experimental error in the independent variable x_i, which as we commented at the beginning of this chapter is usually either under our precise control or may be measured very accurately, and thus that all the experimental uncertainty lies in the y_i values. Fig. 5.3 shows a general plot of

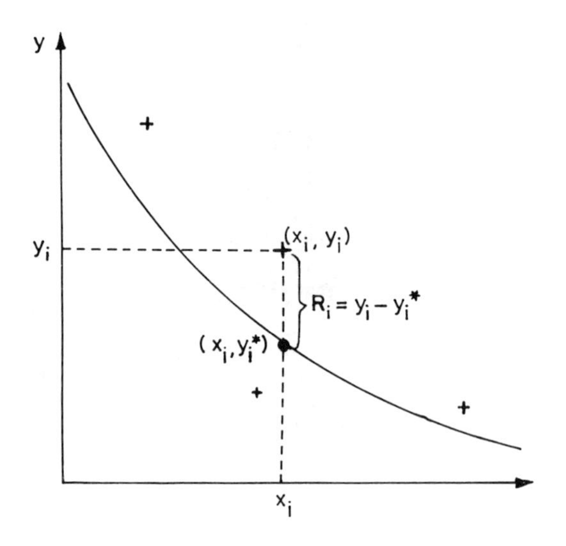

Fig. 5.3 — Plot of experimental (x,y) values together with a possible theoretical curve expressing the underlying physical relationship between y and x. The deviation of an experimental point from the curve is measured by the residual R_i.

experimental points together with a theoretical curve. Any criterion for goodness of fit is clearly going to incorporate the quantities R_i

$$R_i = y_i - y_i^* \qquad i = 1, \ldots, n \tag{5.3}$$

the differences between the measured and predicted ordinate values, and termed the residuals. The simplest possibility for a test of goodness of fit is to say that the points are 'balanced' about the curve if the negative residuals equal the positive residuals, i.e.

$$\sum_{i=1}^{n} R_i = 0 \tag{5.4}$$

However, a consideration of the graph in Fig. 5.4, shows that an infinite number of

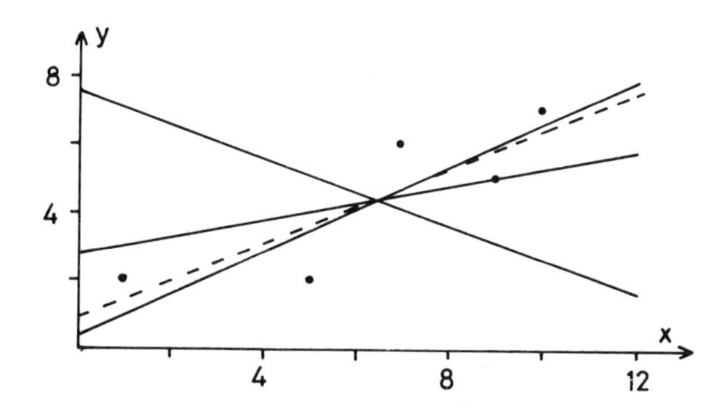

Fig. 5.4 — An infinite number of lines satisfying the criterion $\sum R_i = 0$ may be drawn through experimental data; in this example just three are shown (solid lines) to demonstrate the phenomenon. The dotted line is the 'best fit', computed as described later.

lines could be drawn satisfying this criterion. An attempt to improve this approach by requiring

$$S = \sum_{i=1}^{n} |R_i| \tag{5.5}$$

to be a minimum (since in general it will not be zero) is also unacceptable since minimizing S often requires it to be differentiable; unfortunately, at its minimum point there is a discontinuity in the slope of S.

The expression

$$S = \sum_{i=1}^{n} R_i^2 \tag{5.6}$$

looks more promising since it disregards the sign of the residuals (as did the absolute value function (5.5)), it exhibits no discontinuities in slope, and it only has one turning point which is a minimum point. In fact (5.6) forms the basis of the least squares principle — the best interpretation of an experiment is that which minimizes the sum of the squared errors of individual measurements. A mathematically rigorous derivation of this principle is given in section 6.6.

The expression (5.6) gives equal weight to all the data points, whereas strictly speaking we should minimize the weighted sum of squares function

$$S([p]) = \sum_{i=1}^{n} w_i R_i^2 \tag{5.7}$$

where the m-component vector $[p]$ is simply a convenient way of representing the set of parameters p_j ($j = 1, \ldots, m$). A discussion of the choice of the weights w_i is left until Chapter 6. During the remainder of this chapter we will assume all the weights are identical and equal to unity.

5.4 BEST STRAIGHT LINE

In general, the theoretical relationship we wish to fit to the data will yield a curve, but a commonly encountered variant is a straight line. This is sometimes only obtained after transforming the data and fitting function; the most popular example of this is taking logarithms and we will return to the possible problems inherent in this approach in section 5.6.

5.4.1 Mathematical analysis

The straight line variation of (5.1) is

$$y_i^* = p_1 x_i + p_2$$

or using the conventional nomenclature

$$y_i^* = m x_i + c \tag{5.8}$$

where the variable parameters are the slope of the line, m, and its intercept, c, on the y-axis (Fig. 5.5). The residuals are given by

$$R_i = y_i - m x_i - c \tag{5.9}$$

and the sum-of-squares function by

$$S([p]) = \sum_{i=1}^{n} R_i^2 = \sum_{i=1}^{n} (y_i - m x_i - c)^2$$

We seek those particular values of m and c which cause $S([p])$ to be a minimum. At

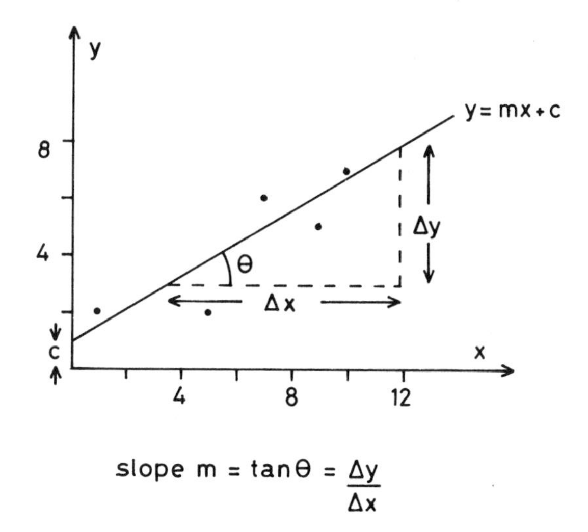

Fig. 5.5 — Graphical description of the parameters m (slope) and c (intercept) defining the straight line $y = mx + c$.

the minimum point the partial derivatives of S with respect to both m and c must be simultaneously zero (Chapter 1). Hence

$$\partial S/\partial m = \sum_{1}^{n} 2(y_i - mx_i - c)(-x_i) = 0$$

$$\partial S/\partial c = \sum_{1}^{n} 2(y_i - mx_i - c)(-1) = 0$$

Gathering terms and simpifying gives us what are called the normal equations for the system

$$c\sum x_i + m\sum x_i^2 = \sum x_i y_i \tag{5.10}$$

$$nc + m\sum x_i = \sum y_i \tag{5.11}$$

where the summations run from 1 to n. These are two simultaneous equations in m and c and may readily be solved using simple manipulation procedures, giving finally

$$m = \frac{n\sum x_i y_i - \sum x_i \sum y_i}{n\sum x_i^2 - \left(\sum x_i\right)^2} \tag{5.12}$$

and

$$c = \frac{\sum x_i^2 \sum y_i - \sum x_i \sum x_i y}{n\sum x_i^2 - \left(\sum x_i\right)^2} \tag{5.13}$$

and the best fit line is thus defined. If we divide (5.11) by n and rearrange we find

$$\left(\sum y_i\right)/n = m\left(\sum x_i\right)/n + c$$

or

$$\bar{y} = m\bar{x} + c$$

where $\bar{x}$ and $\bar{y}$ are the mean values of the independent and dependent variables respectively, showing that the best straight line passes through the centroid of the experimental data points.

The standard deviations σ_m and σ_c of the parameters m and c respectively are given by

$$\sigma_m^2 = n\sigma^2 / \left(n\sum x_i^2 - \left(\sum x_i\right)^2\right) \tag{5.14}$$

and

$$\sigma_c^2 = \sigma^2 \sum x_i^2 / \left(n\sum x_i^2 - \left(\sum x_i\right)^2\right) \tag{5.15}$$

where σ is the standard deviation of the fit (see equation 5.17). A detailed derivation of these expressions is provided by Francis (1984).

The best straight line is sometimes used to predict a value of y for a particular value of x, or conversely a value of x given y. Expressions for the errors in these calculated values may be similarly derived (Lee and Lee, 1982).

5.4.2 Computational procedure

A program to implement the method described above is reasonably straightforward and is given in listing P5.2. A problem arises with (5.12) and (5.13) as formulated since the denominator is the difference difference between two large numbers. If the data is such that this difference is close to the rounding error in the microcomputer (or to be more accurate, the error in the BASIC interpreter) then wildly inaccurate values for m and/or c may be obtained. An instructive example is provided by Lee and Lee (1982).

A formulation of the expressions for m and c which leads to improved accuracy may be derived as follows. Shifting the origin of the coordinate system from (0,0) to the centroid (centre of 'mass' of the points) $(\bar{x},\bar{y})$ and writing $X_i = x_i - \bar{x}$, $Y_i = y_i - \bar{y}$, we obtain via (5.12)

$$m = \sum X_i Y_i / \sum X_i^2 \tag{5.16}$$

c is then calculated from $\bar{y} = m\bar{x} + c$. Whereas the various terms in the original equations (5.12) and (5.13) could be calculated 'on the run' without storing the n (x_i,y_i) values, the improved method requires that $\bar{x}$ and $\bar{y}$ be first computed and then the (X_i, Y_i) values so that all the original data must be stored in memory. This is not likely to cause problems on any but the most primitive microcomputers, but might be an important consideration if the calculation is performed on a pocket calculator with limited storage capacity. It is left as an exercise for the reader to implement this modification to program P5.2.

5.4.3 Some comments on the quality of fit

In the previous analysis the quantity we have used to measure the goodness of fit is the sum of the squares of the errors or residuals, $S([p])$ (equation (5.7)). It must be emphasized that this number is useful only in finding the best-fit values of m and c, and provides no information on the validity of fitting a straight line to the data. The program P5.2 will blindly return optimum values of m and c, even if a plot clearly indicates a more complex trend line. The estimate of the standard deviation of the straight line fit, σ, is given by

$$\sigma^2 = \min(S([p]))/(n-2) \tag{5.17}$$

where $\min(S([p]))$ is evaluated at the minimum point, and σ may be regarded as a quantity which normalizes the sum of squares of the residuals to the number of degrees of freedom, $(n-2)$, and allows the goodness of fit of different sets of data to be compared, providing the magnitude of the y values is comparable. A high standard deviation, i.e. a poor fit, may be due to a bad scatter of points, the solution to which is either to improve the experiment or employ a numerical smoothing technique, or may be due to a bad choice of model. We are reminded that a least square analysis is only as good as the model used.

There are two commonly used methods for testing the reasonableness of a straight line fit: the correlation coefficient and a plot of residuals. Pearson's correlation coefficient, r, is defined by

$$r = \left(\sum X_i Y_i\right) / \left(\sum X_i^2 Y_i^2\right)^{1/2} \tag{5.18}$$

and its value lies in the range -1 to $+1$. If $r = 0$ then there is no correlation between x and y. If $r > 0$ then a direct correlation is indicated (y increasing with x) whereas if $r < 0$ then an inverse correlation (y decreasing with x) is suggested. If $r = +1$ or $r = -1$ then the points are perfectly correlated, but these ideal values can never be achieved in practice. It is a widespread practice to misuse the correlation coefficient as an indicator of the validity of a straight-line fit — a popular approach is to take $r > 0.95$ as the criterion for a linear relationship to exist between the (x,y) data. We emphasize that r only tells one how well the (x_i, y_i) pairs are correlated and conveys no information about whether a straight line or polynomial or more complex model best fits the data (section 5.4.5).

5.4.4 Examining the residuals

Up to this point the residuals have simply been used to construct the sum-of-squares function in order to provide a measure of the goodness of fit. However, they are much more useful than this might imply, and indeed are our best indicator of how well a line, curve or surface (if we have more than one independent variable) fits a set of data points. In this section we widen the discussion to develop ideas relevant to general model-fitting (Chapter 6) but which are of course applicable to the 'best straight line' special case.

A plot of residuals consists of a graph of R_i against x_i, where x_i represents the independent variable which in many cases is time. For a valid fit not only must the residuals be balanced evenly about zero, they must be so distributed randomly — non-linearity in the trend line is reflected in the presence of regions where the residuals are predominantly of the same sign. We assume generally (and correctly) that we will usually only be dealing with one independent variable. Apart from experimental considerations (it is often convenient to have only one independent variable), this is mainly due to the fact that we need an exponentially increasing amount of data to define the surfaces as we increase the number of dimensions (variables) — if 10 points adequately 'cover' some range of one independent variable, then it requires 100 points to give a similar 'coverage' of a rectangular region in two independent variables. In the pencil-and-paper days of numerical analysis this provided a severe barrier. However, it is not too difficult to program a microcomputer to plot residuals in two such variables either by labelling points on a two-dimensional graph with residual value, or to make a pseudo-three-dimensional plot using an isometric plotting program. Although we could plot residuals against each independent variable in turn, this would not reveal any effects where unusual behaviour in the residuals depends on a simultaneous change of the two independent variables.

If we have sufficient residuals then it may be instructive to plot a histogram of number of residuals against deviation from the fitted value, to see if it appears gaussian or not. If it appears significantly non-gaussian then it may be that there is some trend in the data that we have not accounted for; in this case we should look more deeply at our model function and also check that our fitting routine has found a genuine minimum. Alternatively, it may be that the true distribution of error is not gaussian, in which case we should question whether we should be using least-squares as our fitting criterion (Chapter 6).

Another valuable way of displaying the residuals is to plot them against a corresponding fitted value. This shows up any dependence of the standard deviation of the noise on fitted value. This occurs for example when we measure the absorption of light by a solution using a flickering light source. In this case we can express the observed signal as

$$\begin{aligned}\text{signal} &= \text{absorption} \times (\text{lamp mean intensity} + \text{noise}) \\ &= \text{absorption} \times \text{mean intensity} + \text{absorption} \times \text{noise}\end{aligned}$$

and the noise will appear additive but its standard deviation will vary with absorption. Similarly, when a measurement consists of counting particles, for example photoemitted electrons in photoelectron spectroscopy, the noise varies as the square root of the signal (counts). Thus if we are, for example, fitting complex spectral peaks with sums of simpler peak-shapes (Chapter 6) and we find that the noise (residual) intensity increases with increasing fitted value, so that the tops of the peaks are noisier than the baseline, then we should consider the necessity of using weighted least squares.

5.4.5 Some instructive examples

In order to investigate aspects of numerical analysis it frequently proves fruitful to apply the analysis in the first instance to synthesized data, where the true result is known beforehand and may be compared with the estimated one.

In Fig. 5.6(a) are plotted data calculated from $y_i = 1.5x_i + 1.0 + n_i$ where n_i is the error in y_i, derived from the gaussian noise generator (listing P5.1). Also shown is the best straight line computed using program P5.2. The predicted parameter values are $m = 1.501$ and $c = 1.017$, in excellent agreement with the true values. The value of the correlation coefficient is close to unity ($r = 0.995$) and the random distribution of the residuals about a zero mean (Fig. 5.6(b)) indicates that the assumption of a linear relationship is valid (as it must be in this case).

Data derived from $y_i = 1.5\exp(-0.5x_i)$ are plotted in Fig. 5.7(a); the best straight line with $m = -0.467$ and $c = 1.416$ and a correlation coefficient $r = 0.992$ is also shown. The merest visual inspection tells us that the data is significantly non-linear and this is confirmed, if confirmation were really needed, by the residuals plot (Fig. 5.7(b)) which shows clear non-random behaviour. It is worth noting that despite this obvious poor straight-line fit, the correlation coefficient is close to unity. The situation becomes less clear when we add some noise to the data of Fig. 5.7(a); a visual inspection of the fit is no longer to be trusted and we must rely on the residuals.

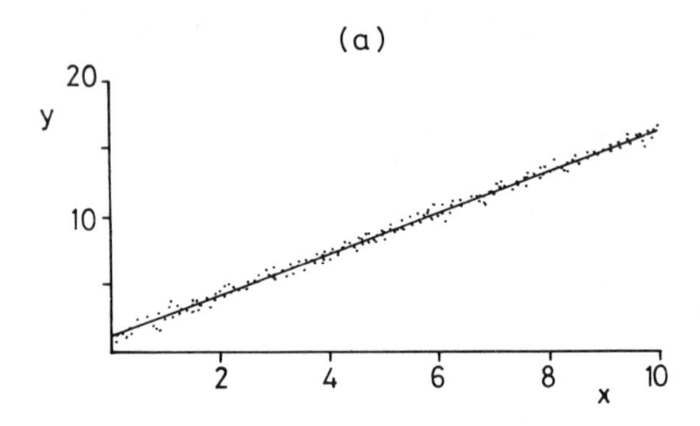

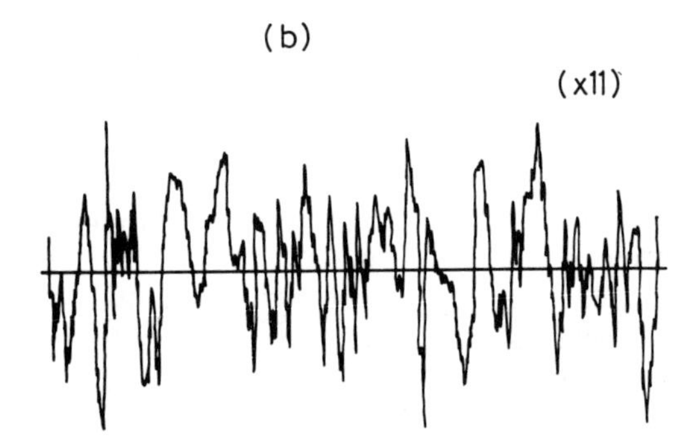

Fig. 5.6 — (a) Plot of data points (x,y) calculated from the equation $y^* = 1.5x + 1.0$, with the addition of gaussian noise. The solid line is the least-squares fit computed using program P5.2. (b) Plot of the residuals $R_i = y_i^* - y_i$ for the fit in (a).

The results of Fig. 5.8 still point clearly to non-linearity, but when the noise level is increased by 50% (Fig. 5.9) there is only a diluted indication in the residuals (Fig. 5.9(b)) that the straight-line fit is not valid. Once again, the correlation coefficients provide little help. This is a pointed reminder that we must get the model right first— we cannot expect the data, especially if it is noisy, to give us an unambiguous indication as to whether the model is valid.

Clearly, we are also limited by relying on a visual examination of the residuals, and would benefit from a more rigorous, programmable procedure. This is pursued in the next section.

5.4.6 Tests for randomness

Non-randomness in the residuals plot is a sure sign of an inadequate model for the data, even in those cases where the error (noise) distribution is non-gaussian and where the least-squares criterion cannot be guaranteed to give good results (although it may in fact do so). A test of the residuals is also applicable in those cases we are

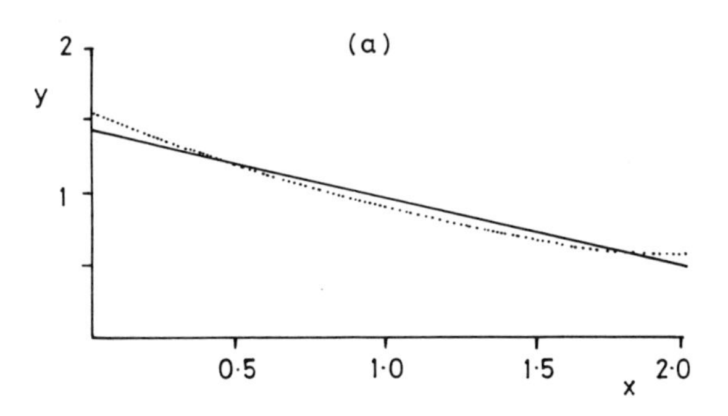

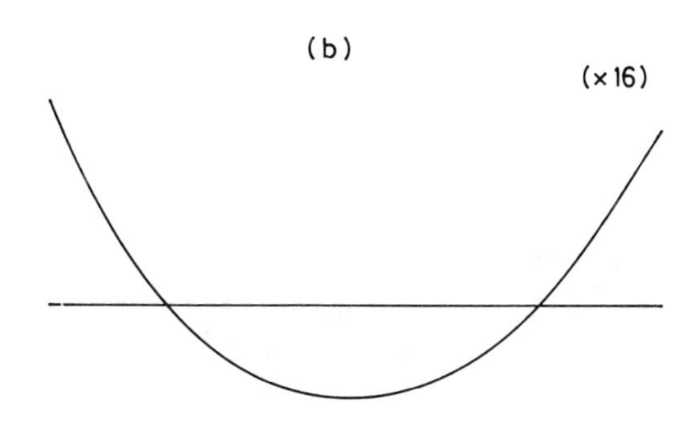

Fig. 5.7 — (a) Data points calculated from $y^* = 1.5\exp(-0.5x)$ with no added noise. The solid line is the least squares best-line fit to this clearly non-linear data. (b) Residual plot for (a), highlighting the inadequacy of the straight line fit.

unable, through a lack of replicate data values, to estimate the standard deviations σ_i for each datum point and so use more sophisticated methods (Chapter 6).

The runs test looks for non-randomness in the residuals by counting the number of runs of positive or negative values. If we have a total of $n_+ = 2$ positive residuals and $n_- = 4$ negative ones, and say three runs such as

$$(+)(----)(+) \quad \text{or} \quad (--)(++)(--)$$

then we pose the question, 'Is this a likely number of runs to have arisen by chance?'. If so we remain happy with our hypothesis of randomness; if not we may choose to reject it.

There are $(n_+ + n_-)!/n_+!n_-! = 6!/2!4! = 15$ possible combinations of two positive and four negative signs (Perrin, 1970). If we were to write all of these out (the reader may care to do this) we would find three arrangements giving five runs, six arrangements with four runs, four with three runs and two with two runs. If we

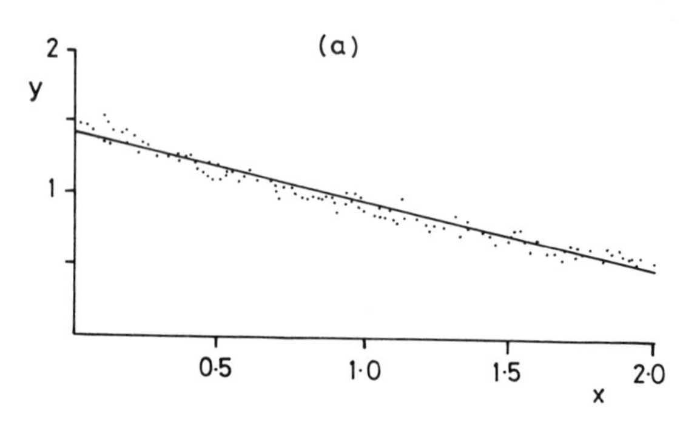

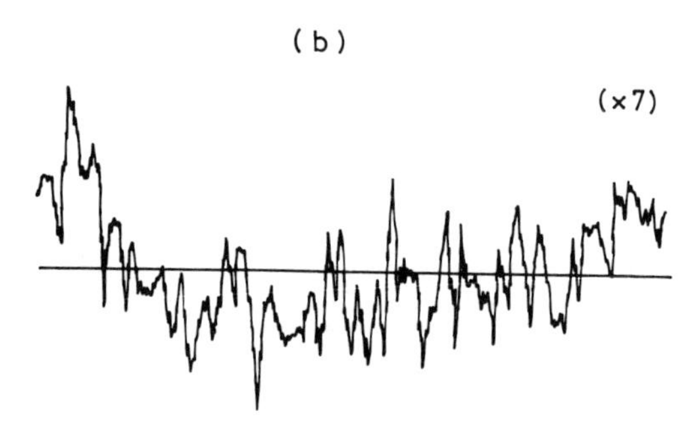

Fig. 5.8 — (a) Data of Fig. 5.7(a) with added noise, re-fitted with a straight line. (b) Residuals plot for (a); the non-linearity of the data is still obvious.

assume that arrangements with a small number of runs (say, three or less) such as $(++)\ (----)$ indicate non-randomness, we can say that the probability of getting three or fewer is

> (number of arrangements with three or fewer runs)/(total number of arrangements)

which equals $6/15 = 0.4$; thus, it is not a rare event. If, however, we had $n_+ = n_- = 8$ and three runs again, there are two ways of getting two runs and 14 ways of getting three runs, out of a total of 12,870 possible arrangements. There is thus a probability of $16/12{,}870 \simeq 0.0012$ of getting three or less runs by chance; this is clearly a rare event and we would be suspicious of our curve fit. We might also be suspicious of the distribution $(+-+-+-+-+-+-+-+-)$ with 16 runs (a very rare event) as showing a strong negative correlation between successive residuals, indicating perhaps some feedback in our instrumentation. Depending on the circumstances we thus choose to consider very large or very small numbers of runs as

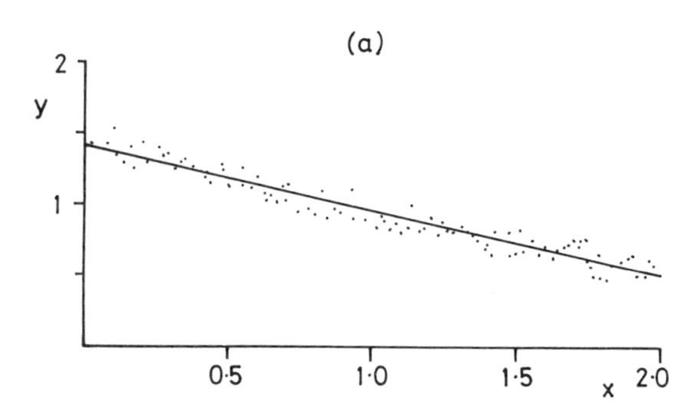

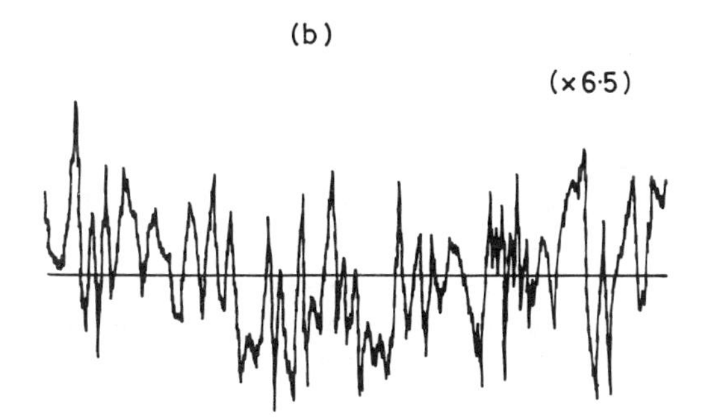

Fig. 5.9 — (a) Data of Fig. 5.8(a) with even more noise incorporated. The best straight line for this data is also shown (solid line). (b) Residuals plot for (a); the non-linearity is only just detectable.

suspicious. Commonly we set a limit of, say, 0.05 and if the probability of obtaining the observed number of runs by chance falls below this value, we reject the hypothesis of randomness. If we also accept a large number of runs as indicative of non-randomness, then we might set half this value, 0.025, at either end of the probability distribution of runs (see later). If the total number of residuals is very small then, since there are few possible alternatives, none of them will be particularly improbable, and a low value of the probability limit will be impossible to apply.

Program P5.3 calculates the probability $P(n,n_1,r)$ of obtaining r runs by chance in a series of n values n_1 of which are positive and the rest (n_2) negative (note that the same procedure applies as a test of randomness to any series of events each of which has two alternative states or values). The mathematical basis of the computation is as follows. If r is odd then $(r+1)/2$ are positive runs and $(r-1)/2$ negative, or vice versa. considering the former case, we can regard the set of n_1 positive values as being divided into $(r+1)/2$ subsets (runs) by $(r+1)/2-1=(r-1)/2$ partitions. For example, if $n_1=10$ and $r=7$ then there are four positive runs with three partitions, such as (+ + / + + + / + + + + / +); each slash represents a run of negative values in practice. Now, there are $(n_1 - 1)$ possible positions between values where the $(r-1)/2$ partitions could be placed. Since the number of ways of distributing q things among p piles is given by

$$ {}^{p}_{q}\mathrm{C} = p!/(p-q)!q! $$

we have ${}^{n_1-1}_{(r-1)/2}\mathrm{C}$ possible sets of $(r+1)/2$ positive runs. Similarly, there are ${}^{n_2-1}_{(r-3)/2}\mathrm{C}$ ways of getting $(r-1)/2$ negative runs. The total number of ways of obtaining $(r+1)/2$ positive and $(r-1)/2$ negative runs is then given by the product of these two numbers $({}^{n_1-1}_{(r-1)/2}\mathrm{C})\ ({}^{n_2-1}_{(r-3)/2}\mathrm{C})$. If instead we had had $(r-1)/2$ positive and $(r+1)/2$ negative runs we would have calculated $({}^{n_2-1}_{(r-1)/2}\mathrm{C})\ ({}^{n_1-1}_{(r-3)/2}\mathrm{C})$; the grand total of arrangements is thus the sum of these two quantities. A similar procedure is used in the case that r is even to give the number of possible arrangements as $2({}^{n_1-1}_{(r-2)/2}\mathrm{C})$ $({}^{n_2-1}_{(r-2)/2}\mathrm{C})$. To get the probability of obtaining r runs by chance we must divide this sum by the number of possible arrangements of n_1 positive and n_2 negative values, which equals ${}^{n_1+n_2}_{n_1}\mathrm{C}$ $(={}^{n}_{n_1}\mathrm{C})$ where $n = n_1 + n_2$. Finally, in order to compute the quantity we are after, the probability of obtaining n_r or fewer runs by chance, we must perform the calculation for $r = 2, \ldots, n_r$ and add all the probabilities together.

A common way to display the probability distribution is to tabulate the cumulative probability $Q(n,n_1,r)$ which is the probability of obtaining r or fewer runs. Since it is a cumulative probability, its value approaches unity as r approaches its maximum value; this is n if $n_1 = n_2$ (and corresponds to alternating positive/negative values), or $2n_1 + 1$ or $2n_2 + 1$, whichever is the smaller, if $n_1 \neq n_2$. Then if our observed value of r gives an accumulated probability close to 0 or 1, we suspect that a factor exists that we have not accounted for.

If both n_1 and n_2 are greater than 10, we can approximate the runs distribution by a normal (gaussian) distribution, which makes the computation easier (see Draper and Smith, 1966). The mean of the normal distribution μ and the standard deviation σ are given by

$$ \mu = 1 + 2n_1n_2/(n_1 + n_2) $$

$$ \sigma^2 = 2n_1n_2(2n_1n_2 - n_1 - n_2)/((n_1 + n_2)^2\ (n_1 + n_2 - 1)) $$

Thus if we subtract the mean from r and divide by σ we obtain the quantity $z(r)$

$$ z(r) = (r - \mu + 1/2)/\sigma $$

which should come from a gaussian distribution with zero mean and unit standard deviation

$$ z(r) = (1/\sqrt{2\pi})\ \exp(-r^2/2) $$

(for an explanation of the factor $+1/2$ see Draper and Smith (1966)). Thus to calculate the probability of obtaining r or less runs we perform a numerical integration of this gaussian distribution from 'minus infinity' (in practice a sufficiently large negative number) up to $z(r)$, using one of the methods discussed in Chapter 3; this is implemented in program P5.4.

If we apply P5.4 to the synthesized data of Figs 5.6, 5.7, 5.8 and 5.9 we obtain probability values of 0.988, 6.1×10^{-21}, 0.026 and 0.963 respectively. Recall that Fig. 5.6 was constructed from a linear model, whereas Figs 5.7–5.9 represent 'data' constructed from an exponential expression, with increasing amounts of noise. The two very low values correspond to situations where the non-linearity is unmistakeable from a visual inspection of the raw residuals (Figs 5.7 and 5.8). It would require the analysis of many more similar data sets to decide on the significance of the difference between the two values corresponding to the linear data (0.988) and the noisy exponential data (0.963); although the runs test provides a convenient and relatively impartial measure of randomness in the residuals, it can probably perform no better than the experienced eye.

5.4.7 Extended linear least squares

In the foregoing least squares analysis we have assumed that we can neglect the experimental error in the x_i and that all uncertainty resides in the y_i values. although this is a good approximation to many experimental situations, where we are often able to control the value of x_i precisely, we must also consider the case where there is significant measurement error in both the x_i and y_i values.

There are two basic procedures for accommodating this complication. The mathematically more rigid, but much more difficult, approach is to reformulate the least squares expression (5.6) replacing the residuals R_i by the distances d_i of the points from the nearest point P_i on the fitting curve — at this point the tangent to the curve is perpendicular to the line joining P_i to the datum point. The reader is referred to Churchhouse (1981) for details of the subsequent mathematical analysis. It suffices to say here that the best straight line still passes through the centroid of the data points, and m and c may be calculated, though not as easily as in the simpler case, In fact, a quadratic in m results, the two roots corresponding to two lines perpendicular to each other: one is the best straight line as required, the other is the worst straight line (that is, with the maximum sum-of-squares value) which also passes through the centroid $(\bar{x},\bar{y})$.

The second method is implemented in two symmetrical stages. The first is to assume that only the y_i are in error and to fit a straight line $y = mx + c$; the second considers that the error lies only in the x_i values and we fit a line $x = m'y + c'$ to the data. The latter is easily achieved using the standard program P5.2 by interchanging the values of x and y. We thus obtain two lines of slope (that is $\mathrm{d}y/\mathrm{d}x$) m and $1/m'$, both lines passing through the centroid $(\bar{x},\bar{y})$. Intuitively, we would expect that the smaller the angle between these two lines, then the better the correlation between the x_i and y_i values. In fact the ratio of the two slopes is equal to r^2 (Francis, 1984), where r is Pearson's correlation coefficient (5.16)

$$m/(1/m') = mm' = r^2 \tag{5.19}$$

A logical prescription, in those cases where the errors in x_i and y_i are of similar significance, is that the best straight line passes through $(\bar{x},\bar{y})$ and has a slope which is the average of m and $1/m'$, so bisecting the angle between the two alternative lines.

5.5 POLYNOMIAL CURVE-FITTING

In general, the theoretical relationship between y and x will not be linear, and a plot of y_i against x_i will indicate a curved trend-line. Although the experimental data may sometimes be transformed to give a straight line relationship, the validity of the subsequent linear least squares analysis must be viewed with a critical eye (section 5.6). We learned in Chapter 1 about the useful properties of polynomials and how we can approximate complicated functions by a power series; it is reasonable therefore to expect that they will provide a basis for extending the least squares method to the non-linear situation.

5.5.1 Simple polynomial curve-fitting

Let us consider, as before, a set of data points $(x_i,y_i;\ i=1,\ldots,n)$ and a theoretical model based on a power series

$$y_i^* = c_0 + c_1x_i + c_2(x_i)^2 + \ldots + c_m(x_i)^m = \sum_{j=0}^{m} c_j(x_i)^j \tag{5.20}$$

where the order of the polynomial m is less than n; if $m = n-1$ then the polynomial corresponds to the interpolating polynomial (Chapter 2). Proceeding as in the linear case, we seek to minimize

$$S = \sum_{i=1}^{n} (y_i - y_i^*)^2 \tag{5.21}$$

At the minimum in S, the first derivatives of S with respect to all the coefficients c_j $(j=0,\ldots,m)$ must be simultaneously equal to zero (Chapter 1). Now we have

$$\partial S/\partial c_k = (\partial/\partial c_k) \sum_{i=1}^{n} (y_i - y_i^*)^2$$

$$= \sum_{i=1}^{n} (\partial/\partial c_k)(y_i - y_i^*)^2$$

which gives us, recalling the rule for differentiating a function of a function,

$$\partial S/\partial c_k = \sum_{i=1}^{n} 2(y_i - y_i^*)(-\partial y_i^*/\partial c_k) \tag{5.22}$$

From (5.20)

$$\partial y_i^*/\partial c_k = (\partial/\partial c_k)(c_0 + c_1 x_i + c_2(x_i)^2 + \ldots + c_m(x_i)^m)$$

and all the terms disappear except the one containing c_k, giving $\partial y_i^*/\partial c_k = (x_i)^k$. Substituting this and the expression for y_i^* in (5.22) leads to

$$\partial S/\partial c_k = \sum_{i=1}^{n} (-2(x_i)^k)(y_i - c_0 - c_1 x_i - \ldots - c_m(x_i)^m)$$

The set of conditions $\partial S/\partial c_k = 0$ $(k = 0, \ldots, m)$ thus becomes

$$\sum_{i=1}^{n} (-2(x_i)^k)(y_i - c_0 - c_1 x_i - \ldots - c_m(x_i)^m) = 0 \qquad k = 0, \ldots, m$$

or

$$\sum_{i=1}^{n} (c_0(x_i)^k + c_1(x_i)^{k+1} + c_2(x_i)^{k+2} + \ldots + c_m(x_i)^{k+m}) = \sum_{i=1}^{n} (x_i)^k y_i$$
$$k = 0, \ldots, m \tag{5.23}$$

which may be written more concisely as

$$\sum_{j=0}^{m} c_j \sum_{i=1}^{n} (x_i)^{k+j} = \sum_{i=1}^{n} (x_i)^k y_i \qquad k = 0, \ldots, m \tag{5.24}$$

This rather unapproachable-looking expression in fact represents a set of $(m+1)$ normal equations in $(m+1)$ coefficients, c_k. As an illustrative example the set of simultaneous normal equations for $m = 2$ (that is, a quadratic polynomial) are, written out in full

$$c_0 \sum 1 \quad + c_1 \sum x_i \quad + c_2 \sum (x_i)^2 = \sum y_i$$

$$c_0\sum x_i \quad + c_1\sum (x_i)^2 + c_2\sum (x_i)^3 = \sum x_i y_i$$

$$c_0\sum (x_i)^2 + c_1\sum (x_i)^3 + c_2\sum (x_i)^4 = \sum (x_i)^2 y_i$$

where the summations run over all the data points, and $\sum 1 = n$. In matrix notation

$$\begin{bmatrix} \sum 1 & \sum x_i & \sum (x_i)^2 \\ \sum x_i & \sum (x_i)^2 & \sum (x_i)^3 \\ \sum (x_i)^2 & \sum (x_i)^3 & \sum (x_i)^4 \end{bmatrix} \begin{bmatrix} c_0 \\ c_1 \\ c_2 \end{bmatrix} \begin{matrix} = \\ = \\ = \end{matrix} \begin{bmatrix} \sum y_i \\ \sum x_i y_i \\ \sum (x_i)^2 y_i \end{bmatrix}$$

which in the general case is written

$$[M][c] = [d] \tag{5.25}$$

with

$$M_{kj} = M_{jk} = \sum_{i=1}^{n} (x_i)^{k+j} \qquad j,k = 0, \ldots, m \tag{5.26}$$

and

$$d_j = \sum_{i=1}^{n} (x_i)^j u_i \qquad j = 0, \ldots, m \tag{5.27}$$

This system of equations may be solved for the component vector $[c]$ (which is shorthand for saying that it may be solved for all the coefficients c_k, $k = 0, \ldots, m$), using one of the methods of Chapter 1. Listing P5.5 shows such a program for polynomial curve-fitting, which utilizes the Gauss–Jordan elimination method (with pivoting) for the solution of (5.25). Given the ill-conditioned nature of the problem (q.v.), it may be desirable to use the LU decomposition technique (Chapter 1). Once the c_k are known, the predicted value of y can be evaluated for any value of x within the range of the original data using (5.20) and Horner's Rule (Chapter 1). For reasons similar to those discussed in Chapter 2 in the case of interpolation polynomials, extrapolation outside the range of the known x_i data is inadvisable.

The standard deviations σ_k of the coefficients c_k $(k=0,\ldots,m)$ are conveniently estimated (Johnson, 1980) in terms of the diagonal elements of the inverse of the coefficient matrix $[M]$

$$\sigma_k^2 = \sigma^2(M^{-1})_{kk} \tag{5.28}$$

where σ^2 is the estimated variance of the fit given by

$$\sigma^2 = \min(S)/(n-m-1) \tag{5.29}$$

The reader is referred to Chapter 6 for a detailed discussion of the estimation of parameter errors.

We would appear now to have a near-perfect solution to the non-linear curve-fitting problem. Since most functions can be approximated by a Taylor polynomial (Chapter 1), then providing we take enough terms in the polynomial representation we ought in principle to be able to cope with most curve-fitting situations. Unfortunately, for $m>7$ the method as formulated here becomes virtually unusable. The problem lies with the coefficient matrix $[M]$ which suffers from being both near-singular and ill-conditioned. The former property manifests itself as a determinant value Carley and Morgan (1989) which is close to zero and a very large range in the magnitude of the elements of $[M]$, both of which symptoms rapidly worsen as m increases; ill-conditioning means that the size of the elements of the inverse matrix $[M]^{-1}$ is very large compared with the magnitude of the elements of $[M]$. Thus small rounding errors in the calculation of the elements of $[M]$ are greatly magnified by the inversion process. The reader is referred to Churchhouse (1981) for further discussion of this point. Whenever a simple polynomial curve-fit is performed, Horner's Rule (Chapter 1) should be used wherever possible to minimize errors in computation, and double-precision arithmetic selected on those microcomputers which offer this option.

In the next section we describe a generalization of the polynomial fitting technique which circumvents these difficulties, and permits the fitting of polynomials of much higher order.

5.5.2 Use of orthogonal polynomials

Let us consider a more general form of fitting function

$$y^* = c_0p_0(x) + c_1p_1(x) + c_2p_2(x) + \ldots + c_mp_m(x) \tag{5.30}$$

where $p_k(x)$ $(k=0,\ldots,m)$ is an arbitrary polynomial of order k, that is the highest power of x it contains is x^k. The polynomial used in section 5.5.1 is thus a particular example of this formulation with $p_k(x)=x^k$. Substituting (5.30) in (5.22) leads, via the conditions $\partial S/\partial c_k = 0$ $(k=0,\ldots,m)$, to the generalized normal equations $[M][c]=[d]$ where now

$$M_{jk} = M_{kj} = \sum_{i=1}^{n} p_k(x_i)p_j(x_i) \qquad j,k = 0, \ldots, m \tag{5.31}$$

and

$$d_k = \sum_{i=1}^{n} p_k(x_i)y(x_i) \qquad k = 0, \ldots, m \tag{5.32}$$

Note that putting $p_k(x_i) = (x_i)^k$ in (5.31) gives (5.26) as it should. The reason for this seemingly more complicated formulation of the problem becomes clear if we consider the simplifications arising from using orthogonal polynomials. A set of polynomials $(p_j(x_i); j = 0, \ldots, m)$ is defined to be orthogonal over the data set $(x_i;$ $i = 1, \ldots, n)$ if, for $j,k = 0, \ldots, m$

$$\begin{aligned} \sum_{i=1}^{n} p_j(x_i)p_k(x_i) &= 0 \qquad j = k \\ &= a_j \qquad j = k \end{aligned}$$

where the a_j are constants. If $a_j = 1$ $(j = 0, \ldots, m)$ then the polynomials are termed orthonormal. Readers should verify for themselves that the set of simple polynomials $(x_i)^k$ $(k = 0, \ldots, m)$ are not orthogonal, using an arbitrary data set. If we are able to find a set of orthogonal polynomials, then the substitution of these into (5.31) give us

$$M_{kj} = M_{jk} = 0 \qquad j \neq k$$

$$M_{jj} = a_j$$

and thus

$$(M^{-1})_{jk} = 0 \qquad j \neq k$$

$$(M^{-1})_{jj} = 1/a_j$$

so that

$$c_j = (1/a_j) \sum_{i=1}^{n} p_j(x_i)y(x_i)$$

$$= \sum_{i=1}^{n} p_j(x_i) y(x_i) / \sum_{i=1}^{n} p_j(x_i) p_j(x_i) \tag{5.33}$$

and there are no problems due to ill-conditioning. Remember that the c_j are now coefficients of (as yet unknown) polynomials $p_j(x)$ rather than coefficients of powers of x. Forsythe (1957) derived a set of rules for constructing an orthogonal set of polynomials from the data x. They are defined as follows

$$\begin{aligned}
&p_0(x) = 1\\
&p_1(x) = (x - \alpha_1)p_0(x)\\
&p_2(x) = (x - \alpha_2)p_1(x) - \beta_1 p_0(x)\\
&\vdots\\
&p_{j+1}(x) = (x - \alpha_{j+1})p_j(x) - \beta_j p_{j-1}(x)
\end{aligned} \tag{5.34}$$

where the quantities α_{j+1} and β_j are given by

$$\alpha_{j+1} = \sum x_i p_j^2(x_i) / \sum p_j^2(x_i) \tag{5.35}$$

$$\beta_j = \sum p_j^2(x_i) / \sum p_{j-1}^2(x_i) \tag{5.36}$$

where as before the summations are over all the data and superscripts represent powers. Since the α and β values depend on the values of all previously calculated polynomials evaluated at each x_i, the computation must be organized carefully. The reader is strongly advised to scrutinize listing P5.6, where Forsythe's method is implemented, until satisfied about the structure of this segment. The values of $\sum p_j^2(x_i)$ and $\sum p_j(x_i)y(x_i)$ are stored as the computation proceeds so that the coefficients c_j are easily calculated at the end of the computation. Knowing the α and β values, the predicted value $y^*(x)$ for any value of x in the range of the x_i may be evaluated according to (5.30) and (5.34) without having to express the polynomial sum (5.30) as an explicit power series (5.20). The latter may be achieved, if desired, in a relatively straightforward fashion (Johnson, 1980): if the coefficients of the m-order power series are c'_k $(k = 0, \ldots, m)$ and the coefficients of the Forsythe polynomial sum are $c_k (k = 0, \ldots, m)$ then

$$c'_k = \sum_{j=k}^{m} c_j b_{kj} \qquad k = 0, \ldots, m \tag{5.37}$$

where

$$
\begin{aligned}
b_{jk} &= 0 && \text{for } k<0 \text{ or } k>j \\
&= 1 && \text{for } k=j \\
&= b_{k-1,j-1} - \alpha_j b_{k,j-1} - \beta_{j-1} b_{k,j-2} && \text{for } 0 \leqslant k < j
\end{aligned}
$$

this rather tricky piece of programming is shown in listing P5.7.

Many other sets of orthogonal polynomials have been discovered, amongst the most well-known being the Chebyshev polynomials, $T_n(x)$, defined by

$$T_n(x) = \cos(n \arccos(x))$$

but more conveniently represented by the relation

$$
\begin{aligned}
&T_0(x) = 1 \\
&T_1(x) = x \\
&\vdots \\
&T_{n+1}(x) = 2xT_n(x) - T_{n-1}(x) \qquad (5.38)
\end{aligned}
$$

The use of Chebyshev polynomials in curve-fitting is discussed by Southworth and Deleeuw (1965); they are not as straightforward to use as Forsythe's polynomials.

5.5.3 Polynomial smoothing

One consequence of fitting a polynomial to a set of experimental data points (x_i, y_i; $i = 1, \ldots, m$) is that the predicted curve is 'smoother' than the experimental distribution of points. The 'degree of smoothing' decreases as the order of the polynomial, m, increases, and for $m = n - 1$ the predicted curve is the interpolation polynomial which corresponds to no smoothing. Savitzky and Golay (1964) have exploited this result and developed a simple but elegant method for smoothing noisy experimental data, which has found particular application to the suppression of noise in spectral lineshapes.

The procedure is an N-point algorithms, with N odd and less than n, and one runs through the data set point by point. At each point the $(N-1)/2$ points on either side are taken with it to form the current smoothing set containing N points. A simple polynomial of pre-chosen order (usually less than six) is fitted to this smoothing set by the least squares approach and the central ordinate replaced by the predicted (that is, smoothed) value. We then move to the next point in the original n-point data set, dropping a point from one end of the N-point smoothing set and picking one up at the other end, repeating the procedure for the rest of the experimental points. Since $(N-1)/2$ points are required on either side of the point being smoothed, then clearly we must begin and terminate the processing at the $(1 + (N-1)/2)$th point from each end of the experimental data set. The smoothed data set thus contains $(N-1)$ fewer points than the original set.

The procedure as stated would be rather time-consuming and less widely used if it were not for the fact that Savitzky and Golay (1964) showed that for equally spaced data the polynomial fitting could be simplified to a weighted summation of the N-point smoothing data segment, followed by division by a normalizing factor. In order to illustrate this, if we perform five-point quadratic ($m=2$) smoothing and y_0 is the current central ordinate in the smoothing set $(y_{-2}, y_{-1}, y_0, y_1, y_2)$ then the corrected (smoothed) value of the central ordinate is

$$y_S = (-3y_{-2} + 12y_{-1} + 17y_0 + 12y_1 - 3y_2)/35 \tag{5.39}$$

and all one needs to do is step through the data set point by point, updating the smoothing subset at each step. Savitzky and Golay (1964) have tabulated the (integer) weights and normalizing factors for a range of values of N and polynomial order (some corrections were published subsequently by Steiner *et al.*, 1972). Listing P5.8 is a program for implementing N-point quadratic smoothing of a data set; rather than store the tables in the program, the N weights and the normalizing factor are calculated as required (Bromba and Ziegler, 1979).

The benefit of data smoothing is illustrated in Fig. 5.10(a) where we show an example of a X-ray photoelectron spectrum before and after 11-point quadratic smoothing. The enhancement of the signal-to-noise, and consequent accessibility of the spectral information, is clear. The residuals (differences between the raw and smoothed ordinates) are plotted in Fig. 5.10(b) and show a random distribution about zero. Like all numerical methods for spectral analysis, polynomial smoothing must be used with care: over-smoothing (using too large a value for N) results in lineshape distortion (Fig. 5.11), whereas under-smoothing (N too small) gives poor noise-suppression. Experience of its systematic application to particular types of data is the only way to determine the optimum operating parameters.

5.6 FITTING NON-POLYNOMIAL MODELS

Real physical models do not usually generate a polynomial relationship between the dependent and independent variables, not least because polynomicals become infinite as the independent variable approaches positive or negative infinity. There are two simple ways we might adopt to deal with this. The first is to transform the data so that the model transforms to a linear relationship between y and x, which may be accommodated by the method of section 5.4, and the second is to approximate the theoretical function by a power series in x and use the analysis of section 5.5.

As an example, consider a physical system where the rate of consumption of some quantity A is proportional to the amount of A, $a(t)$, remaining at time t

$$\mathrm{d}a(t)/\mathrm{d}t = -c_2 t$$

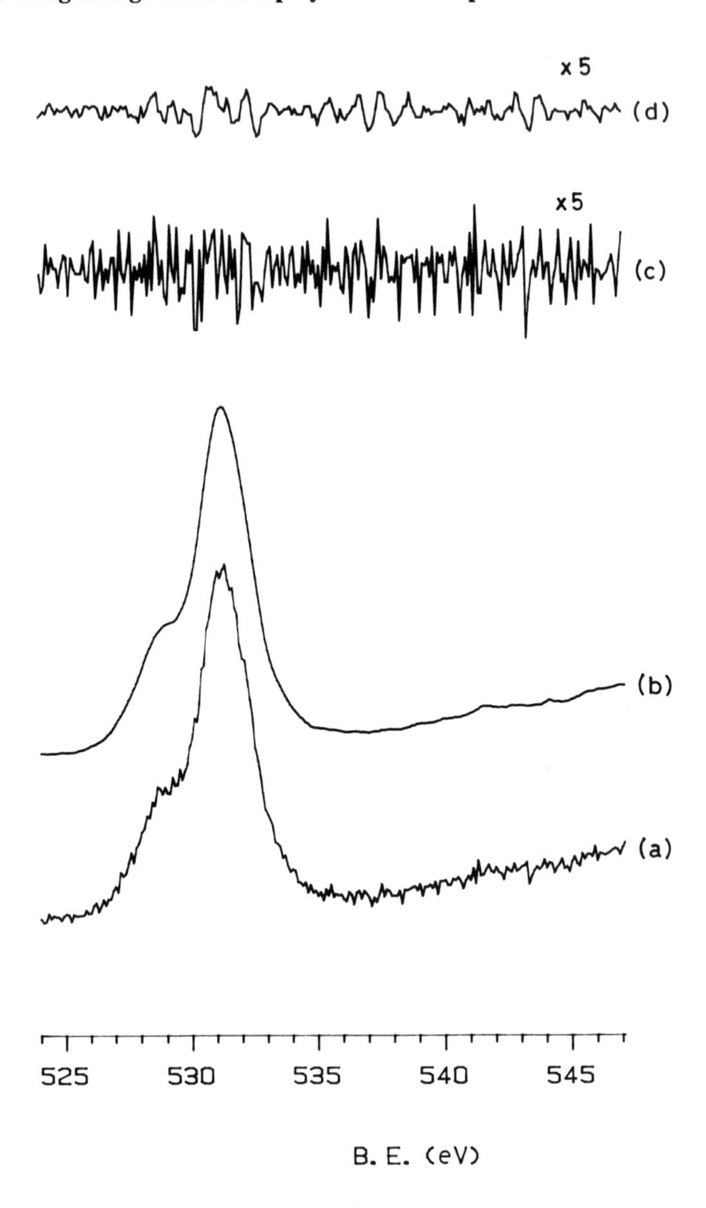

Fig. 5.10 — Smoothing of experimental spectral data. (a) O(1s) X-ray photoemission peak for a sample of the high critical-temperature superconductor $YBa_2Cu_3O_{7-x}$ Carley *et al.*, 1989). (b) Spectrum (a) after 17-point quadratic smoothing. (c) Residuals obtained by subtraction of (b) from (a). (d) Residuals of (c) after 9-point smoothing to highlight any non-randomness. The deviations in the region of the experimental peak position are no larger than the general noise amplitude in the residuals.

This situation is frequently encountered in science, for instance in the first-order decay of a chemical species A to product B, $A \rightarrow B$ with no back-reaction, and in simple radioactive decay. The solution is of the form

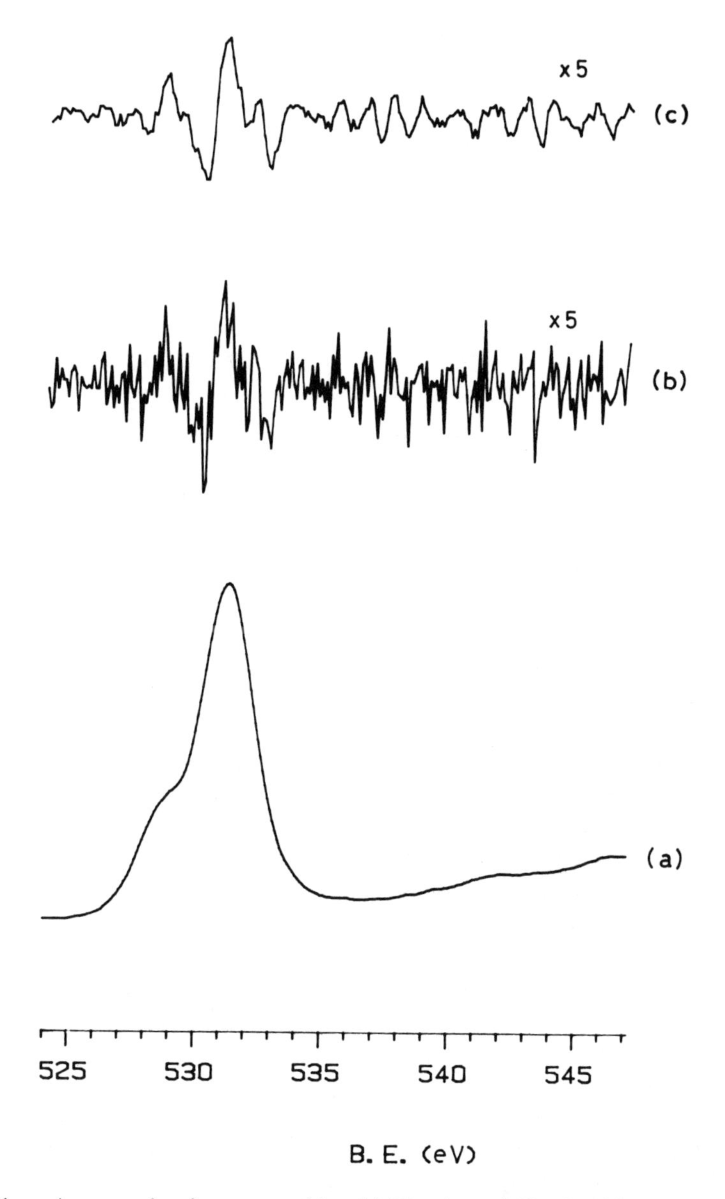

Fig. 5.11 — An example of over-smoothing (a) The data of Fig. 5.10(a) after 27-point smoothing. (b), (c) Plot of residuals and 9-point smoothed residuals. The deviations in the region of 530 eV are now considerable, and indicate significant distortion of the peak-shape.

$$a^*(t) = c_1 \exp(-c_2 t) \tag{5.40}$$

where in the case of the first-order decay example, we associate c_2 with the rate constant for $A \rightarrow B$, and c_1 with the initial quantity of A at $t = 0$. If we derive the normal equations for this fitting function as before we find (the reader should confirm this)

$$c_1\sum_{i=1}^{n} \exp(-c_2t_i) = \sum_{i=1}^{n} y_i$$

$$c_1\sum_{i=1}^{n} t_i \exp(-c_2t_i) = \sum_{i=1}^{n} t_iy_i$$

These cannot be solved explicitly for c_1 and c_2, no natter how we rearrange them. General methods for dealing with non-linear curve-fitting are discussed in Chapter 6.

We can linearize (5.40) by taking the natural logarithm of each side so that

$$y^* = \ln(a^*(t)) = \ln(c_1) - c_2t \tag{5.41}$$

Hence, if we measure various $(t,\ a(t))$ values, then a plot of $\ln(a(t))$ against t is predicted to be a straight line, notwithstanding scatter due to experimental error, and we can perform a linear least squares analysis to find the slope $(-c_2)$ and the intercept (c_1) of the best straight line. However, we must not forget that the least squares analysis is based on the premise that the experimental noise obeys a gaussian distribution (section 5.3); whilst this is likely to be true for the original $(t,\ a(t))$ measurements, it is by no means certain that this will remain so in the transformed data $(t,\ \ln(a(t)))$. We may investigate this by fitting synthesized data sets, where we know what the correct answer is. Fig. 5.12(a) shows a plot of the function $y=2\exp(-2.0x)$ to which some gaussian noise (section 5.2 and program P5.1) has been added; also shown is a plot of $\ln(y)$ against x (Fig. 5.12(b)). The best straight-line fit to the transformed data gives $c_1=2.003$ and $c_2=2.012$, which compares well with the true values. The residuals are plotted in Fig. 5.12(c); their appearance is rather unusual, increasing in size with increasing x_i, owing to the non-linear effect of the transformation (taking logs) on the data. However, the runs test (section 5.4.6) gives a probability value of 0.988, identical with that obtained for 'real' linear data. Thus it seems that, for a simple exponential, linearization by transformation does work, and a very popular procedure is vindicated.

The second method for dealing with a fitting function $y=c_1\exp(-c_2t)$ is to expand it as a Taylor polynomial of order m (Chapter 1)

$$y=c_1(1+c_2t+(c_2t)^2/2!+\ldots+(c_2t)^m/m!)$$

and to compare this term by term with the best-fit polynomial of order m, converted if necessary to power series form, in order to derive values for c_1 and c_2. The problems inherent in using too large a value for m have already been stated.

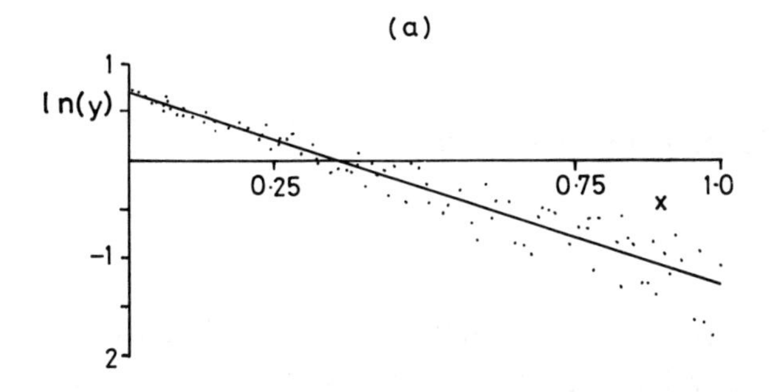

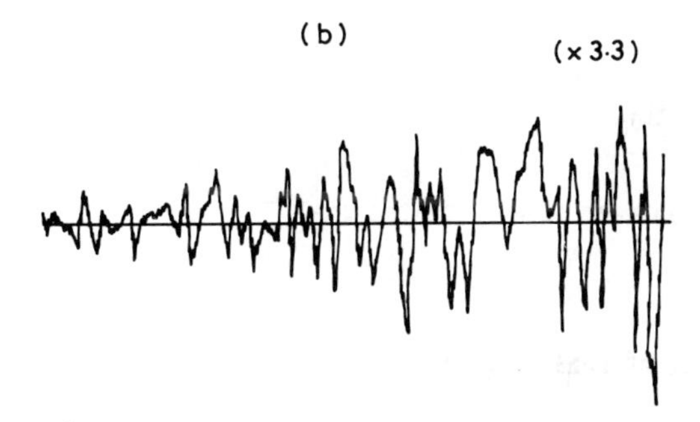

Fig. 5.12 — (a) Plot of transformed data ln(y) against x, derived from $y = 2\exp(-2x) + n$, where n is a gaussian noise contribution. Also shown is the least squares fit to a straight line (solid line). (b) Residuals plot for the fit in (a).

In the next chapter we shall meet general methods for fitting functions to data, allowing us to use functions which better represent the physical model under investigation.

```
    0 REM P5.1
   10 :
   20 REM Gaussian pseudorandom number generator
   30 :
   40 : GOTO 10000
   50 :
   60 REM use built-in random number generator with uniform distribution
   70 REM to compute Normally distributed random numbers using the fact
   80 REM that uniform distribution on interval (0,1) has SD=1/12 and SD's add
   90 REM If you want a new series of random numbers at each call of the
  100 REM you may have to "reseed" the system RND function.
  110 :
 7000 X=0
 7010 FOR I=1 TO 12
 7020           X=X + RND(1)
 7030 NEXT I
 7040 X=X - 6 : REM subtract mean to centre on 0
 7050 RETURN
 7060 :
10000 REM main segment of some program
10010 GOSUB 7000
10020 :
10030 REM  X is a random number from a population which approximates a
10040 REM a Normal distribution with unit standard deviation (SD) over the
10050 REM range (-6,6).
10060 :
10070 REM If Y is a quantity to which we may wish to add Gaussian "noise"
10080 REM then we might write YN=Y*(1 + F*X) where F is the estimated SD
10090 REM of the noise
```

```
    0 REM P5.2
   10 :
   20 REM best straight line by least squares
   30 :
   40 DIM X(100),Y(100),R(100)
   50 GOTO 10000
   60 :
  500 REM subroutine to read datafile containing experimental pairs
  510 REM X(I),Y(I) I=1,2,...,N  N>2 else est. SD residuals is infinite
  520 :
 7000 REM calculate slope (M), intercept (C) and correlation coefficient for
 7010 REM best line
 7020 S1=0 : S2=0 : S3=0 : S4=0 : S5=0 : REM zero before accumulating sums
 7030 FOR I=1 TO N
 7040         S1=S1 + X(I) : REM calculation of sum of X values
 7050         S2=S2 + X(I)^2 : REM calculate sum of squares
 7060         S3=S3 + Y(I)
 7070         S4=S4 + X(I)*Y(I) : REM sum of "cross products"
 7080         S5=S5 + Y(I)^2
 7090 NEXT I
 7100 :
 7110 N2=N*S2 - S1^2 : IF ABS(N2)<1E-30 THEN PRINT "ALL X VALUES IDENTICAL" : END
 7120 M=(N*S4 - S1*S3)/N2
 7130 C=(S2*S3 - S1*S4)/N2
 7140 T1=S4 - S1*S3/N : T2=S2 - (S1^2)/N : T3=S5 - (S3^2)/N
 7150 R=T1/SQR(T2*T3)
 7160 RETURN
 7170 :
10000 REM main segment
10010 GOSUB 500 : REM read datafile
10020 :
10030 REM take logs if required
10040 PRINT : PRINT "LOG TRANSFORM (Y/N)? ";
10050 GET Z$ : IF Z$<>"Y" AND Z$<>"N" THEN 10050
10060 IF Z$="N" THEN 10120
10070 FOR I=1 TO N
10080         IF Y(I)<0 THEN PRINT "NEGATIVE Y-VALUE" : END
10090         Y(I)=LOG( Y(I) )
10100 NEXT I
```

```
10110 :
10120 GOSUB 7000
10130 :
10140 REM calculate residuals for fit
10150 FOR I=1 TO N : R(I)=Y(I) - M*X(I) - C : NEXT I
10160 :
10170 REM compute best estimate of standard deviation of noise in data
10180 S=0 : FOR I=1 TO N : S=S + R(I)^2 : NEXT I : S=SQR(S/(N-2)) : REM N>2
10190 REM could put in formulae for estimated error in M,C and output these
10200 REM together with M,C and S as required
10210 :
10220 REM and plot residuals
10230 MI=1E38 : MX=-1E38 : REM set "impossible" values to max and min
10240 FOR I=1 TO N
10250         IF R(I)>MX THEN MX=R(I) : REM find max value
10260         IF R(I)<MI THEN MI=R(I) : REM find min value
10270 NEXT I
10280 FOR I=1 TO N : R(I)=R(I) - MI : NEXT I : REM shift residuals up
10290 MX=MX - MI : IF MX<1E-30 THEN PRINT "ZERO RANGE" : END : REM calc range
10300 SF= 159 /MX : REM scale factor - 160 points over vertical plot range
10310 HGR : HCOLOR=3
10320 ZE=ABS(MI)*SF
10330 HPLOT 0,159 - ZE TO 279,159 - ZE
10340 HPLOT 0,159 - R(1)*SF
10350 FOR I=2 TO N
10360         X=(I-1)* 279 /(N-1)
10370         HPLOT TO X,159 - R(I)*SF
10380 NEXT I
10390 :
10400 END
```

```
   0 REM P5.3
  10 :
  20 REM Runs test for short series of data values
  30 :
  40 DIM RR(15)
  50 GOTO 10000
  60 :
 500 REM get N data  values and calculate number of runs, R, and the number
 510 REM of positive, N1, and negative, N2, values
 520 INPUT "NUMBER OF VALUES = ";N
 530 FOR II=1 TO N : PRINT "VALUE # ";II : INPUT " = ";RR(II) : NEXT II
 540 :
 550 N2=0 : R=1
 560 FL=RR(1)<0 : REM negative flag
 570 N2=N2 + FL
 580 FOR II=2 TO N
 590         Z=RR(II)<0
 600         IF Z<>FL THEN R=R + 1 : FL=Z : REM if sign change count a run
 610         N2=N2 + Z : REM  only increment N2 if RR(II)<0
 620 NEXT II
 630 N1=N-N2
 640 RETURN
 650 :
7100 REM compute NU!
7110 FA=1
7120 IF NU<2 THEN RETURN : REM 1!=0!=1
7130 FOR II=2 TO NU : FA=FA*II : NEXT II
7140 RETURN
7150 :
7200 REM compute pCq - number combinations of P things Q at a time
7210 NU=P : GOSUB 7100 : CO=FA
7220 NU=P-Q : IF NU<0 THEN CO=0 : RETURN
7230 GOSUB 7100 : CO=CO/FA
7240 NU=Q : GOSUB 7100 : CO=CO/FA
7250 RETURN
7260 :
7300 REM compute number of arrangements, NA, of runs
7310 REM different cases for odd/even numbers of runs controlled by Q1,Q2
7320 P=N1-1 : Q=Q1 : GOSUB 7200 : Z1=CO
```

```
7330 P=N2-1 : Q=Q2 : GOSUB 7200 : Z1=Z1*CO
7340 P=N2-1 : Q=Q1 : GOSUB 7200 : Z2=CO
7350 P=N1-1 : Q=Q2 : GOSUB 7200 : Z2=Z2*CO
7360 NA=Z1 + Z2
7370 RETURN
7380 :
10000 REM main segment
10010 GOSUB 500 : REM get data and N,N1,N2,R
10020 PRINT N1;" POSITIVE VALUES"
10030 PRINT N2;" NEGATIVE VALUES"
10040 PRINT R;" RUNS"
10050 REM calculate total possible number of possible arrangements
10060 REM of N1 positive and N2 negative values
10070 :
10080 P=N : Q=N1 : GOSUB 7200 : NT=CO
10090 :
10100 REM loop to compute cumulative probability, CP, of obtaining
10110 REM  R or less runs by chance
10120 :
10130 CA=0
10140 FOR I=1 TO R
10150           IF N1*N2=0 AND I=1 THEN NA=1 : GOTO 10200
10160           IF I=1 THEN 10210 : REM deal with special cases first
10170           Q1=(I - 1)/2 : Q2=(I -3)/2 : REM assume I odd
10180           IF 2*INT(I/2)=I THEN Q1=(I -2)/2 : Q2=Q1 : REM if even
10190           GOSUB 7300
10200           CA=CA + NA
10210 NEXT I
10220 :
10230 CP=CA/NT
10240 PRINT : PRINT "PROBABILITY OF OBTAINING BY CHANCE"
10250 PRINT "LESS =< R RUNS IS"
10260 PRINT CP
10270 END
```

```
   0 REM P5.4
  10 :
  20 REM  Adaptive Simpson's Rule (P3.2) applied to the Runs Test
  30 REM  for large run numbers
  40 :
  50 F=SQR( 8*ATN(1) ) : F=1/F : REM TAN(PI/4)=1  - saves long decimal
  60 :
  70 DIM RR(250),AL(35),U(35),X(35)
  80 GOTO 10000
 100 U=F*EXP(-X*X/2) : RETURN : REM Gaussian distribution of runs for N>20
 110 :
 500 REM  subroutine to read N residuals data from datafile
 510 REM  compute quantities for runs test
 520 INPUT "NUMBER OF VALUES";N : FOR I= 1 TO N : INPUT RR(I) : NEXT I
 530 :
 540 N2=0 : R=1
 550 FL=RR(1)<0 : REM negative flag
 560 N2=N2 + FL
 570 FOR II=2 TO N
 580         Z=RR(II)<0
 590         IF Z<>FL THEN R=R + 1 : FL=Z : REM if sign change count a run
 600         N2=N2+Z : REM only increment N2 if RR(II)<0
 610 NEXT II
 620 N1=N - N2
 630 RETURN
 640 :
5100 REM  Simpson's Rule with interval halving
5110 ES=0 : IL=0 : S=0 : REM initialize total error estimate, integral and S
5120 X(1)=XL : X(5)=XR : REM set end points
5130 X=X(1) : GOSUB 100 : U(1)=U
5140 X=X(5) : GOSUB 100 : U(5)=U
5150 :
5210 X=( X(5) + X(1) )/2 : GOSUB 100 : U(3)=U : X(3)=X
5220 A=( X(5) - X(1) )*( U(1) + 4*U(3) + U(5) )/6 : REM H is 2*( X(5) - X(1) )
5230 :
5240 H=( X(S+5) - X(S+1) )/4 : REM return to here unless finished
5250 X=X(S+1) + H : GOSUB 100 : U(S+2)=U : X(S+2)=X : REM calc. new
5260 X=X(S+5) - H : GOSUB 100 : U(S+4)=U : X(S+4)=X : REM intermediate values
5270 AL(S/2)=H*( U(S+1) + 4*U(S+2) + U(S+3) )/3 : REM store left hand area
```

```
5280 AR=H*( U(S+3) + 4*U(S+4) + U(S+5) )/3 : REM calculate right hand area
5290 AA=AL(S/2) + AR : REM  calculate better area estimate with half step size
5300 ER=(A - AA)/15 : REM error est. if 4th deriv. U is const. in subinterval
5310 IF ABS(ER)<TE*( X(S+5) - X(S+1) )/(XR - XL) THEN 5340 : REM interval OK
5320 S=S+2 : IF S>2*LM THEN PRINT "LEVEL LIMIT EXCEEDED" : END
5330 A=AR : GOTO 5380
5340 IL=IL + AA : ES=ES + ER : REM update integral value and error estimate
5350 IF S=0 THEN RETURN : REM calc. completed or H too small
5360 S=S-2 : A=AL(S/2) : REM try larger subinterval
5370 :
5380 U(S+5)=U(S+3) : U(S+3)=U(S+2) : REM renumber values
5390 X(S+5)=X(S+3) : X(S+3)=X(S+2)
5400 GOTO 5240
5410 :
8000 REM alternative approx. of integral of Gaussian function usable -4<A<4
8010 REM 1 - X + (X^2)/2! - (X^3)/3! + ... is series for EXP(-X)
8020 REM Substitute (X^2)/2 for X and integrate powers of X term by term
8030 REM put in limits 0 and A  - T is current term in power series for IL
8040 REM  RT is ratio of (NT+1)th term to (NT)th
8050 IF A>4 THEN PRINT "TOO CLOSE TO UNITY TO CALCULATE" : IL=-.5 : RETURN
8060 IF A<-4 THEN PRINT "TOO CLOSE TO ZERO TO CALCULATE" : IL=.5 : RETURN
8070 REM for such values, first few terms are large and round-off dominates
8080 IL=XL : NT=0 : T=XL : AA=A*A : F=1/SQR( 8*ATN(1) ) : REM initialize (see 1. 70 for
8090 RT=-AA*(2*NT + 1)/( (2*NT + 3)*2*(NT + 1) )
8100 T=RT*T : IF ABS(T)<TE THEN IL= - IL*F : RETURN : REM change sign, mult. F
8110 IL=IL + T : NT=NT + 1 : GOTO 8090
8120 :
10000 REM main segment
10010 GOSUB 500 : REM get data and N1, N2, and R
10020 TE=1E-6 : LM=15 : REM see P3.2
10030 :
10040 REM Gaussian approximation for N>15
10050 NN=2*N1*N2 : IF NN<1E-30 THEN PRINT "ONLY ONE RUN" : END
10060 MU=1 + NN/N : SG=NN*(NN-N)/(N*N*(N - 1))
10070 XR=(R - MU + .5) /SQR(SG) : REM addition of .5 is a continuity correction
10080 ML$="LESS" : IF XR>0 THEN XR=-XR : ML$="MORE"
10090 REM integrating just the tail of the Gaussian more accurate for XR>>0
10100 XL=-MU : GOSUB 5100 : XL set to what is, in effect, negative infinity
10110 PRINT "PROBABILITY OF GETTING ";R;" OR ";ML$;" RUNS BY CHANCE IS ";IL
```

```
10120 PRINT "MEAN NUMBER OF RUNS FOR ";N1; "+ AND ";N2;"- VALUES IS";MU
10130 END
```

```
    0 REM P5.5
   10 :
   20 REM least squares fit to polynomial
   30 REM polynomial of form coeff0 + coeff1*X + coeff2*X^2 + ... + coeffM*X^M
   40 :
   50 DIM X(100),Y(100),T(100)
   60 GOTO 10000
   70 :
  500 REM subroutine to read datafile - X(I),Y(I) I=1,2, .....,N
  510 REM or input from keyboard e.g. :-
  520 INPUT "NUMBER OF DATA POINTS";N
  530 FOR I=1 TO N : INPUT "X-VALUE, Y-VALUE";X(I),Y(I) : NEXT I
  540 RETURN
  550 :
 7200 REM solve system of normal equations using Gauss-Jordan method (Chap 1)
 7210 FOR I=0 TO M : MT(I,M1)=DM(I) : NEXT I : REM augment matrix with RHS
 7220 :
 7230 FOR C=0 TO M
 7240         MD=0
 7250         FOR CC=C TO M : REM search down rest of column C
 7260                 DD=MT(CC,C) : AD=ABS(DD)
```

```
7270                IF AD>MD THEN MD=DD : MC=CC : REM find largest element
7280          NEXT CC
7290          :
7300          IF AD<TL THEN PRINT "ZERO PIVOT" : END
7310              :
7320          IF MC=C THEN 7380 : REM largest element in current row - no swap
7330          FOR CC=C TO M1
7340                MT=MT(C,CC) : MT(C,CC)=MT(MC,CC) : MT(MC,CC)=MT
7350                REM MT and MT( , ) are different variables
7360          NEXT CC
7370          :
7380          FOR J=C+1 TO M1 : MT(C,J)=MT(C,J)/MD : NEXT J
7390          REM  skip elements in current and previous columns
7400          FOR I=0 TO M
7410                IF I=C THEN 7440 : REM current row already processed
7420                MU=MT(I,C)
7430                FOR J=C+1 TO M1 : MT(I,J)=MT(I,J) - MU*MT(C,J) : NEXT J
7440          NEXT I
7450 NEXT C
7460 RETURN
7470 :
10000 REM main segment
10010 PRINT : INPUT "ORDER = ";M
10020 TL=1E-30 : GOSUB 500 : REM set threshold for pivotting and get data
10030 :
10040 REM set up RHS of system
10050 DM(0)=0
10060 FOR I=1 TO N : DM(0)=DM(0) + Y(I) : NEXT I
10070 FOR J=1 TO M
10080          DM=0
10090          FOR I=1 TO N : DM=DM + Y(I)*X(I)^J : NEXT I
10100          DM(J)=DM
10110 NEXT J
10120 :
10130 REM set up elements of coefficient matrix
10140 T(0)=N
10150 FOR J=1 TO 2*M
10160          T=0
10170          FOR I=1 TO N : T=T + X(I)^J : NEXT I
```

```
10180         T(J)=T : REM using T saves searching array T( )
10190 NEXT J
10200 :
10210 FOR J=0 TO M : FOR K= 0 TO M : MT(J,K)=T(J + K) : NEXT K : NEXT J
10220 :
10230 REM solve for coefficients in polynomial which are returned in MT(I,M1)
10240 M1=M + 2 : GOSUB 7200 : REM augmented matrix is M1 columns wide
10250 FOR I=0 TO M : PRINT "COEFFICIENT #";I;" = ";MT(I,M1) : NEXT I
10260 REM can add a routine to calc. best fit values from X() and coefficients
10270 REM also residuals, Y(I)- best fit values, and estimate of noise SD
10280 END
```

```
   0 REM P5.6
  10 :
  20 REM data fitting using Forsythe polynomials
  30 :
  40 DIM X(100),Y(100),P(10,100)
  50 GOTO 10000
  60 :
 500 REM routine to read datafile or to input data from keyboard e.g.
 510 INPUT "NUMBER OF DATA POINTS":N
 520 FOR I=1 TO N : INPUT "X-VALUE, Y-VALUE";X(I),Y(I) : NEXT I
 530 RETURN
 540 :
7300 REM compute alpha and beta values - A() and B()
7310 REM Forsythe polynomials and derived quantities
7320 REM calculate zero and first order values
7330 A(1)=0
7340 FOR I=1 TO N : A(1)=A(1) + X(I) : NEXT I
7350 A(1)=A(1)/N : REM mean of data X-values
7360 :
7370 M1=0 : V0=0 : V1=0
7380 FOR I=1 TO N
```

```
7390        P(0,I)=1 : REM zero order has unit value at all I
7400        P=X(I) - A(1) : Y=Y(I)
7410        V0=V0 + Y
7420        V1=V1 + P*Y
7430        M1=M1 + P*P
7440        P(1,I)=P
7450 NEXT I
7460 M(1)=M1 : V(0)=V0 : V(1)=V1
7470 A(0)=0 : B(0)=0 : M(0)=N
7480 :
7490 REM calculate values for second order and above
7500 FOR J=2 TO M
7510        D1=0 : D2=0 : N1=0 : N2=0
7520        FOR I=1 TO N
7530                P1=P(J-1,I) : P2=P(J-2,I) : X=X(I)
7540                PP=P1*P1 : PQ=P1*P2 :QQ=P2*P2
7550                D1=D1 + PP : D2=D2 + QQ
7560                N1=N1 + X*PP : N2=N2 + X*PQ
7570        NEXT I
7580        A(J)=N1/D1 : B(J-1)=N2/D2
7590        :
7600        MJ=0 : VJ=0
7610        FOR I=1 TO N
7620                PJ=( X(I) - A(J) )*P(J-1,I) - B(J-1)*P(J-2,I)
7630                MJ=MJ + PJ*PJ : REM normalizing factor (Jth polynomial)
7640                VJ=VJ + PJ*Y(I) : REM Jth polynomial contribution to data
7650                P(J,I)=PJ
7660        NEXT I
7670        M(J)=MJ : V(J)=VJ
7680 NEXT J
7690 RETURN
7700 :
10000 REM main segment
10010 INPUT "ORDER";M
10020 GOSUB 500 : REM get data
10030 :
10040 GOSUB 7300 : REM calculate polynomials and derived summations
10050 :
10060 REM calculate Forsythe polynomial coefficients
```

```
10070 FOR J=0 TO M : C(J)=V(J)/M(J) : NEXT J
10080 :
10090 REM can now use contents of A(),B(),C() for other calculations
10100 REM see P5.7 and text
10110 END
```

```
    0 REM  P5-7
   10 :
   20 REM Compute power series coefficients
   30 REM  from Forsythe polynomial coefficients
   40 :
   50 GOTO 10000
   60 :
 7400 REM subroutine either to calculate alpha and beta values, A(),B(),
 7410 REM and the Forsythe coefficients, C(), or to input them
 7420 REM order of method is M - dimension arrays if M>10
 7430 :
 7700 REM caLculate power series coefficients D()
 7710 BB(0,0)=1 : BB(0,1)= -A(1) : BB(1,0)=0 : BB(1,1)=1
 7720 FOR K=2 TO M : BB(K,0)=0 : BB(K,1)=0 : NEXT K
 7730 FOR K=0 TO M
 7740        FOR J=2 TO M
 7750                IF K>J THEN BB(K,J)=0
 7760                IF K=J THEN BB(K,J)=1
 7770                IF K<J AND K>0 THEN GOSUB 7900
 7780                IF K<J AND K=0 THEN GOSUB 7920
 7790        NEXT J
```

```
7800 NEXT K
7810 :
7820 FOR K=0 TO M
7830        D(K)=0
7840        FOR J=K TO M : D(K)=D(K) + C(J)*BB(K,J) : NEXT J
7850 NEXT K
7860 RETURN
7870 :
7900 BB(K,J)=BB(K-1,J-1) - A(J)*BB(K,J-1) - B(J-1)*BB(K,J-2) : RETURN
7910 :
7920 BB(K,J)=-A(J)*BB(K,J-1) - B(J-1)*BB(K,J-2) : RETURN
7930 :
10000 REM main segment
10010 GOSUB 7400 : REM get Forsythe coefficients
10020 :
10030 GOSUB 7700 : REM calculate power series coefficients
10040 END
```

```
    0 REM  P5.8
   10 :
   20 REM  M-point quadratic smoothing
   30 :
   40 DIM W(100),Y(100),Q(20)
   50 GOTO 10000
   60 :
  500 REM read datafile or input from keyboard e.g.
  510 FOR I=1 TO N : INPUT "ORDINATE VALUE";Y(I) : NEXT I
  520 RETURN
  530 :
  600 REM routine to display W(), write it to disc or print e.g.
  610 FOR I=1 TO N : PRINT "VALUE";I;" = ";Y(I) : NEXT I
  620 RETURN
  630 :
 7800 REM calculate Savitsky-Golay smoothing weights from formula - see text
 7810 MM=INT(M/2) : REM this number of points either side of central point
 7820 D=(2*MM - 1)*(2*MM + 1)*(2*MM + 3)/3
 7830 FOR I=0 TO MM
 7840         Q(I)=(3*MM*MM + 3*MM - 1 - 5*I*I)
 7850 NEXT I
 7860 :
 7870 REM smooth the data in Y() by taking a weighted sum of ordinates
 7880 N1=N1 + MM : N2=N2 - MM : REM remember ends of smoothed array
 7890 FOR K=N1 TO N2 : REM lose MM points at each end
 7900         W=Y(K)*Q(0)
 7910         FOR I=1 TO MM : REM Q() symmetric about 0
 7920                 W=W + ( Y(K + I) + Y(K - I) )*Q(I)
 7930         NEXT I
 7940         W(K)=W/D
 7950 NEXT K
 7960 :
 7970 FOR I=N1 TO N2 : Y(I)=W(I) : NEXT I
 7980 FOR I=1 TO N1-1 : Y(I)=0 : NEXT I
 7990 FOR I=N2+1 TO N : Y(I)=0 : NEXT I : REM emphasise points lost by zeroing
 8000 RETURN
 8010 :
10000 REM main segment
10010 INPUT "NUMBER OF POINTS IN SMOOTHING WINDOW (MUST BE ODD)";M
```

```
10020 IF INT(M/2)=M/2 THEN PRINT "YOU HAVE INPUT AN EVEN NUMBER" : GOTO 10010
10030 :
10040 INPUT "NUMBER OF EQUISPACED ORDINATE VALUES";N
10050 :
10060 GOSUB 500 : REM input data
10070 :
10080 GOSUB 600 : REM show data
10090 :
10100 N1=1 : N2=N : REM initialize start and end of array
10110 GOSUB 7800 : REM calculate coefficients and smooth
10120
10130 IF N2-N1 < M THEN PRINT "TOO FEW ORDINATES" : END
10140 GOSUB 600 : REM print, store or show data
10150 :
10160 PRINT "VALUES";N1;"TO";N2;"REMAIN"
10170 INPUT "SMOOTH AGAIN (Y/N)";S$
10180 IF S$="Y" OR S$="y" THEN GOSUB 7870 : GOTO 10130 : REM only smooth
10190 END
```

6

Optimization methods and non-linear least squares minimization

6.1 INTRODUCTION

The use of polynomial functions (of which the straight line is a particular example) to fit experimental data according to the least squares criterion, has the great advantage of being a single step method which reduces to the inversion of a matrix of coefficients. However, a polynomial function is not always the most appropriate, especially if one wishes to test the validity of a theoretical model, which is most unlikely to generate a polynomial expression. In section 5.6 we discussed two methods for reducing the general non-linear problem to a linear or polynomial fitting exercise. However, the ideal solution is to develop methods which can use directly the true fitting function, no matter how complex. The distinctive feature of such generalized minimization methods is that they are iterative in operation as opposed to the one-step procedures of Chapter 5. Although the discussion of such methods below is often made in the context of least squares minimization, they apply to any situation where the aim is to find the minimum value of a function of several variables; in fact the methods are applicable to the more general task of optimization since a search for the maximum value of a function $f(x,y,\ldots)$ may be reformulated as a search for the minimum value of $-f$.

6.2 THE MINIMIZATION PROBLEM

The problem may be stated thus: to find values of the variables (parameters) p_j $(j=1,\ldots,m)$ which minimize the value of the function $f(p_1,\ldots,p_m)$. In the case of least squares minimization, the function is the sum-of-squares function which also depends on the experimental data x_i,y_i $(i=1,\ldots,n)$. However, these are predetermined and may be regarded as constants as far as the minimization process is concerned and will be disregarded except where explicitly required. In order to illustrate the fundamental features of the problem will consider the simple case of a function of two variables p_1 and p_2, and to provide a visual interpretation we will represent the function on a 'contour' graph. Fig. 6.1 shows a typical such plot where

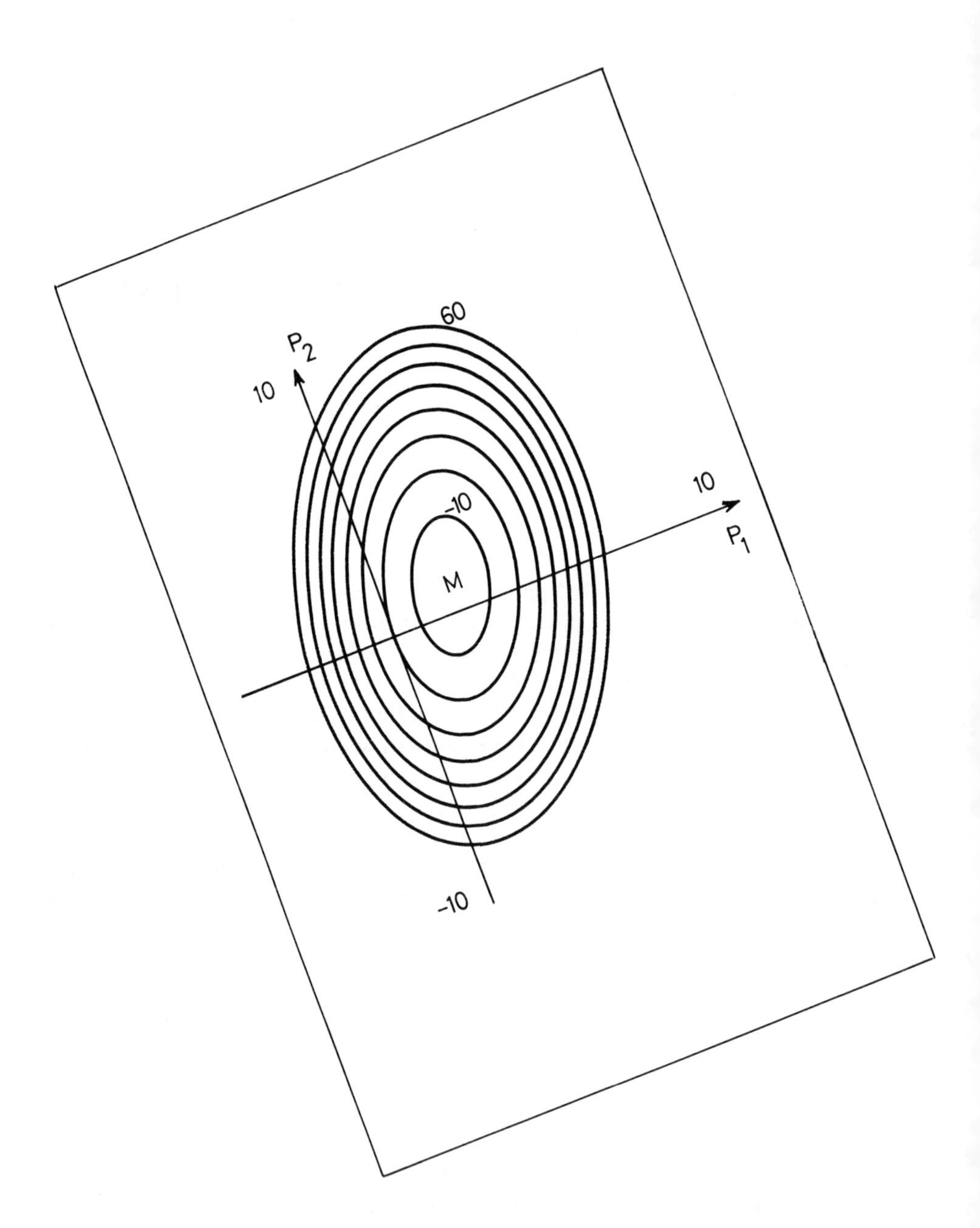

Fig. 6.1 — Example of a contour plot of a function of two variables (parameters) $f(p_1,p_2)$. The closed curves (ellipses in this example) connect points (p_1,p_2) of equal function value.

the closed curves join points of constant functional value, in much the same way as contours on a map represent lines of equal elevation. The contours in our simple example (Fig. 6.1) may be thought of as successive cross-sections through a bowl-shaped valley, the bottom of the valley being the minimum point which we seek, and labelled M in the diagram. For a function which is quadratic in the parameters p_1 and p_2 (that is only contains a constant term and terms in p_1, p_2, p_1^2, p_2^2 and p_1p_2) the contours are exactly elliptical; for more complex functions it is generally assumed (usually correctly in practical problems) that this is approximately true near the minimum point. In the case of least squares minimization, this corresponds to the contours being elliptical if the fitting function is linear in parameters, the degree of distortion depending on the degree of non-linearity in the fitting function.

All the methods reduce to choosing a starting point $(p_1^{(0)},p_2^{(0)})$ and then generating, according to a formula or a set of rules, a sequence of points $(p_1^{(k)},p_2^{(k)}; k=1,2,\ldots)$ which 'home in' on the minimum point, M of the function $f(p_1,p_2)$ (Fig. 6.1). In the general case, where we may be minimizing a very complex function, one may visualize the exercise as a journey through a landscape of hills, passes and valleys, the end of the journey being the bottom of the lowest valley in our hypothetical world. The important feature of such a search is the capability embodied in the updating algorithm for using local knowledge of the 'terrain' to decide where to move next. There are two main classes of method: direct search methods and gradient techniques. The latter method uses explicitly the derivatives (slopes) of the function with respect to the parameters (variables) in order to find the next point in the search sequence. In contrast, direct search methods use the local functional information in a less formalized fashion, and are especially useful in those situations where it is not possible or not practicable to calculate the derivatives.

Our aim is not to give an exhaustive compilation of all the minimization techniques and their variations which are available, rather to discuss in some depth those more frequently used methods which the reader may find applicable to his own problems, and to also cover some less useful approaches where it is instructive to do so. For a more comprehensive coverage, particularly of the more advanced methods, the reader is directed to other sources. In particular, the proceedings of the 'Optimization' symposium held at the University of Keele (Fletcher, 1969) comprise a unique source-book of optimization techniques, discussed by many of the great names in the field. Other useful texts include those by Walsh (1975) and Adby and Dempster (1974).

6.3 DIRECT SEARCH METHODS

6.3.1 Grid search

The simplest direct method is to divide the search region up into a regular pattern of grid points and to evaluate the function at each point. This is readily visualized in the two-dimensional case (Fig. 6.2), where the function depends on two parameters. The smallest value so obtained is taken to be the minimum value of the function and the coordinates of the corresponding grid point are the minimizing parameter values. In order to improve the calculation, one could then construct a finer grid in a smaller region centred on the approximate minimum point just found, and repeat the process. The main drawback with this scheme is the prohibitively large number of

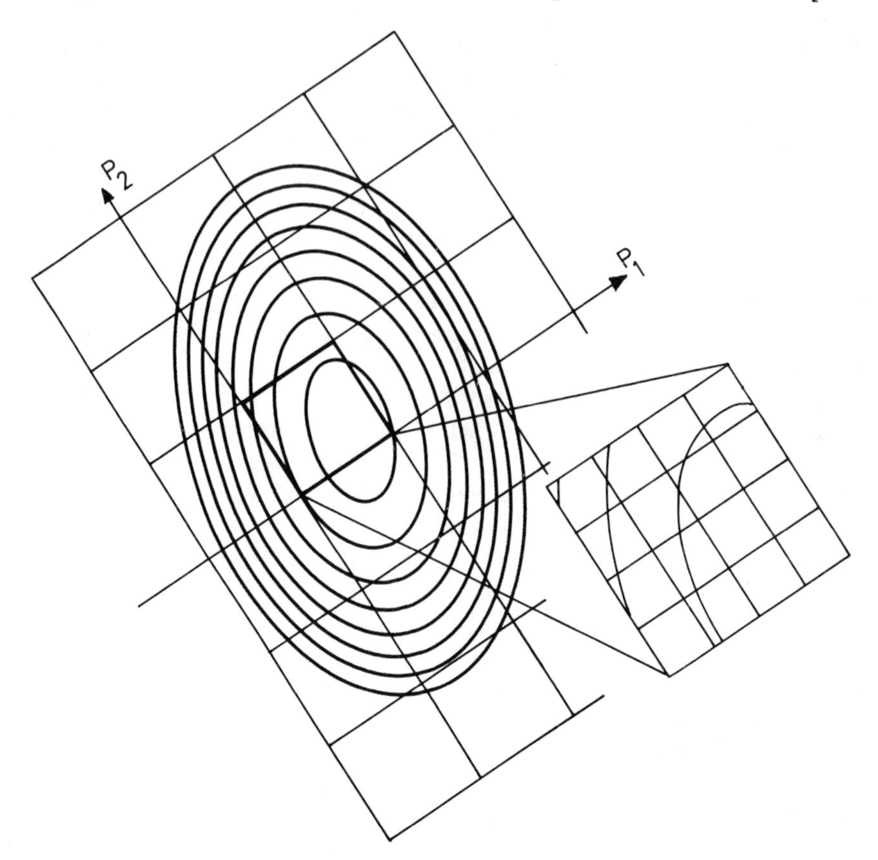

Fig. 6.2 — Optimization via a grid search. The function is evaluated at each grid point, and then a finer grid search is undertaken in the region of smallest function value. The process is repeated until a suitable accuracy (fineness of grid mesh) has been reached.

function evaluations required (most of them unwanted), unless one is able to concentrate on a very small region near the minimum point. Furthermore, the information acquired during the search about the behaviour of the function is not used to speed up the progress of the search.

A more practical alternative to the global grid-search involves exploring only in the neighbourhood of a current search point. We start by choosing a point $(p_1^{(0)},p_2^{(0)},\ldots,p_m^{(0)})$ which we shall represent by the column matrix or vector $[p]^{(0)}$. (We thus write $f([p])$ as shorthand for $f(p_1,p_2,\ldots,p_m)$.) We proceed by evaluating $f([p])$ at $[p]^{(0)}$ and all surrounding grid points, e.g. in two-dimensions ($m=2$) we consider the nine points

$$\begin{bmatrix} p_1^{(0)} \\ p_2^{(0)} \end{bmatrix}, \begin{bmatrix} p_1^{(0)}+h \\ p_2^{(0)} \end{bmatrix}, \begin{bmatrix} p_1^{(0)}-h \\ p_2^{(0)} \end{bmatrix}, \begin{bmatrix} p_1^{(0)} \\ p_2^{(0)}+h \end{bmatrix}, \begin{bmatrix} p_1^{(0)} \\ p_2^{(0)}-h \end{bmatrix},$$

$$\begin{bmatrix} p_1^{(0)} + h \\ p_2^{(0)} + h \end{bmatrix}, \begin{bmatrix} p_1^{(0)} - h \\ p_2^{(0)} + h \end{bmatrix}, \begin{bmatrix} p_1^{(0)} - h \\ p_2^{(0)} - h \end{bmatrix}, \begin{bmatrix} p_1^{(0)} + h \\ p_2^{(0)} - h \end{bmatrix}$$

which lie on the edges and corners of a square surrounding $[p]^{(0)}$ (Fig. 6.3(a)). (We

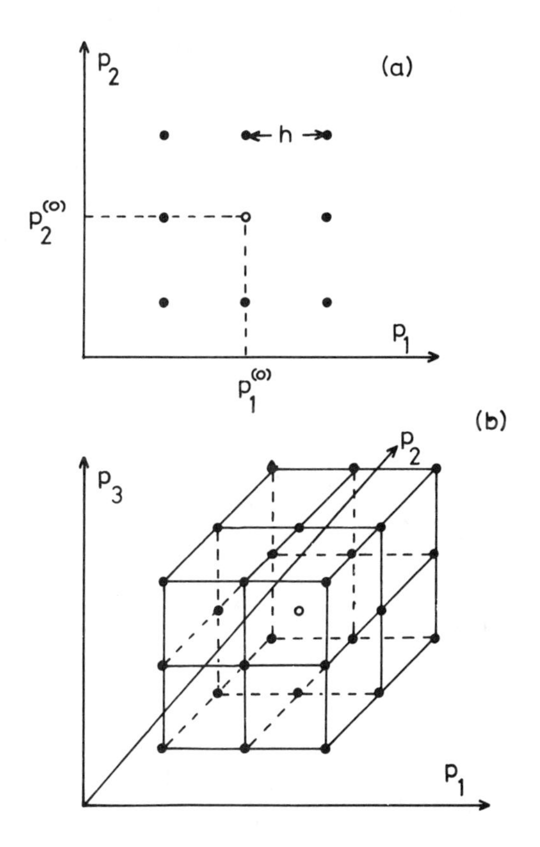

Fig. 6.3 — Points used in a local grid search for the minumum of a function of (a) two and (b) three parameters. The current search point is indicated by an open circle.

have assumed implicitly that the grid size is the same for each parameter. This need not be so, in which case the mesh units are rectangular, not square.) In three dimensions the points lie on the corners, edges and faces of a cube with $[p]^{(0)}$ at its centre, and are 27 in number (Fig. 6.3(b)). In the general case, we must evaluate the function of 3^m points during each local search.

If $[p]^{(1)}$ is the point which corresponds to the smallest value of the function after this procedure then $[p]^{(0)}$ is replaced by $[p]^{(1)}$ and the process repeated. If it turns out that $[p]^{(0)}$ is the best point then the grid spacing is reduced and the search continued. The search is terminated when the minimum point has been found to the required accuracy; the criterion for this may be either that the change in 'best' function value between cycles is below a pre-chosen limit, that the changes in values of the parameters corresponding to the approximation to the minimal point are small enough, or that the grid spacing has been reduced below a pre-set value. Although a considerable improvement on the global grid search, it is still not a practicable method unless m is small.

The requirement on the user to choose a starting point from which to search is a feature of all the methods discussed in this chapter, and clearly this choice should be made with care, utilising all the prior information at one's disposal in order that $[p]^{(0)}$ be as close as possible to the minimum point.

6.3.2 'One-at-a-time' minimization

More correctly referred to as single-parameter variation, this long-known method seeks a minimum with respect to each parameter in turn. In our particular example this means searching from the starting point along a direction parallel to the p_1 axis until a minimum point is found, then moving along the p_2 direction until a minimum with respect to p_1 is discovered (Fig. 6.4). The problem is that after the second

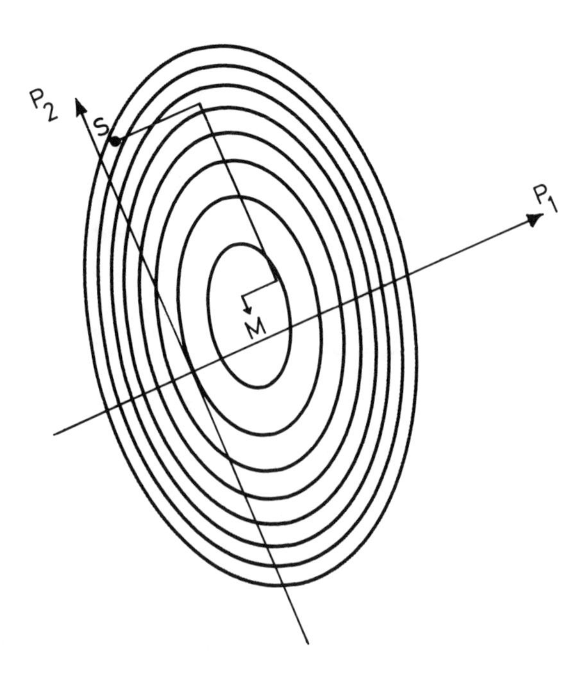

Fig. 6.4 — 'One-at-a-time' minimization. Starting at S we move in the direction of each parameter axis in turn, changing direction when we find a minimum point in the current direction. M is the sought after minimum point.

minimization we will generally no longer be at a minimum with respect to p_1 and the whole process must be repeated (iterated) until it converges, which it usually does. The rate of convergence, however, is often painfully slow, for reasons which are illustrated in Fig. 6.5 where the method is applied to a situation in which the contour plot forms a narrow valley.

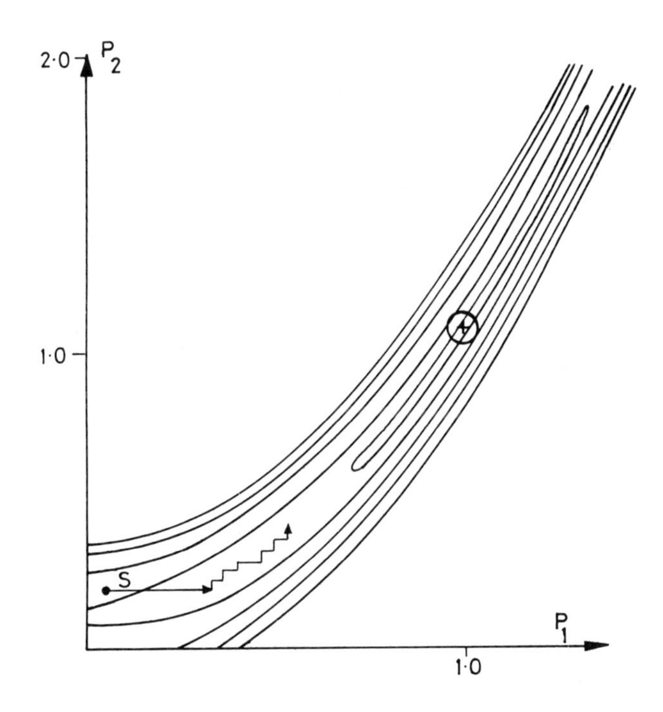

Fig. 6.5 — Example of a contour plot where 'one-at-a-time' minimization performs poorly — the narrow valley problem.

The search for the minimum point along the axial directions may be performed by the one-dimensional analogue of the grid search described in the previous section, but a much more efficient approach is to use the technique of quadratic interpolation. As this will form an integral part of the gradient methods to be described later, an explanation will be postponed until then.

A method which suffers less from the 'narrow valley problem' by virtue of its having some 'intelligence' built in is described next.

6.3.3 Hooke and Jeeves' method

The method of Hooke and Jeeves is based on combinations of two basic 'moves': exploratory and pattern or leapfrog moves (Walsh, 1975). It incorporates a higher level of 'intelligence' (where the term is used in the sense of adaptability to the environment) to speed up the search. The two moves are defined as follows:

(1) An exploratory move around a point $[p]$ may be regarded as 'one-at-a-time' minimization applied using a fixed step size (which may be different for each parameter). Thus, for $j=1$ to m we evaluate $f([p_+])$ where $[p_+]=[p_1,p_2,\ldots,p_j+h_j,\ldots,p_m]$. If $f([p_+])<f([p])$ then replace $[p]$ by $[p_+]$ and deal with the next parameter. Otherwise, evaluate $f([p_-])$ with $[p_-]=[p_1,p_2,\ldots,p_j-h_j,\ldots,p_m]$. If $f([p_-])<f([p])$ then replace $[p]$ by $[p_-]$ and continue to the next parameter. If both moves fail then retain $[p]$ and try the next parameter. Let $[p]'$ denote the new point found after cycling through all the parameters once.

(2) A pattern move is only performed after an exploratory move. Its purpose is to increase the efficiency of the optimization process by taking steps along a line which approximates the local 'downhill' direction. It is at this point that we make use of the previously acquired information about $f([p])$: if the function value decreases in going from $[p]$ to $[p]'$ then a logical direction in which to move next from $[p]'$ is defined by the vector $([p]'-[p])$, which points from $[p]$ to $[p]'$. A pattern move is thus from $[p]'$ to $[p]'+([p]'-[p])=2[p]'-[p]$, and is always followed by an exploratory move. Denote the point reached after this combined pattern/exploratory move by $[p]''$. Note that the intermediate point reached by the pattern move itself must also be remembered, since it may be required as part of the input to another combination move.

The method of Hooke and Jeeves uses these moves in a systematic fashion:

(a) Explore about the current point $[p]$ and find a new point $[p]'$.
(b) If $[p]'=[p]$ then halve all step sizes h_j. Go back to (a) unless the step sizes are now below some prescribed limit, in which case stop.
(c) Make a combination move ((2) above) from $[p]'$ to $[p]''$.
(d) If $f([p]'')<f[p]')$ then let $[p]'=[p]''$ and go to (c), otherwise let $[p]=[p]'$ and go to (a).

Note that it is only if the combined pattern/exploratory move is successful (i.e. the function value decreases) that a new point is accepted. If the combination does not succeed then the pattern move is discarded and an exploratory move performed. In either case a new combination move is then attempted. The process is continued until, for example, the lower limits on the step lengths has been reached, although other 'stopping rules' can be used.

The program in listing P6.1 shows how Hooke and Jeeves' method may be implemented. Worked examples for this and many other minimization methods, not all of which are covered in this chapter, have been given by Walsh (1975) and are recommended reading for those who experience any difficulties in following the various steps and their interplay.

The method of Hooke and Jeeves may be easily adapted to include ideas discussed in the previous section, such as a modified pattern move involving quadratic interpolation along the pattern move direction or indeed single parameter variation as the exploratory step followed by a modified pattern move. Rosenbrock's method (1960) employs a refinement of the latter approach.

6.3.4 Simplex method

For our purposes a simplex may be defined as a geometrical figure which has one more vertex than the dimension of the space in which it exists. A simplex in two dimensions is thus a triangle and in three dimensions a tetrahedron. For the purpose of geometrical illustration we will consider the two-dimensional (i.e. two-parameter) case but the arguments and results apply with equal validity to the general situation. The aim of the simplex method is to distort and move the simplex around in a 'downhill' direction in parameter space according to a set of well-defined rules, until it surrounds the sought-after minimum point and then to shrink it about this point (Fig. 6.6). The process is terminated when a sufficient accuracy is achieved, i.e. the shrunken simplex is small enough, or in other words the function values defined at all the vertices are in close enough agreement. A major difference between this method and others is that since each simplex defines $(m + 1)$ vertices (parameter vectors), we generate a sequence of sets of $(m + 1)$ vectors as we proceed, rather than a sequence of single vectors.

Spendley, Hext and Himsworth (1962) developed a method based on a regular simplex (e.g. in two dimensions this would be an equilateral triangle), but we shall base our discussion on the improved, more efficient formulation due to Nelder and Mead (1965). In this version the simplices are free to become irregular, and thus take better advantage of favourable search directions. Not unexpectedly, one begins by making an intelligent estimate of the optimizing parameters; the coordinates (parameters) of the other vertices of the simplex are then derived by means of an algorithm which results in an approximately regular (in other words, evenly spread) initial simplex. The simplex is moved by replacing the vertex with the highest function value (in our case, the sum of the squares) by a new vertex arising from one of the operations reflection, expansion or contraction (see below). This is repeated until a better vertex cannot be so found, when the simplex is shrunk towards the vertex with the smallest function value. The whole process is repeated until sufficient accuracy is obtained. The flowchart in Fig. 6.7 details the steps involved in the algorithm, which is a slightly simplified form of that due to Nelder and Mead (1965). They considered also the vertex with the second highest function value, a complicating refinement which is not essential to the efficient operation of the method (Caceci and Cacheris, 1984). The movement operations and shrinkage step are explained below and illustrated in Fig. 6.8. The worst vertex is labelled W, the best (i.e. lowest function value) is called B, the reflected one R, the result of an expansion is termed E, and that of a contraction C.

(1) Reflection: let X be the centroid of all the vertices excluding the worst one. For the two-dimensional case (triangular simplex) the centroid is the midpoint of the line joining the other two vertices. The reflected vertex R lies on the continuation of the line joining the worst vertex W to the centroid X, and the same distance from X as W (Fig. 6.8(a)). Nelder and Mead's more general method used a variable reflection factor so that the distances WX and XR were not necessarily equal. However, their recommended value for this factor corresponds to WX = XR.

    If R is better than W (i.e. has a lower function value associated with it) and is worse than B then it is accepted and W is replaced by B — in this case we can

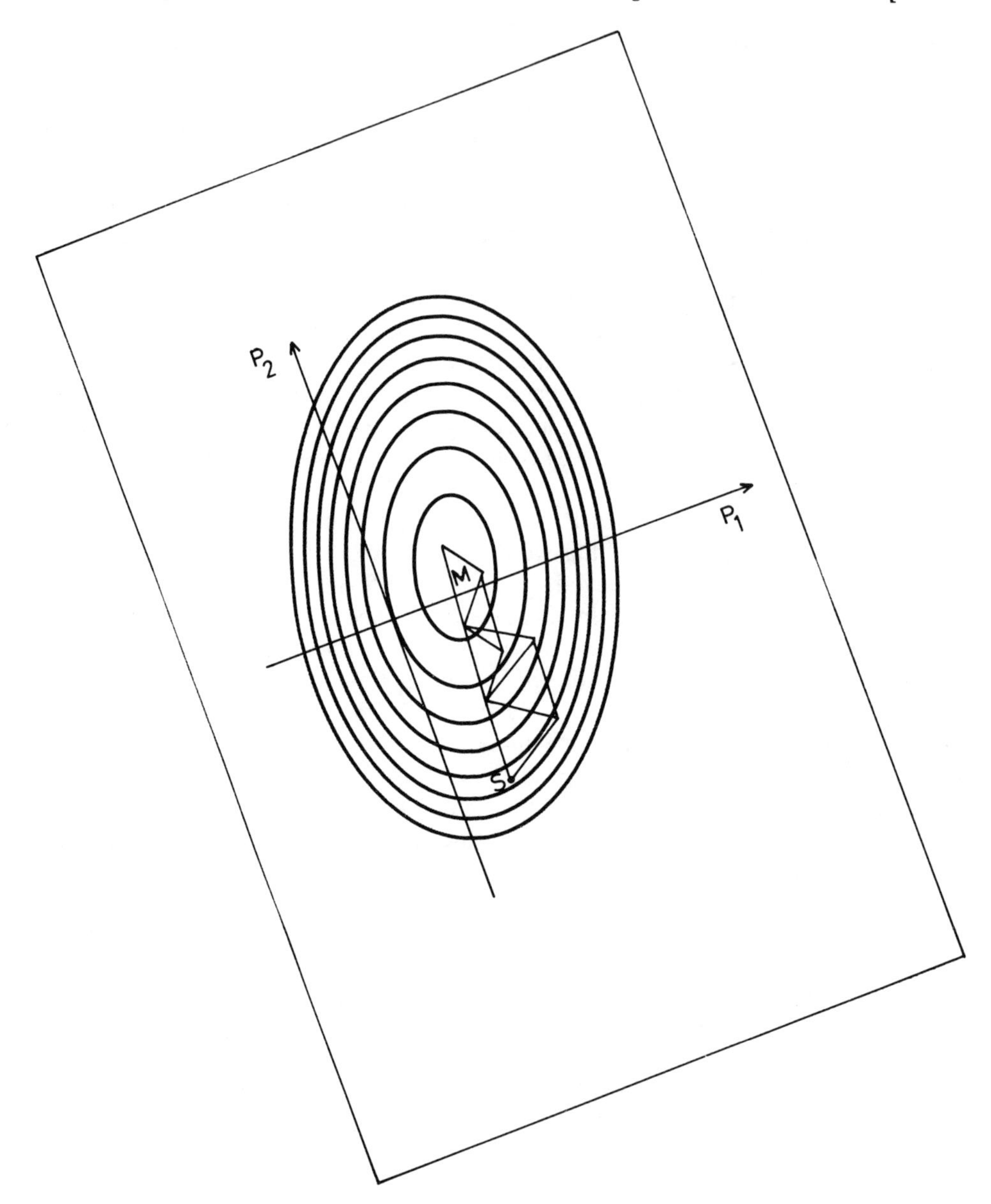

Fig. 6.6 — Simplex method illustrated in two dimensions. The simplex (triangle in this case) starts with a vertex at S and moves and distorts according to a set of rules, until it finds and contracts around the minimum point M.

think of B and W bracketing R in terms of function value. However, if R is better than B then an expansion is attempted in order to take advantage of a promising search direction, and if R is worse than W then possibly the current simplex straddles the valley and a contraction is tested.

(2) Expansion: this operation is only performed when a reflection has resulted in a

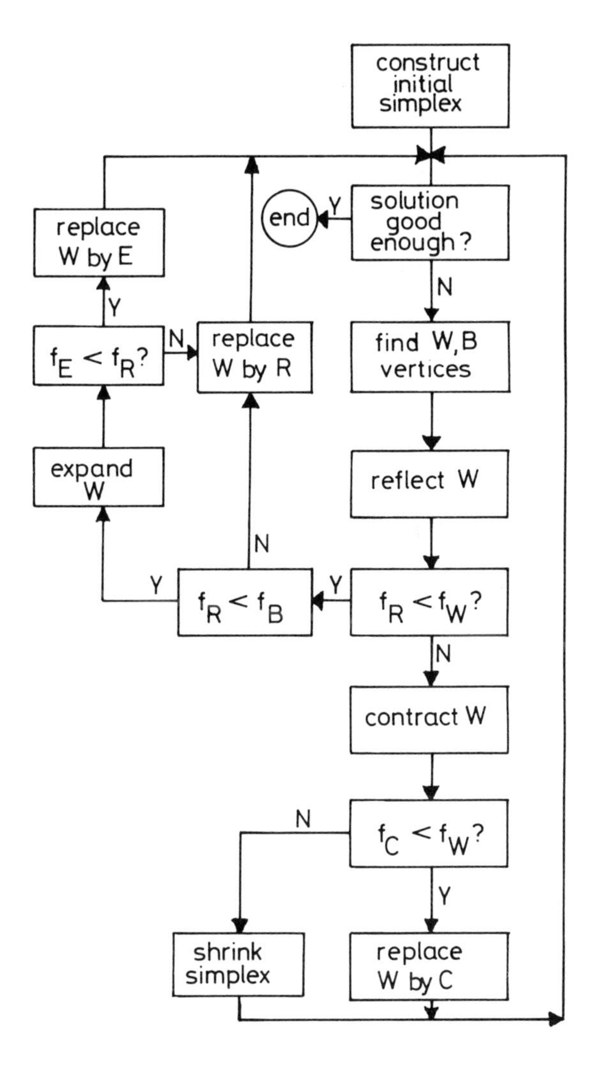

Fig. 6.7 — Flowchart of the Simplex method. For an explanation of the symbols and terminology, consult the text and Fig. 6.8.

vertex which is better than the previous best vertex, and seeks to use acquired information to accelerate the movement of the simplex in favourable search directions. In this sense it is akin to the pattern move of Hooke and Jeeves (section 6.3.3). An expansion is like a reflection but involves a larger step in the direction WX. Using the recommended expansion factor of Nelder and Mead (1965) the expanded point E is such that XE = 2WX (Fig. 6.8(b)). If E is better then R then it is accepted, if not then R is used to replace W.

(3) Contraction: this is tried if R is worse then W. The contracted vertex C lies on WX, its position relative to W being determined by a contraction factor. Nelder

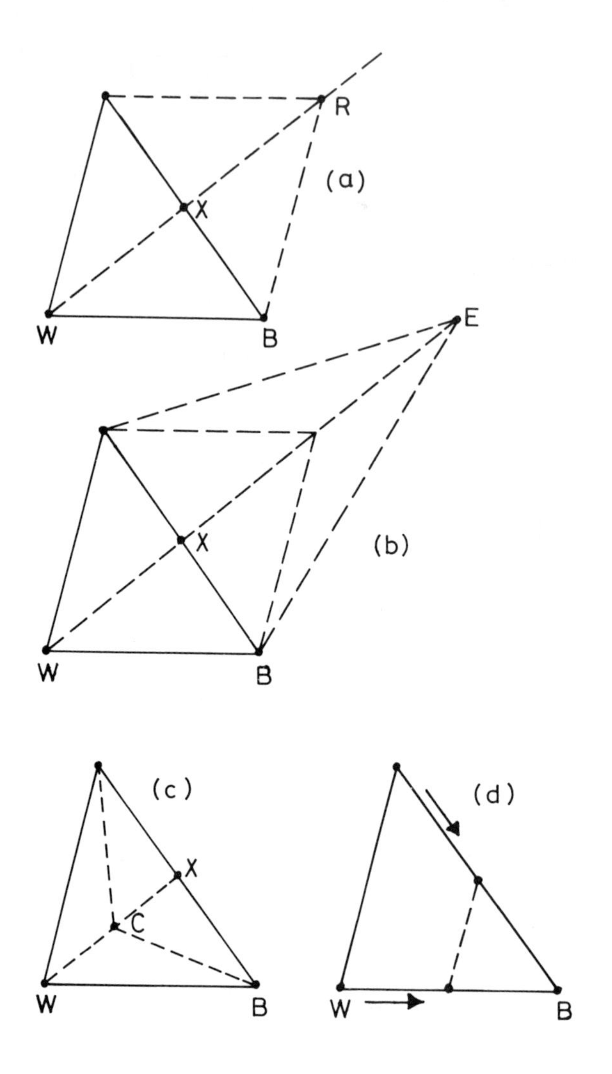

Fig. 6.8 — Geometrical representation of operations applied to a two-dimensional simplex (triangle): (a) reflection; (b) expansion; (c) contraction; (d) shrinkage.

and Mead (1965) suggest a factor which places C midway between W and X (Fig. 6.8(c)). If C is better than W then it is accepted and replaces W.

If none of these operations leads to an improvement in W then a shrinkage is implemented: all the vertices except B are moved toward B by an amount equal to half their original distances from B (Fig. 6.8(d)). Another reflection is now tested and the cycle repeated until an adequate solution has been found. Fig. 6.6 shows schematically the evolution of a simplex in two dimensions, and listing P6.2 shows the program to implement the simplex algorithm.

The simplex method is the best of the pattern-type approaches and offers several advantages:

(a) it never diverges, although it may not always converge properly, especially if the initial simplex is poorly constructed (that is, with the function values at all the vertices very close to each other);
(b) it requires relatively few function evaluations;
(c) it does not require derivatives of the function (this is true of all direct search methods);
(d) the arithmetic is straightforward.

These aspects are discussed in more detail by Caceci and Cacheris (1984). A brief review of the simplex technique and a historical perspective on its application has been given by Golden and Deming (1984); this paper also provides a valuable list of references to practical uses of the method.

6.3.5 Terminating the simplex search

In the case of those minimization methods where a simple sequence of parameter vectors $[p]^{(k)}$ is generated, the convergence criteria are straightforward — the process is terminated when the change in function value between iteration cycles is less than a pre-chosen amount and/or the change in each component of $[p]^{(k)}$ between cycles is small enough.

In the simplex method, however, we generate a sequence of sets of parameter vectors, each set containing $(m+1)$ vectors corresponding to the vertices of a simplex. Furthermore, each member of the set has a function value (e.g. sum of squares) associated with it. We cannot simply base the termination decision on the closeness of all the $(m+1)$ function values since we can easily envisage a situation where the vertices of the simplex lie on the same contour (and are thus identical in function value) but some distance from the minimum point. Although such an occurrence is unlikely in practice, we may avoid it altogether by examining the closeness of the $(m+1)$ parameter sets, but also by considering the parameter and function values at the centroid of the simplex — if there is an unreasonable discrepancy between the centroid and the vertices, then we cannot be close enough to the minimum point.

Walsh (1975) sugests that the search be terminated when the standard deviation of the $(m+1)$ function values at the vertices, defined by

$$\sigma^2 = \sum_{k=1}^{m+1} (f_k - \bar{f})^2/m$$

where $\bar{f}$ is the mean of the function values, is less than some prescribed value. We also advise that one should test that the function value at the centroid is within the prescribed tolerance of $\bar{f}$, and that the parameters be analysed in a similar manner. Listing P6.2 illustrates how this may be achieved. The termination criterion used by

Caceci and Cacheris (1984) was that the percentage difference between the extreme values of each parameter was below some prescribed level.

A discussion of parameter errors will be deferred until section 6.6.

6.4 GRADIENT METHODS

At the most fundamental level, all of the more sophisticated minimization methods are similar — choose a starting point (or points in the simplex case) in parameter space and then move in a sequence of steps in directions determined, at least partially, by the local geography so as to converge on the neighbourhood of the point of minimum function value. The basic difference between direct search and gradient methods is that the search direction in the latter case is calculated by making explicit use of the derivatives of the function with respect to each of the dependent variables (parameters). Naturally, such methods only offer potential advantages over, for example, the simplex technique when it is both feasible and practicable to calculate all the derivatives required.

6.4.1 Basic algorithm

The general form of gradient methods is:

(1) Choose a starting point $(p_1^{(0)}, p_2^{(0)},\ldots,p_m^{(0)})$ which again we represent by the vector $[p]^{(0)}$.
(2) Calculate a search direction represented by the vector $[d]^{(k)}$ and based on the partial derivatives of the function $f([p]^{(k)})$ with respect to each parameter p_j $(j=1,\ldots,m)$.
(3) Evaluate

$$[p]^{(k+1)} = [p]^{(k)} + h[d]^{(k)} \tag{6.1}$$

where the step length h along the direction $[d]^{(k)}$ is obtained by a linear search (usually by quadratic interpolation) in the direction $[d]^{(k)}$.
(4) If sufficient convergence has been obtained (i.e. if $f([p]^{(k+1)}) - f([p]^{(k)})$ is small enough and/or the parameter changes in the current cycle are small enough) then stop; otherwise replace $[p]^{(k)}$ by $[p]^{(k+1)}$ and go back to step (2).

In some variations the step length in (3) is taken to be fixed (often $h=1$) and the quadratic interpolation stage is omitted. There are many variants of the general gradient approach, of differing degrees of complexity, but a discussion will be deferred until after a description of Powell's quadratic interpolation method.

6.4.2 Quadratic interpolation

This method (which was introduced in Chapter 2), due to Powell (1964), seeks to minimize the function $f([p])$ along the line $[p]=[p]^{(k)} + h[d]^{(k)}$, where the nomenclature is as above. In other words, for what value of h (h_m say) is $f([p]^{(k)} + h[d]^{(k)})$ a minimum? Once found, then clearly $[p]^{(k+1)} = [p]^{(k)} + h_m[d]^{(k)}$ is a better estimate of the parameter vector which minimizes $f([p])$ than is $[p]^{(k)}$. The basis of the method is to approximate the variation of $f([p])$ with $[p]$ along $[p]^{(k)} + h[d]^{(k)}$ by a quadratic function, the minimum point of which may be predicted analytically. The algorithm, which is iterative, is as follows,

(1) Choose three points on the line $[p]^{(k)} + h[d]^{(k)}$. Let the corresponding values of h be h_1, h_2, h_3 and the function values be designated f_1, f_2, f_3 for convenience.
(2) Construct the (unique) quadratic $y(h)$ which passes through the three points. Calculate the value of h for which this curve has a turning point (call it h_m) and check whether it is a minimum or maximum point.
(3) If the point $h = h_m$ corresponds to a maximum in $y(h)$ then discard the point which is furthest from the maximum point and replace it with a new point obtained by moving a distance H (chosen by the user) away from the maximum. This step is made from the point which is furthest from the turning point. Go to step (2).
(4) If the point $h = h_m$ corresponds to a minimum point which is at a distance greater than H from the nearest of the current three points then replace the point which is furthest from the minimum point by one obtained by taking a step H towards the minimum. This time the step is made from the current point which is nearest to the turning point. This stage is necessary since otherwise there would be a danger that the predicted minimum point would be obtained by excessive extrapolation. Go to step (2).
(5) If the point $h = h_m$ corresponds to a minimum of $y(h)$ and if also it lies within a small (prescribed) distance of the nearest of the three current points, say $h = h_2$, then take the smaller of $f([p]^{(k)} + h_m[d]^{(k)})$ and $f([p]^{(k)} + h_2[d]^{(k)})$ as the sought-after minimum of $f([p])$. Replace $[p]^{(k)}$ by $[p]^{(k)} + h[d]^{(k)}$ and continue with the next step of the gradient method.
(6) If $h = h_m$ corresponds to a minimum point which satisfies neither (4) nor (5) then replace the point with highest function value by $[p]^{(k)} + h_m[d]^{(k)}$, and go to (2).

The reasons why the process has to be iterated rather than being performed in one step are twofold: firstly to avoid excessive extrapolation, and secondly because the quadratic approximation $y(h)$ to $f([p])$ is only valid close to the starting point $f([p]^{(k)})$. The initial choice of three points on the search line may be simply made. Decide on a step length h_0 and evaluate $f([p]^{(k)})$ and $f([p]^{(k)} + h_0[d]^{(k)})$. If $f([p]^{(k)}) < f([p]^{(k)} + h_0[d]^{(k)})$ then calculate $f([p]^{(k)} - h_0[d]^{(k)})$, otherwise evaluate $f([p]^{(k)} + 2h_0[d]^{(k)})$, yielding the three required points. The formula for calculating h_m is

$$h_m = \tfrac{1}{2}\left\{\frac{(h_2^2 - h_3^2)f_1 + (h_3^2 - h_1^2)f_2 + (h_1^2 - h_2^2)f_3}{(h_2 - h_3)f_1 + (h_3 - h_1)f_2 + (h_1 - h_2)f_3}\right\} \tag{6.2}$$

and the condition for a minimum is

$$\frac{(h_2 - h_3)f_1 + (h_3 - h_1)f_2 + (h_1 - h_2)f_3}{(h_1 - h_2)(h_2 - h_3)(h_3 - h_1)} < 0 \tag{6.3}$$

Their derivation was outlined in Chapter 2 and may also be found in Walsh (1975).

Although this method may appear complicated it is well-suited to implementation on a microcomputer (listing P6.3). Walsh (1975) has given a modification of the methods which reduces the number of function evaluations which are needed if more than one iteration of the algorithm is required.

As we commented earlier, a fixed step length is substituted quite often for the superior quadratic interpolation stage — this omission seems to be tolerable in many least-squares minimization problems, but would probably be unacceptable in the general minimization case.

A cubic interpolation method was introduced by Davidon. The formulae derived are more complicated than in Powell's method but fewer iterations are usually required. A detailed analysis has been given by Walsh (1975).

6.4.3 Method of steepest descent

Although of little practical application, because of its slow convergence near the minimum point and susceptibility to the 'narrow valley syndrome', this simple gradient method is worthy of examination since it can be used as a basis for the development of more sophisticated approaches.

The aim of the method is to choose the search direction which maximizes the initial (local) rate of decrease of the function value. If $[d]$ is a general direction, then the rate of decrease of $f[p]$) in this direction is approximated by

$$\mathrm{d}f/\mathrm{d}h = -(f([p]+h[d]) - f([p]))/h \tag{6.4}$$

where h is a small step length. In order to proceed further we need to know how the function $f([p])$ varies near the starting point, so that we can write a suitable approximate expression for $f([p]+h[d])$.

In Chapter 1 we saw how a function of one variable, for example $g(x)$, could be approximated in the region of $x = x_0$ by a polynomial called a Taylor series, and we could make this approximation as good as we liked by choosing enough terms. The m-order Taylor series approximation is

$$g(x) = g(x_0) + (x - x_0)(\mathrm{d}g/\mathrm{d}x) + (1/2!)(x - x_0)^2\ (\mathrm{d}^2g/\mathrm{d}x^2) + \ldots + (1/m!)(x - x_0)^m\ (\mathrm{d}^m g/\mathrm{d}x^m)$$

There are analogous but more complex series for functions of more than one variable. For a function of two variables, say $g(x,y)$, the first terms are

$$g(x,y) = g(x_0,y_0) + (x - x_0)(\partial g/\partial x) + (y - y_0)(\partial g/\partial y) \tag{6.5a}$$

or

$$g(x_0 + \Delta x,\ y_0 + \Delta y) = g(x_0,y_0) + \Delta x(\partial g/\partial x) + \Delta y(\partial g/\partial y) \tag{6.5b}$$

where $\Delta x = x - x_0$ and $\Delta y = y - y_0$. We can thus estimate the function value at (x,y)

from a knowledge of the function value and its partial derivatives at (x_0, y_0). We can simplify and at the same time generalize the expression by introducing matrix notation and letting

$$[\upsilon_0] = \begin{bmatrix} x_0 \\ y_0 \end{bmatrix} \text{ and } \Delta\upsilon = \begin{bmatrix} \Delta x \\ \Delta y \end{bmatrix}$$

Hence

$$g([\upsilon_0] + [\Delta\upsilon]) = g([\upsilon_0]) + [(\partial g/\partial x)\ (\partial g/\partial y)] \begin{bmatrix} \Delta x \\ \Delta y \end{bmatrix}$$

and we have written the second term on the right-hand side as the scalar product of $[\Delta\upsilon]$ with a vector whose elements are the partial derivatives of $g(x,y)$. This vector occurs frequently in vector analysis and is given the special name [grad(g)]; it will often be seen written as ∇g. Thus

$$g([\upsilon_0] + [\Delta\upsilon]) = g([\upsilon_o]) + [\text{grad}(g)]^{\text{T}}[\Delta\upsilon] \tag{6.6}$$

This equation holds for the general m-variable situation, and returning to our original problem we have

$$f([p] + h[d]) = f([p]) + h[\text{grad}(f)]^{\text{T}}[d]$$

and finally

$$\mathrm{d}f/\mathrm{d}h = -[\text{grad}(f)]^{\text{T}}[d] \tag{6.7}$$

Now, the scalar product of two vectors $[\upsilon_1]$ and $[\upsilon_2]$ is $\upsilon_1\,\upsilon_2 \cos\theta$ where θ is the angle between $[\upsilon_1]$ and $[\upsilon_2]$. Clearly, this has a maximum value when $\cos\theta = 1$, that is when the vectors are parallel ($\theta = 0$) or in other words are pointing in the same direction. Thus $\mathrm{d}f/\mathrm{d}h$ is a maximum when $[d] = -[\text{grad}(f)]$.

Hence the direction of steepest descent is $-[\text{grad}(f)]$, and the core of the updating algorithm is

$$[p]^{(k+1)} = [p]^{(k)} - h[\text{grad(f)}] \tag{6.8}$$

where either h is fixed (often we take $h = 1$) or it is determined each iteration by quadratic interpolation.

6.4.4 Gauss–Newton method

The Gauss–Newton method, or simply Newton's method, provides an improvement in the rate of convergence over the method of steepest descent by making use of the second derivatives of $f([p])$, which describe the curvature of the function. Its basis lies in approximating $f([p])$ by a quadratic, which is the simplest polynomial with a minimum. We then take the turning point of this quadratic as the next point in the search sequence.

We begin by returning to the Taylor series expansion of $g(x,y)$ we encountered in the previous section (equation (6.5a)). If we let $g(x,y) = \partial f(x,y)/\partial x$ then (6.5b) becomes

$$\partial f/\partial x = (\partial f/\partial x)_0 + \Delta x(\partial^2 f/\partial x^2)_0 + \Delta y(\partial^2 f/\partial x \partial y)_0 \tag{6.9}$$

where the subscript indicates that the derivatives on the right-hand side are evaluated at (x_0, y_0). This expression assumes $\partial f/\partial x$ is linear in x, which means $f(x,y)$ is quadratic in x. Similarly

$$\partial f/\partial y = (\partial f/\partial y)_0 + \Delta x(\partial^2 f/\partial y \partial x)_0 + \Delta y(\partial^2 f/\partial y^2)_0 \tag{6.10}$$

Equations (6.9) and (6.10) estimate the derivatives of $f(x,y)$ at a point displaced by $(\Delta x, \Delta y)$ from (x_0, y_0). At the minimum point $\partial f/\partial x = \partial f/\partial y = 0$, and thus the displacements needed to reach this point are given by the solution of the set of simultaneous equations

$$(\partial^2 f/\partial x^2)\Delta x + (\partial^2 f/\partial x/\partial y)\Delta y = -\partial f/\partial x$$
$$(\partial^2 f/\partial x \partial y)\Delta x + (\partial^2 f/\partial y^2)\Delta y = -\partial f/\partial y$$

where for clarity we have dropped the zero subscript from the derivatives. In matrix terms, and recalling the definitions of the previous section, we have

$$[H][\Delta v] = -[\mathrm{grad}(f)]$$

and thus

$$[\Delta v] = -[H]^{-1}[\mathrm{grad}(f)] \tag{6.11}$$

The matrix $[H]$ of all partial second derivatives is called the Hessian matrix. For the general case, $[H]$ is $m \times m$ and has elements

$$H_{ij} = \partial^2 f([p])/\partial p_i \partial p_j \qquad (i,j = 1,\ldots,m) \tag{6.12}$$

Applying (6.11) to our original problem we have

$$[\Delta p] = -[H]^{-1}[\text{grad}(f)]$$

where $[\Delta p]$ is the 'correction vector' we add to the current parameter vector to provide a better estimate of the minimizing parameter vector. Since $f([p])$ is unlikely to be exactly quadratic in the parameters, the corrected parameter vector is not the exact solution, and the process must be repeated. The iterative algorithm for updating the estimate to the solution vector is thus

$$[p]^{(k+1)} = [p]^{(k)} - h^{(k)}\,([H]^{(k)})^{-1}\,[\text{grad}(f)] \tag{6.13}$$

where $h^{(k)}$ may be found by a linear search in the direction $[d]^{(k)} = -([H]^{(k)})^{-1}$ $[\text{grad}(f)]$ or may be fixed.

In the special case of least squares minimization. the formula for calculating the elements of $[H]$ (equation (6.12)) reduces to a form which involves first derivatives only. An explanation of this result and some examples will be given later in this chapter. A program for Gauss–Newton minimization is given in listing P6.4. In this implementation we re-formulate the problem as

$$[H]^{(k)}([p]^{(k+1)} - [p]^{(k)}) = -[\text{grad}(f)]$$

and solve for the parameter adjustments $([p]^{(k+1)} - [p]^{(k)})$, using one of the methods discussed in Chapter 1. This avoids the necessity for computing the inverse of the Hessian matrix, but if this is required (for example, for error analysis, see section 6.6) then its computation may be incorporated into the solution procedure (Chapter 1).

6.4.5 Quasi-Newton methods

The updating formula in gradient methods may be generalized as

$$[p]^{(k+1)} = [p]^{(k)} - h^{(k)}[M]^{(k)}[\text{grad}(f)] \tag{6.14}$$

where $[M]^{(k)}$ is an $m \times m$ matrix. For the method of steepest descent $[M]^{(k)}$ is the unit matrix $[I]$, whereas for the Gauss–Newton method it is the inverse of the Hessian matrix. The use of other forms of $[M]$ gives rise to a class of methods termed quasi-Newton methods, in which the inverse Hessian matrix is replaced by $[M]^{(k)}$ which is updated each cycle according to a particular set of rules, without the need for explicit matrix inversion.

A popular method of this type is the Davidon–Fletcher–Powell (DFP) method which starts with $[M]^{(0)}$ equal to $[I]$, the unit matrix but updates $[M]$ in such a way that as the minimum point is approached $[M]^{(k)}$ becomes a closer and closer approximation to the inverse Hessian matrix, i.e. the method starts out as steepest descent (with its initially fast decrease in $f([p]^{(k)})$ and finishes as Gauss–Newton (with its rapid convergence near the minimum point). The method thus proceeds in the standard fashion for gradient methods, as outlined earlier, except that at the end of each cycle when $[p]^{(k+1)}$ has been evaluated, the matrix $[M]^{(k)}$ must be updated. Two alternative formulae are in common use, the original due to DFP and the Complementary DFP due to Fletcher (1970). The latter is superior and is the one quoted here; the derivation is to be found in Walsh (1975).

$$[M]^{(k+1)} = [M]^{(k)} + \frac{\beta^{(k)}[s]^{(k)}([s]^{(k)})^{\mathrm{T}} - [s]^{(k)}([g]^{(k)})^{\mathrm{T}}[M]^{(k)} - [M]^{(k)}[g]^{(k)}([s]^{(k)})^{\mathrm{T}}}{([g]^{(k)})^{T}[s]^{(k)}} \tag{6.15}$$

where $[s]^{(k)} = [p]^{(k+1)} - [p]^{(k)}$, $[g]^{(k)} = [\mathrm{grad}(f([p]^{(k+1)}))] - [\mathrm{grad}(f([p]^{(k)}))]$ and

$$\beta^{(k)} = 1 + \frac{([g]^{(k)})^{\mathrm{T}}[M]^{(k)}[g]^{(k)}}{([g]^{(k)})^{\mathrm{T}}[s]^{(k)}}$$

The superscript T indicates the transpose of a vector; note the outer product terms, e.g. $[g]^{(k)}([s]^{(k)})^{\mathrm{T}}$ (see Chapter 1). Clearly, $[s]^{(k)}$ is the change in position (i.e. parameter) vector during the current cycle whilst $[g]^{(k)}$ is the change in gradient vector. The DFP algorithm is coded in listing P6.5.

It is tempting to assume that the more complex the optimization method, the more powerful it will be. The possible dangers in this blind faith are highlighted by considering an example. Walsh (1975) gives the function $f([p]) = (p_1)^4 - 3p_1p_2 + (p_2 + 2)^2$ which he shows cannot be minimized using Newton's method starting from the point (0,0) — the Newton search direction is uphill. However, the pattern search technique of Hooke and Jeeves solves the problem easily, yielding a minimum point $(-1.5, -4.25)$ with a minimum function value of -9. Although such pathological examples are not commonly met in practice, especially where least squares minimization is concerned, it is worthwhile always to bear in mind the less sophisticated optimization methods. Even the simple direct grid search has successfully minimized 'difficult' standard test functions (Becsey *et al.*, 1968).

As we mentioned in Chapter 3, Stewart (1967) modified the DFP method to accept difference approximations to derivatives. It may thus in a sense be regarded as a direct search method since only function evaluations are required. This approach allows the DFP method to be used to minimize complex functions where derivatives are not easily calculated.

6.4.6 Errors

Once the minimum point has been located with the desired precision, it is often of interest to be able to estimate the errors in the parameters defining this minimum. Intuitively, one would expect that the errors would depend on the curvature of the function near the minimum point — if the valley bottom is gently curved in one direction then the corresponding parameter is less well-defined (changes in the parameter value have a small effect on the function value) and vice versa. This is illustrated in the one-parameter case in Fig. 6.9. Mathematically this is described by the covariance matrix which is simply the inverse of the Hessian matrix and must be calculated explicitly in the Gauss–Newton method. The DFP method yields a good approximation to it, as discussed earlier. The diagonal elements of the covariance

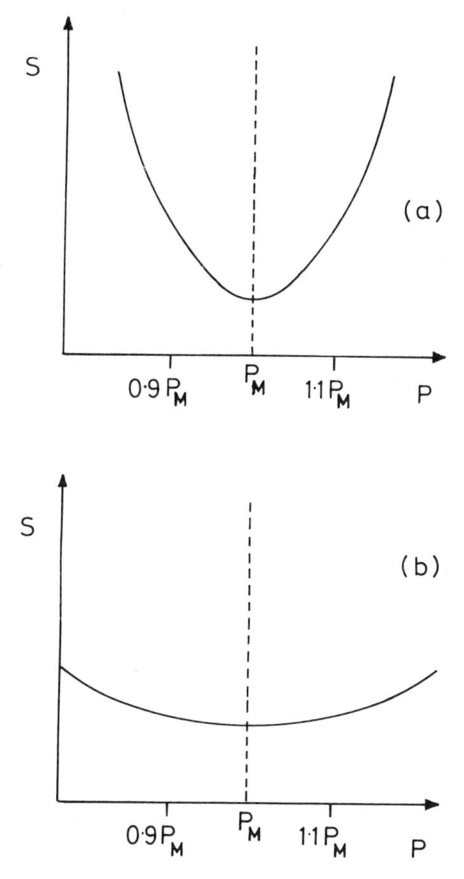

Fig. 6.9 — Illustration of the relationship between the curvature of the sum-of-squares function S at the minimum point M of a function $f(p)$ and the likely error in the parameter value p_M at that point. S is much more sensitive to a change in p_M in example (a) than it is in (b); p_M is thus much better defined (more accurately known) in the former case.

matrix are related to the squares of the errors in the parameter values, and the off-diagonal elements provide information on the correlation between parameters — the smaller these are, the less interdependent are the parameters. In the case of search methods, such as the simplex method, the Hessian matrix may be calculated and inverted once an acceptable minimum has been found. The estimation and interpretation of errors is discussed in greater detail in section 6.6.

6.5 LEAST SQUARES MINIMIZATION

For the majority of applied scientists the most useful application of the optimization methods discussed above is the fitting of a model $F([p]), x)$ to experimental data. This is achieved through minimization of the (weighted) sum of the squares of the difference between the predicted and measured data points (see previous chapter). Thus the function $f([p])$ defined above is given by

$$f([p]) = \sum_{i=1}^{n} w_i \,(F([p], x_i) - y(x_i))^2 \tag{6.16}$$

$$= \sum_{i=1}^{n} w_i (R_i)^2$$

where the fitting function is $F([p], x)$ and thus $F([p], x_i)$ is the predicted datum corresponding to a value x_i of the independent variable, $y(x_i)$ is the actual measured value and R_i is the residual at x_i. The weights w_i are often, either explicitly or implicitly, assumed to be equal and discarded. Whilst this may be justifiable in many instances (although the justification is frequently not discussed), a properly normalized sum-of-squares function (sometimes referred to as a chi-square function) is essential in an analysis of parameter errors. The two commonly used forms for w_i are (Johnson, 1980)

$$w_i = 1/y_i \tag{6.17}$$

known as statistical weighting, and

$$w_i = 1/\sigma_i^2 \tag{6.18}$$

which is termed instrumental weighting, where σ_i is the standard deviation in y_i and $y_i = y(x_i)$. For experimental situations where the measurements consist of accumulating counts into independent channels

$$\text{noise}_i \propto (y_i)^{\frac{1}{2}}$$

and the two formulations are identical, the choice then being made on the grounds of convenience. Strictly speaking one should measure a value for at each point by making repeated measurements of y_i, but it is often possible to simplify the procedure by assuming the σ_i are all equal. In this case the value of the standard deviation is only required for error analysis (see below).

The parameters represented by the vector $[p]$ are thus adjusted to find the minimum value of $f([p])$. Any of the methods discussed earlier may be used — the Simplex procedure (P6.2) is generally accepted to be the best of the search-type methods, although a Newton, combined Simplex–Newton or quasi–Newton method (DFP) is to be recommended where derivative calculations are feasible.

6.5.1 Newton's method and least squares

Major simplifications obtain when Newton's method is applied to least squares minimization. The general element of the Hessian $H_{jk} = \partial^2 f/\partial p_j \partial p_k$ which, substituting for $f([p])$ from (6.16), becomes (taking $w_i = 1$)

$$
\begin{aligned}
H_{jk} &= (\partial^2/\partial p_j \partial p_k) \sum_{i=1}^{n} (R_i)^2 \\
&= \sum_{i=1}^{n} (\partial/\partial p_j) 2R_i(\partial R_i/\partial p_k) \\
&= \sum_{i=1}^{n} \{2(\partial R_i/\partial p_j)(\partial R_i/\partial p_k) + 2R_i(\partial^2 R_i/\partial p_j \partial p_k)\}
\end{aligned}
\tag{6.19}
$$

It is usually assumed that the term in second derivatives on the right-hand side may be neglected compared with the first derivative term — this is called linearization (in linear least squares the second term is identically zero). If the second term is included, the procedure is sometimes referred to as the Newton–Raphson method. The vector of first derivatives [grad(f)] is given by

$$
(\text{grad}(f))_j = 2\sum_{i=1}^{n} \partial R_i/\partial p_j \tag{6.20}
$$

Note that $\partial R_i/\partial p_j = \partial F_i/\partial p_j$ since the y_i are independent of the p_j. Program P6.4 may be easily modified to incorporate expressions (6.16), (6.19) and (6.20); the result is the least squares minimization program P6.4A.

We have used Newton's method successfully for several years to decompose overlapping bands in multicomponent photoelectron spectra. A gaussian model function, defined by the three parameters peak height (p_1), peak position (p_2) and full-width at half-maximum (p_3), is used to represent the shape of the individual peaks

$$
G([p], x) = p_1 \exp(-\ln(2)(2(x - p_2)/p_3)^2)
$$

and the fitting function $F([p], x)$ is constructed as the algebraic sum of several such gaussians. An example of a typical fit using three component bands is shown in Fig. 6.10. Newton's method is used in its simplest form, with a fixed step length of unity and no linear search between iterations. Broyden (in Fletcher, 1969) found that when using a pseudo-Newton method in least-squares minimization, a linear search each cycle actually inhibited convergence.

The method is relatively problem-free, although it may diverge if the experimental data is very noisy, if the initial estimate of the minimum is very poor or if the

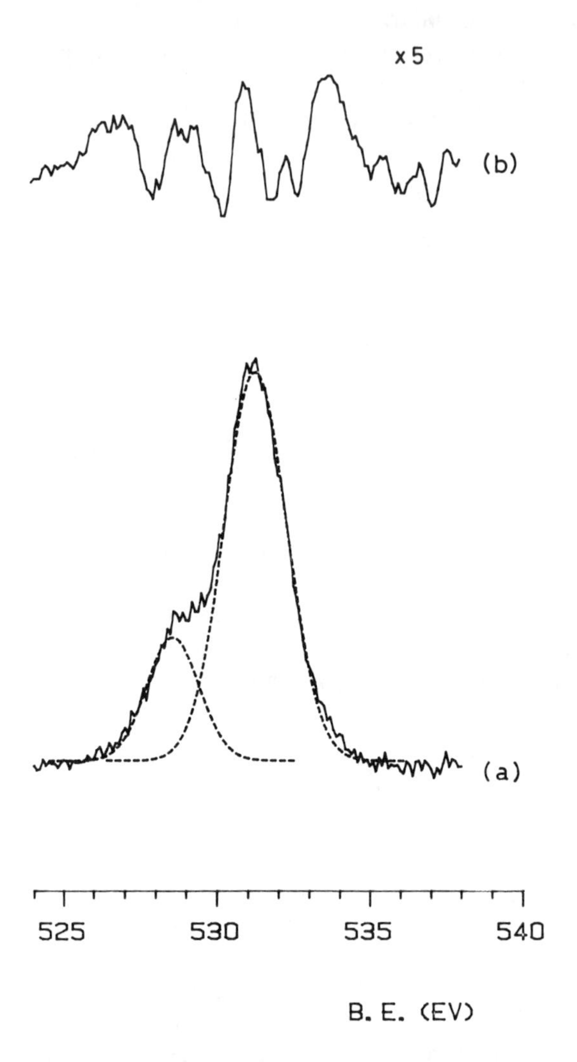

Fig. 6.10 — An example of curve-fitting from X-ray photoelectron spectroscopy. (a) O(1s) photoemission spectrum from a sample of the high critical temperature superconductor $YBa_2Cu_3O_{7-x}$ (Carley *et al.*, 1989), fitted with two gaussian components. For clarity the calculated envelope is not shown. (b) Residuals of the fit, i.e. the difference between the raw data and fitted envelope. The residuals have been subjected to nine-point smoothing.

overlap of components in the spectrum is severe. These effects generally worsen as the number of fitted gaussian bands increases. The divergence may be controlled by means of a modification to the standard Gauss–Newton method due to Levenberg (1944), which was subsequently developed by Marquardt (1963) and later extended by Meiron (1965); essentially a damping factor is introduced which unfortunately may also result in a slowing-down in the rate of convergence to the minimum (Fraser and Suzuki, in Blackburn (1970)).

6.5.2 Levenberg and Marquardt's method

The least-squares form of the Gauss–Newton method may be written in a more convenient form (Adby and Dempster, 1974). If we define a vector of residuals $[r]$ and a matrix $[A]$ by

$$r_i = F_i - y_i \qquad i = 1,\ldots,n \tag{6.21}$$

and

$$A_{ij} = \partial r_i / \partial p_j \qquad i = 1,\ldots,n \qquad j = 1,\ldots,m \tag{6.22}$$

then we find (Adby and Dempster, 1974)

$$[\text{grad}(f)] = 2[A]^{\text{T}}[r] \tag{6.23}$$

(where $[A]^{\text{T}}$ means the transpose of $[A]$) and

$$[H] = 2[A]^{\text{T}}[A] \tag{6.24}$$

Hence

$$[\Delta p] = -([A]^{\text{T}}[A])^{-1}[A]^{\text{T}}[r] \tag{6.25}$$

In 1944 Levenberg proposed a modification to (6.25) which allowed it to follow the basic strategy of variable metric or quasi-Newton methods mentioned earlier, that is to switch from steepest descent behaviour to Gauss–Newton as the minimum is approached:

$$[\Delta p] = -(\lambda[I] + [A]^{\text{T}}[A])^{-1}[A]^{\text{T}}[r] \tag{6.26}$$

where $[I]$ is the unit matrix, and λ the damping factor referred to earlier. For small λ, the method clearly reduces to the ordinary Newton method, whereas if λ is large $[\Delta p]$ becomes the steepest descent displacement. The method used by Levenberg to find the optimum value of λ in each cycle involved a time-consuming linear search; subsequently Marquardt (1963) proposed a more tractable procedure which was later improved by Fletcher. A detailed analysis of these procedures is given by Adby and Dempster (1974). The improvement due to Meiron (1965) replaces the unit matrix $[I]$ in (6.26) by a matrix $[C]$ which is formed from $[A]^{\text{T}}[A]$ by setting all off-diagonal elements to zero; the bracketed term thus reduces in this case to multiplying the diagonal elements of $[A]^{\text{T}}[A]$ by $(1 + \lambda)$. Pitha and Jones (1966) compared seven different optimization methods for the fitting of complex infrared band envelopes and, interestingly, found Meiron's method provided the speediest convergence; it should be noted that the value of λ was optimized each iteration cycle.

A good example of the Meiron 'damping factor' in action, suppressing divergence due to poor initial parameter estimates, is to be found in the article by Fraser and Suzuki (in Blackburn, 1970). The approach may be successfully used in an empirical, interactive fashion, allowing the operator to interrupt the fitting program when problems of divergence occur and increase the magnitude of the damping factor. The value can then be decreased progressively as the convergence of the process stabilizes, and the minimum is approached.

6.6 ERROR ESTIMATION

Having performed a least-squares curve-fit, we will naturally be interested in the reliability of the parameter values so obtained. We consider next two basic methods for estimating the errors in the best-fit parameters.

6.6.1 Variation of the sum-of-squares function with parameter value

As we mentioned earlier (section 6.4.6 and Fig. 6.9), on no other grounds than intuition, parameter reliability is related to the curvature of the sum-of-squares function and this in turn is related to the Hessian matrix containing the second derivatives of the sum-of-squares function. In this section we develop the mathematical basis for this relationship.

We must first return to basics and derive mathematically the least squares principle, for which we have provided only empirical justification (Chapter 5). We assume that the 'real' values of the quantity we are measuring fall along the curve $y^* = F(p,x)$ where for the purposes of illustration we use only a single parameter, p. We further assume that the observed values y_i are displaced about the y_i^* according to the known probability distribution which we are often justified in taking to be gaussian. Hence

$$\text{prob}(y_i) = (1/\sigma_i\sqrt{2\pi})\ \exp(-0.5(y_i^* - y_i)^2/\sigma_i^2) \tag{6.27}$$

and the probability of simultaneously finding a particular set of data values $y_1, y_2, \ldots, y_n$ (that is, the probability of obtaining a particular experimental result) is the product of the individual probabilities; thus

$$P = \text{prob}(y_{1,2,\ldots,n}) = \left\{ \prod_i (1/\sigma_i\sqrt{2\pi})\ \exp(-0.5 \sum_i (y_i^* - y_i)^2/\sigma_i^2) \right. \tag{6.28}$$

using the fact that

$$\prod_i \exp(x_i) = \exp\left(\sum_i x_i\right) .$$

We now invoke the 'principle of maximum likelihood' which states that we should choose p (and hence $y_i^* = F(p,x_i)$) so as to make our data set the most likely one, or in other words makes $P = \text{prob}(y_{1,2,...,n})$ a maximum. If we maximize P then we also maximize $L = \ln(P)$ since if any quantity reaches a maximum, any continuously increasing function of that quantity reaches a maximum at the same time. This proves convenient since

$$L = \text{constant} - 0.5 \sum_i (y_i^* - y_i)^2/\sigma_i^2$$

so that maximizing L (and thus P) is the same as minimizing the weighted sum of squared differences between y_i^* and y_i — the principle of least squares. Each difference is weighted with the standard deviation so that if σ_i is large and therefore y_i unreliable, then y_i has little effect on the sum-of-squares. Very often our data points are equally reliable, or approximately so, and the σ_i are all equal so we can consider σ_i as a constant and the procedure reduces to a simple least squares minimization.

Suppose we have found p_B, the best estimate of the parameter p. We can express L as a Taylor series around p_B, ignoring terms in p_B^3 and higher

$$L = L_B + (\mathrm{d}L/\mathrm{d}p)_{p_B}\,(p - p_B) + (\mathrm{d}^2L/\mathrm{d}p^2)_{p_B}\,(p - p_B)^2/2$$

However, $(\mathrm{d}L/\mathrm{d}p)_{p_B} = 0$ since p_B is at the bottom of the 'bowl', and thus

$$L = L_B + (\mathrm{d}^2L/\mathrm{d}p^2)_{p_B}\,(p - p_B)^2/2$$

Taking antilogs, we obtain $P = \text{prob}(y_{1,2,...,n})$ from L as

$$\begin{aligned} P &= \text{constant } \{\exp((\mathrm{d}^2L/\mathrm{d}p^2)_{p_B}\,(p - p_B)^2/2)\} \\ &= \text{constant } \{\exp(-0.5(p - p_B)^2/(-(\mathrm{d}^2L/\mathrm{d}p^2)_{p_B}^{-1}))\} \end{aligned}$$

This probability distribution tells us what the chances are of finding a particular value of p different from p_B; the variance (square of standard deviation) is given by $-(\mathrm{d}^2L/\mathrm{d}p^2)_{p_B}$. In practice we minimize the weighted sum-of-squares S_w rather than maximize L and since $L = \text{constant} - 0.5S_w$ we have $\mathrm{d}^2L/\mathrm{d}p^2 = -0.5\ d^2S_w/dp^2$. Thus, the standard deviation of the parameter p is given by

$$\sigma_p = \sqrt{\{2(\mathrm{d}^2S_w/\mathrm{d}p^2)^{-1}\}}$$

If we assume all the σ_i are equal then $S_w = S/\sigma^2$ where S is the simple sum-of-squares function and further, we may estimate σ from the minimum sum-of-squares S_B by

$$\sigma = \sqrt{\{S_B/(n-1)\}}$$

and the standard deviation of the parameter p is finally given by

$$\sigma_p = \sqrt{\{2(S_B/(n-1))(d^2S/dp^2)^{-1}\}}$$

In the multiparameter case (with m parameters) the terms d^2S_w/dp^2 and d^2S/dp^2 become the Hessian matrices $[H_w]$ and $[H]$ defined by

$$(H_w)_{ij} = \partial^2 S_w/\partial p_i \, \partial p_j$$

and

$$H_{ij} = \partial^2 S/\partial p_i \partial p_j$$

Furthermore, instead of reciprocals we must use inverse matrices and the standard deviation σ_k of the kth parameter p_k is given by

$$\sigma_k = \sqrt{\{2(S_B/(n-m))(H^{-1})_{kk}\}} \qquad (6.29)$$

where $(H^{-1})_{kk}$ is the kth diagonal element of the inverse of the Hessian matrix, and $S_B/(n-m)$ is the estimated variance of the noise.

Note that this is not the same as using the square root of the reciprocal of the diagonal element of the Hessian matrix, unless the Hessian matrix is diagonal, in which case the parameters are uncorrelated. Strong correlation between parameters is reflected in large off-diagonal elements in $[H]^{-1}$. It means that the sum-of-squares function can remain essentially unchanged if the parameters are changed in a concerted fashion, or conversely that a combined change in the parameters can produce a much larger change in the sum of squares than can any of the parameters separately. Clearly in such a case the parameter values are poorly defined, and may be misleading. Conway (1970) has extended the theory of confidence limits to take account of such correlations.

The Hessian matrix is calculated explicitly at each step in the Gauss–Newton procedure (section 6.4.4) and may be inverted once the minimum in the sum-of-squares function has been determined. This is best achieved by taking the unit matrix and applying to it the elimination/row exchange operations required to solve for the parameter adjustments. In quasi-Newton methods an approximation to the inverse Hessian matrix is generated directly (section 6.4.5). Direct search methods, such as the Simplex method (section 6.3.4), require that the Hessian matrix must be constructed and inverted once the minimization has been achieved. In the special case of polynomial curve-fitting (section 5.5) the coefficient matrix $[M]$ (equations (5.25) and (5.26)) is related to the Hessian matrix by $[M] = (1/2)[H]$ (the reader should confirm this) and thus

$$\sigma_k = \sigma\sqrt{\{(M^{-1})_{kk}\}}$$

6.6.2 Monte Carlo approach

The basis of this method lies in fitting simulated sets of experimental data, and looking at the standard deviation in the resulting set of parameter estimates. In practice, we construct the simulated sets by taking the best fit curve and adding gaussian noise (Chapter 5) having the same standard deviation as that of the original data. If necessary we can estimate the standard deviation of the noise from the residuals (section 6.6.1) as

$$\sigma = \sqrt{(S/(n-m))}$$

where n is the number of points, m is the number of parameters and S is the sum of squares of the residuals for the best fit. We use the factor $(n-m)$ rather than n because we have 'used up' m of our sources of variability in fitting the curve, and we only have $(n-m)$ 'degrees of freedom' left. For example, if we take a fitting function $f(t) = A + B\exp(-Ct)$ with $m = 3$, then we could make it pass exactly, through three of our data points by a suitable choice of A, B and C, even though these points contain error.

We use the pseudo-data sets as if they were real data and fit using our original best-fit estimates of the parameters as starting values; clearly these fits will be very fast since the pseudo-fit will be close to the 'true' best fit. The resulting fitted parameters we will term 'pseudo-values', and assume we shall make n_f such pseudo-fits. In order to obtain some idea of the standard deviations of the parameters, we can calculate the error matrix $[E]$ according to

$$[E] = (1/n_f)[P]^T[P]$$

where $[E]$ is $(m \times m)$ and $[P]$ is $(n_f \times m)$ and is the matrix of deviations of the parameter estimates from the original best fit values. That is,

$$P_{ji} = (\text{pseudo-value of the } i\text{th parameter and } j\text{th trial}) - \\ - (\text{original 'true' estimate of the } i\text{th parameter})$$

In fact, $[E]$ is the covariance matrix of the parameter estimates, and the diagonal elements are the variances of the parameter estimates, giving $\sigma_k = \sqrt{(E_{kk})}$.

Chandler *et al.* (1972) suggest that a measure of our confidence in the fit can be provided by the ratio (number of fits with sum-of-squares greater than the original best fit)/(number of pseudo-fits). Thus, for example, if this ratio is 0.9, and if we regard our Monte Carlo simulations as being representative of all possible occurrences of random errors about the true value, then we deduce that 90% of these would be worse fits. Obviously, the higher this number the more confidence we have in our fit.

Another procedure for generating pseudo-data sets is used in the 'jack-knife' method, where subsets of the original data are selected. For example, one might miss out one point of the original data at a time and refit. For more details the reader should consult Duncan (1978).

6.6.3 An example

As an illustration, we consider 'experimental' data constructed according to

$$y_i = 1.5 \exp(-0.5x_i) + n_i \tag{6.30}$$

where n_i is a gaussian noise contribution (Chapter 5). (Fig. 6.11(a) shows a plot of

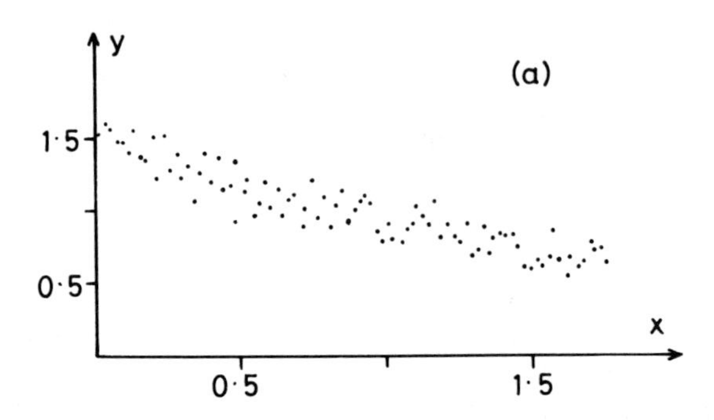

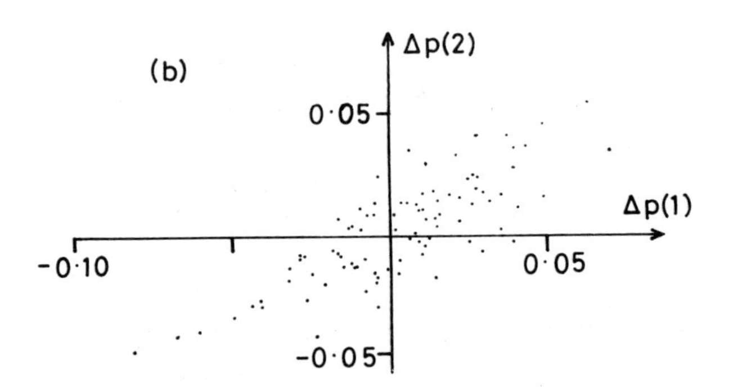

Fig. 6.11 — (a) Data synthesized from the function $y = 1.5 \exp(-0.5x)$, with added noise. The standard deviation of the noise is 0.1. (b) Plot of $\Delta p(1)$ versus $\Delta p(2)$ for 'pseudo-fits' of $y = p(1) \exp(-p(2)x)$ to 100 data sets derived from the least-squares fit to Fig. 6.11(a). $\Delta p(1)$ and $\Delta p(2)$ are the differences between the parameters refitted to the pseudo-data sets, and the original best-fit values. The origin thus corresponds to the best-fit values of p(1) and p(2).

'experimental' data created according to (6.30) with noise standard deviation of 0.1. Next, we fit the data with the model function $y = p(1)\exp(-p(2)x)$, and derive best-fit values for $p(1)$ and $p(2)$ (Table 6.1). At this stage we can estimate the standard deviations in $p(1)$ and $p(2)$ via the Hessian matrix (Table 6.1); remember

Table 6.1 — Error analysis in curve-fitting, using the synthesized 'experimental' data of Fig. 6.11(a) as the example

Best-fit values		Estimated errors			
		From Hessian		Monte Carlo	
$p(1)$	$p(2)$	$p(1)$	$p(2)$	$p(1)$	$p(2)$
1.506	0.504	0.025	0.019	0.027	0.021

that for least squares minimization the Hessian matrix can be calculated using only the first derivatives of the fitting function (section 6.5.2). In order to apply the Monte Carlo analysis, we create 100 simulated data sets using the best-fit parameters and adding noise with the same standard deviation as in the original 'data'. Fig. 6.11(b) shows a plot of $p(1)$ against $p(2)$ for the simulated data sets corresponding to Fig. 6.11(a); the origin has been shifted to coincide with the best-fit values of the parameters. The plot indicates strong correlation between the two parameters since the points lie in a band rather than being clustered randomly around the best-fit point (origin). This may be confirmed by an inspection of either the error matrix or inverse Hessian matrix. The error matrix for this example is

$$\begin{bmatrix} 7.28\times10^{-4} & 4.51\times10^{-4} \\ 4.51\times10^{-4} & 4.34\times10^{-4} \end{bmatrix}$$

and we observe that the off-diagonal elements are of comparable magnitude to the diagonal elements, confirming the strong correlation between parameters. Parameter errors derived from the error matrices are also shown in Table 6.1.

6.7 CONSTRAINED MINIMIZATION

The minimization methods which we discussed earlier are known as unconstrained methods since the parameters are allowed to vary freely during the search for the minimum point. Although this may be acceptable if one is simply interested in finding the minimum value of mathematical function, in the 'real world' one may be obliged to subject some of the parameters to constraints in order to obtain physically meaningful results. In the example of gaussian curve-fitting presented earlier, an unconstrained minimization might lead to parameter values which are in conflict with prior knowledge, past experience or common sense (for example, negative peak heights or unexpectedly large or small FWHM values). The origins of such problems may lie in an inadequate model or high levels of noise in the experimental data.

The most general form of parameter constraint usually required is

$$l_j \leqslant p_j \leqslant u_j \tag{6.31}$$

where l_j and u_j are lower and upper bounds on p_j. Special cases may be accommodated by a suitable choice of bounds:

(1) Positivity — l_j is equated to a small positive number whereas u_j is very large and positive.
(2) Parameter fixing — set l_j equal to u_j. An easily implemented means of parameter fixing for use with Newton's method is described by Fraser and Suzuki (in Blackburn, 1970).
(3) Unconstrained parameters — l_j is set to a large negative number and u_j to a large positive value.

These constraints are frequently expressed as a set of inequality relationships.

$$h_i \leqslant 0 \qquad \text{for } i = 1 \ldots l \tag{6.32}$$

e.g. $p_2 - l_2 \leqslant 0$.

Constrained minimization is usually implemented as an unconstrained minimization of a restated problem. It is possible to remove constraints of the type mentioned above by a change of variable (parameter). The restriction $i_j \leqslant p_j \leqslant u_j$ is equivalent to

$$p_j = l_j + (u_j - l_j)\{\sin(p_j')\}^2 \tag{6.33}$$

with p_j' now unconstrained in the transformed problem, i.e. with p_j replaced by the above expression wherever it appears. This straightforward approach is unfortunately only useful in a few single cases (Walsh, 1975).

In the case of direct search methods parameter constraints are most easily achieved by a modification to the routine which calculates the sum-of-squares *SS* (or indeed any objective function). The aim is to present the search from straying into a non-feasible region, i.e. a region where any of the parameter constraints are violated. A program section of the form

```
FOR I = 1 TO M
IF P(I) < L(I) OR P(I) > U(I) THEN SS = SS + Q
NEXT I
```

where Q is a large positive number, results in the return of a large (unfavourable) value for *SS* if any of the parameter constraints are violated, and the current search direction is thus rejected. For obvious reasons this is known as a barrier function technique, since we have essentially constructed a wall or barrier around the feasible region. This approach has the great merits of simplicity of implementation and flexibility; parameter constraints of great complexity may be easily included, such as a fixed relationship between certain of the parameters. Unfortunately, this technique cannot be used in this simple form with gradient methods of minimization, where the modified objective function must be differentiable, or at least possess derivatives. The most common way of doing this is to use penalty functions, q_i, to construct a new objective function

$$f' = f + \sum_i w_i q_i \tag{6.34}$$

where the w_i are weights and the q_i are (penalty) functions of the h_j. The penalty term causes the search to be deflected back into the feasible region if the constraint boundary is approached. A more rapidly convergent extension of this technique is known as sequential unconstrained minimization. In this procedure a sequence of modified object functions is generated, each one of which is minimized using a standard unconstrained method. The sequence of functions is so constructed that the sequence of unconstrained minima found converges to the constrained minimum of the original function (see Adby and Dempster, 1974).

Penalty function methods are of two kinds: interior point methods, where the search starts in the feasible region and is prevented from leaving by the penalty term which approaches infinity at the constraint boundaries (the barrier function approach already mentioned), or exterior point methods, where the minimization may start outside the feasible region but the penalty term forces a constrained minimum to be reached.

The use of penalty function methods with both inequality and equality constraints and the choice of penalty function are discussed in detail by Adby and Dempster (1974) and Walsh (1975). These references should also be consulted for a discussion of more advanced methods of constrained optimization, which are beyond the scope of this book.

We must bear in mind that we are now minimizing a different function than the original one, into which artefacts may have been introduced by the penalty function(s), and also that there may be problems with derivatives on or near the boundary.

6.8 LEAST SQUARES AND SYSTEMS OF EQUATIONS

In Chapter 1 we tackled the problem of analysing multicomponent spectra and determining how much of each of several known pure component spectra were present. This involved solving the system of simultaneous equations represented by the matrix equation

$$[K][c] = [a]$$

for the mixing coefficients (concentrations of pure species) $[c]$ where $[a]$ represents the measured ordinates in the mixture spectrum. We also saw how the presence of noise adversely affected the reliability of the solution especially when the overlap was severe. The methods used demanded that the matrix $[K]$ be square $(N \times N)$ with columns given by the ordinates of the N pure spectra. Clearly we can then only use N ordinates for the measured spectrum, and $[a]$ must be an N-component vector. In this section we will see how to make use of more measured values (thus making $[K]$ non-square; the set of equations then is called over-determined) by incorporating the criterion of least squares into the procedure.

Assume we have made M measurements, with $M > N$, so that $[K]$ is now an $M \times N$ matrix — that is, with M rows and N columns — and $[a]$ is now an M-element vector. The vector $[c]$ will still, of course, contain N elements. In the case of perfect, noise-free data we could write, as in $N \times N$ case

$$[K][c] = [a] \tag{6.35}$$

For real, noisy, measurements, however, the system cannot be satisfied exactly — we cannot find N values c_j $(j = 1,\ldots,N)$ which satisfy simultaneously the M equations represented by (6.35).

Now, for any $[c]$ we can write

$$[K][c] - [a] = [r] \tag{6.36}$$

where $[r]$ is the vector of residuals, a measure of how poor the 'solution' $[c]$ is. We invoke the least squares principle by choosing as the 'best' solution that which minimizes $[r]^{\mathrm{T}}[\mathrm{r}] = \mathrm{r}^2$.

An obvious way to implement this is to apply one of the direct search methods discussed in section 6.3. Indeed, this was one of the original problems for which the method of Hooke and Jeeves was developed. The 'function' we minimize is $[r]^{\mathrm{T}}[r]$, or in other words the sum of the squares of the M residuals corresponding to the M equations. Instead of a simple function call, we must now use a subroutine call to evaluate the 'function' value during the search.

We leave the investigation of this approach as a project for the interested reader. We shall instead tackle the problem from a very different viewpoint. We assume first of all (and shall prove shortly) that we can find an orthonormal matrix $[Q]$ that reduced $[K]$ to upper triangular form. This statement requires some elaboration. An orthonormal matrix $[Q]$ is one which has the property that its transpose is its inverse, that is

$$[Q]^{\mathrm{T}}[Q] = [Q][Q]^{\mathrm{T}} = [\mathrm{I}] \tag{6.37}$$

Another way to look at it is that the scalar product of any pair of rows or pair of columns is zero, and the scalar product of any row or column with itself is unity. Why we need $[Q]$ to be orthonormal will be made apparent later. We discussed the reduction of matrices to upper triangular form in Chapter 1 (section 1.8), but in the context of square matrices. However, the process can be applied to any matrix, square or not — in the case of a 7×4 matrix, for instance, we represent it schematically as

$$
\begin{matrix}
* & * & * & * \\
* & * & * & * \\
* & * & * & * \\
* & * & * & * \\
* & * & * & * \\
* & * & * & * \\
* & * & * & *
\end{matrix}
\rightarrow
\begin{matrix}
* & * & * & * \\
0 & * & * & * \\
0 & 0 & * & * \\
0 & 0 & 0 & * \\
0 & 0 & 0 & 0 \\
0 & 0 & 0 & 0 \\
0 & 0 & 0 & 0
\end{matrix}
$$

$$[K] \rightarrow \begin{bmatrix} \mathrm{U} \\ 0 \end{bmatrix}$$

where $[U]$ is a $N \times N$ (4×4) upper triangular matrix and $[0]$ is the $(M-N) \times N$ (3×4) zero matrix. Given such a $[Q]$ we proceed

$$
\begin{aligned}
[Q][r] &= [Q]([K)[c] - [a]) \\
&= [Q][K][c] - [Q][a] \\
&= \begin{bmatrix} U \\ 0 \end{bmatrix} [c] - \begin{bmatrix} f \\ g \end{bmatrix} \\
&= \begin{bmatrix} U & c & - & f \\ & & - & g \end{bmatrix}
\end{aligned}
$$

where we have partitioned $[Q][a]$ into two segments $[f]$ with M elements and $[g]$ with $(M-N)$ elements, to correspond with the obvious partitioning of the reduced matrix into $[U]$ and $[0]$. Now since $[Q]$ is orthonormal, the length of $[Q][r]$ is the same as the length of $[r]$, and thus

$$
\begin{aligned}
[r]^{\mathrm{T}}[r] &= ([Q][r])^{\mathrm{T}} \, ([Q][r]) \\
&= ([U][c] - [f])^{\mathrm{T}} \, ([U][c] - [f]) + [g]^{\mathrm{T}}[g]
\end{aligned}
$$

Since the first term on the right-hand side (representing the length of a vector) cannot be negative, and the second term is predetermined by the measurements we have made, the quantity $[r]^{\mathrm{T}}[r]$ is a minimum when the first term is equal to zero, the least squares solution of (6.36) is given by the solution of

$$[U][c] = [f] \tag{6.38}$$

which we solve by back substitution (section 1.8) to obtain $[c]$.

Clearly, it now remains to find $[Q]$ or, as it turns out, a sequence of matrix operations to reduce $[K]$ to upper triangular form. We make use of a class of orthonormal matrices known as rotation matrices. The simplest rotation matrix is

$$\begin{bmatrix} \cos\theta & \sin\theta \\ -\sin\theta & \cos\theta \end{bmatrix}$$

which operates on a vector to rotate it through an angle θ without changing its length (since the matrix is orthonormal). The general rotation matrix we denote by $[Q(i,j)]$. It is constructed by taking the identity matrix and changing four elements according to

$$Q(i,j)_{ii} = \cos\theta \qquad Q(i,j)_{ij} = \sin\theta$$
$$Q(i,j)_{ji} = -\sin\theta \qquad Q(i,j)_{jj} = \cos\theta$$

This notation may at first sight appear confusing to the reader, with its mixture of bracketed and subscripted indices i and j. The subscripts are the usual identifiers for the elements of a matrix. The indices within the brackets denote the rows and columns whose intersections identify where the trigonometric elements are situated. For example, $Q(2,4)$ might be a 5×5 rotation matrix given by

$$\begin{array}{c} \\ \\ \\ 2\rightarrow \\ \\ 4\rightarrow \\ \\ \end{array} \begin{array}{c} \quad\; 2 \qquad\quad 4 \\ \quad\;\downarrow \qquad\quad \downarrow \\ \begin{bmatrix} 1 & 0 & 0 & 0 & 0 \\ 0 & c & 0 & s & 0 \\ 0 & 0 & 1 & 0 & 0 \\ 0 & -s & 0 & c & 0 \\ 0 & 0 & 0 & 0 & 1 \end{bmatrix} \end{array}$$

where $c = \cos\theta$ and $s = \sin\theta$. For the more mathematically inclined reader, $[Q(i,j)]$ represents a rotation in the i,j plane. However, our interest in this type of matrix lies in the fact that for a suitable choice of θ the operation of $[Q(i,j)]$ on a matrix $[K]$ results in a matrix $[K]'$ with $K'_{ji} = 0$. We find, since most of the multiplications are by zeros.

$$K'_{ji} = -sK_{ii} + cK_{ji}$$

Requiring $K'_{ji} = 0$ and recalling that $c^2 + s^2 = 1$ gives us

$$s^2K_{ii}^2 = (1 - s^2)\ K_{ji}^2$$

Thus

$$s^2(K_{ii}^2 + K_{ji}^2) = K_{ji}^2$$

and finally

$$s = K_{ji}/(K_{ii}^2 + K_{ji}^2)^{\frac{1}{2}} \tag{6.39}$$
$$c = K_{ii}/(K_{ii}^2 + K_{ji}^2)^{\frac{1}{2}}$$

Thus, the repeated application of such matrices can be used to make all the subdiagonal elements of $[K]$ equal to zero. We proceed by considering each column in turn, sweeping through the column starting with the element immediately below the diagonal. In this way elements previously set to zero are not affected by subsequent operations.

Program P6.6 applies the method to the least squares solution of (6.35). The $(M \times N)$ system matrix $[K]$ is first reduced to upper triangular form, and the resulting N-equation system solved by back-substitution. The number of arithmetical operations may be minimized by taking into account the fact that the multiplication of a matrix $[K]$ by $[Q(i,j)]$ only affects rows i and j of $[K]$, thus avoiding a complete matrix multiplication.

As we have noted, the columns of $[K]$ in our example consist of M measured values from the N pure absorption spectra, and the vector $[a]$ is made up of M measurements on the mixture at identical wavelengths. The data input routines in P6.6 show how the matrix and vector may be loaded using an Apple II+ microcomputer with data stored on magnetic disk. Table 6.2 shows the result of applying this program to the noisy data of Fig. 1.20(d); recall that the analysis of this mixture 'spectrum' by the solution of N $(=4)$ simultaneous equations failed miserably (Table 1.5). In contrast the least squares solution using 100 values $(M = 100)$ gives a reasonably accurate solution. For convenience we have taken the M points equally spaced throughout the spectra. In order to minimize the number of points needed and also to improve the accuracy of the answer, it might be preferable to select points in the region of the peak and to disregard values from the baseline region, although one must guard against introducing subjective bias by so doing.

Table 6.2 — Effect of applying a least squares approach to the analysis of the mixture spectrum in Fig. 1.20(d). For details of the 'simple analysis' method, see sections 1.8.2. *et seq*.

	Concentrations			
	1	2	3	4
'Simple analysis'	0.817	2.400	−0.046	2.374
Least squares $M=25$	0.873	2.133	−0.151	2.263
$M=100$	1.051	1.420	0.476	1.995
(True values	1.000	1.500	0.500	2.000)

```
    0 REM  P6.1
   10 :
   20 REM Hooke and Jeeves' optimization method
   30 :
   40 DIM OP(20),CP(20),NP(20),H(20),MH(20),TP(20)
   50 GOTO 10000
  492 :
  494 REM example from Walsh (1975) p 77
  495 REM NF is the function value, NP(I) (I=1, ... ,M) the parameters
  497 :
  500 NF=3*NP(1)*NP(1) - 2*NP(1)*NP(2) + NP(2)*NP(2) + 4*NP(1) + 3*NP(2)
  510 RETURN
 7992 :
 7994 REM reduce step size and check if too small
 7996 :
 8000 FL=1
 8010 FOR K=1 TO M
 8020   IF H(K)>=MH(K) THEN H(K)=H(K)/2
 8030   FL=FL*( H(K) < MH(K) )
 8040   REM ( H(K) < MH(K) ) has value of 1 if true and 0 if false
 8050 NEXT K
 8092 :
 8100 REM exploratory move from TP( )
 8110 FOR J=1 TO M : NP(J)=TP(J) : NEXT J
 8120 FOR J=1 TO M
 8130   NP(J)=TP(J)+H(J) : GOSUB 500
 8140   IF NF>=TF THEN NP(J)=TP(J)-H(J) : GOSUB 500
 8145   REM if one direction unsuccessful try opposite one
 8150   IF NF>=TF THEN NP(J)=TP(J) : GOSUB 500
 8160   IF NF<TF THEN TP(J)=NP(J) : REM move base point
 8170 NEXT J
 8180 RETURN
 8190 :
10000 REM main segment - start data input
10010 INPUT "NUMBER OF PARAMETERS = " ;M : PRINT
10020 FOR J=1 TO M
10030   PRINT "PARAMETER #" ;J; : PRINT
10040   INPUT "STARTING VALUE = " : CP(J)
10050   INPUT "INITIAL STEP SIZE = " : H(J)
```

```
10060   INPUT "MINIMUM STEP SIZE = " : MH(J)
10070   NP(J)=CP(J)
10080 NEXT J
10090 GOSUB 500 : REM starting value of function
10100 CF=NF
10110 :
10120 REM search starts now
10130 :
10140 FOR L=0 TO 1 : REM simulates a DO ........ UNTIL type of loop
10150   FOR J=1 TO M : TP(J)=CP(J) : NEXT J
10160   TF=CF : GOSUB 8100 : REM explore
10170   IF NF>=CF THEN GOSUB 8000 : REM if no improvement then reduce step size
10180   :
10190   REM only exit if step size too small or we have moved downhill
10200   L=( NF < CF ) OR ( FL = 1 ) : REM L=1 if either condition is true
10210 NEXT L
10220 :
10230 IF FL=1 THEN END : REM end if step size below minimum specified
10240 :
10250 FOR J=1 TO M : OP(J)=CP(J) : CP(J)=NP(J) : NEXT J
10260 CF=NF
10270 :
10280 REM try pattern move
10290 FOR J=1 TO M
10300   PP(J)=2*CP(J)-OP(J)
10305   REM try a leap twice as far in the same successful direction
10310   TP(J)=PP(J)
10320 NEXT J
10330 TF=CF : GOSUB 8100 : REM explore
10340 IF NF>=CF THEN 10120 : REM pattern move failed
10350 :
10360 FOR J=1 TO M : OP(J)=PP(J) : CP(J) = NP(J) : NEXT J
10370 CF=NF
10380 GOTO 10280 : REM repeat the pattern/exploratory move
```

```
   0 REM P6.2
  10 :
  20 REM simplex optimization
  30 REM example from Walsh (1975) p 79
  31 REM if you want to use this routine for fitting observed data then
  32 REM add GOSUB 500 to line 10020, change GOSUB 8200 to GOSUB 7000 in
  33 REM lines 8680, 9040, 10130, 10430, 10570, 10690
  34 REM fitting function goes in at line 7000 - PR() are parameters
  35 REM observed values in YV(), independent variable values in X1()
  36 REM more than one independent variable needs arrays X2(),X3() etc.
  37 :
  40 REM dimension ST(),P(),Q(),S(,),NE(),PR(),HI(),LO(),CE() if M>9
  45 :
  50 AL=1 : BE=0.5 : GA=2 : REM  alpha, beta and gamma in the text
  60 :
  70 GOTO 10000
  80 :
 500 REM input data on observations
 510 DIM X1(50),YV(50)
 520 INPUT "NUMBER OF OBSERVATIONS":NB
 530 FOR I=1 TO NB : INPUT "X-VALUE, Y-VALUE":X1(I),YV(I) : NEXT I
 540 RETURN
 550 :
7000 REM calculate function as sum of squared differences
7010 SS=0
7020 FOR IK=1 TO NB
7030      SS=SS + ( PR(1) + PR(2)*EXP( - PR(3)*X1(IK) ) - YV(IK) )^2
7040 NEXT IK
7050 RETURN
7060 :
8000 REM construct initial simplex
8010 FOR K=1 TO M
8020      P(K)=ST(K)*( SQR(M1)+M-1 )/( M*SQR(2) )
8030      Q(K)=ST(K)*( SQR(M1)-1 )/( M*SQR(2) )
8040 NEXT K
8050 :
8060 FOR II=2 TO M1
8070      FOR JJ=1 TO M
8080           S(II,JJ)=S(1,JJ)+Q(JJ)
```

```
8090       NEXT JJ
8100       S(II,II-1)=S(1,II-1)+P(II-1)
8110 NEXT II
8120 RETURN
8130 :
8200 REM calculate function value for any vertex
8210 SS=3*PR(1)^2 - 3*PR(1)*PR(2) + PR(2)^2 + 4*PR(1) + 3*PR(2)
8220 RETURN
8230 :
8300 REM find largest and smallest of all parameters
8310 FOR JJ=1 TO M1
8320       FOR II=1 TO M1
8330               IF S(II,JJ) < S(LO(JJ),JJ) THEN LO(JJ)=II
8340               IF S(II,JJ) > S(HI(JJ),JJ) THEN HI(JJ)=II
8350       NEXT II
8360 NEXT JJ
8370 RETURN
8392 :
8400 REM replace worst vertex by new vertex
8410 FOR II=1 TO M1 : S(HI(M1),II)=NE(II) : NEXT II
8420 RETURN
8494 :
8500 REM check for convergence
8510 SM=0 : Z$=""
8520 FOR I=1 TO M1 : SM=SM+S(I,M1) : NEXT I
8530 SM=SM/M1 : REM mean function value over vertices
8540 SD=0
8550 FOR I=1 TO M1 : SD=SD+( S(I,M1)-SM )^2 : NEXT I
8560 SD=SQR(SD/M) : REM standard deviation of function values
8570 PRINT "MEAN ":SM,"STANDARD DEVIATION":SD
8580 :
8600 FOR J=1 TO M
8610       S=0
8620       FOR I=1 TO M1 : S=S+S(I,J) : NEXT I
8630       S=S/M1 : S(M1+1,J)=S
8640 NEXT J
8650 :
8660 IS=M1+1
8670 FOR K=1 TO M : PR(K)=S(IS,K) : NEXT K
```

```
 8680 GOSUB 8200 : REM SS is difference of mean and centroid function values
 8690 S(IS,M1)=SS
 8700 IF SD < TL AND ABS(SS-SM) < TL THEN Z$="OK"
 8710 RETURN
 8720 :
 9000 REM simplex expansion : PRINT "EXPANSION"
 9010 GOSUB 8400
 9020 FOR I=1 TO M : NE(I)=GA*S(HI(M1),I)+(1-GA)*CE(I) : NEXT I
 9030 FOR K=1 TO M : PR(K)=NE(K) : NEXT K
 9040 GOSUB 8200
 9050 NE(M1)=SS
 9060 IF NE(M1)<=S(LO(M1),M1) THEN GOSUB 8400
 9070 GOTO 10480
 9980 :
10000 REM main segment - input data values
10010 INPUT "NUMBER OF PARAMETERS = ";M
10020 M1=M+1 : REM if fitting to data insert GOSUB 500 here
10030 FOR I=1 TO M
10040       PRINT "PARAMETER # " ;I; " = "; : INPUT "";S(1,I)
10050       INPUT "STEP = ";ST(I) : PRINT
10060 NEXT I
10070 INPUT "TOLERANCE = ";TL
10080 PRINT
10090 :
10100 GOSUB 8000 : REM construct initial simplex
10110 FOR IS=1 TO M1
10120       FOR K=1 TO M : PR(K)=S(IS,K) : NEXT K
10130       GOSUB 8200 : REM calculate function values for all vertices
10140       S(IS,M1)=SS
10150 NEXT IS
10160 :
10170 FOR I=1 TO M1 : LO(I)=1 : HI(I)=1 : NEXT I
10180 CY=0
10190 REM always jump back to here
10200 :
10210 GOSUB 8500 : IF Z$ = "OK" THEN 11000 : REM convergence is achieved
10220 :
10230 FL=0 : REM this is an IF flag
10240 GOSUB 8300 : REM find largest and smallest of parameter values
```

```
10250 CY=CY+1
10260 FOR I=1 TO M1 : CE(I)=0 : NEXT I
10270 :
10280 REM  CE( ) will be the centroids of the vertices
10290 FOR I=1 TO M1
10300       IF I<>HI(M1) THEN FOR J=1 TO M : CE(J)=CE(J)+S(I,J) : NEXT J
10310       REM leave out the "worst" vertex
10320 NEXT I
10330 :
10340 REM reflection of worst vertex
10350 PRINT "REFLECTION"
10360 FOR I=1 TO M1
10370       CE(I)=CE(I)/M : REM calculate centroid of "base" of simplex
10380       NE(I)=(1+AL)*CE(I)-AL*S(HI(M1),I) : REM reflect worst point
10390 NEXT I
10400 :
10410 FOR K=1 TO M : PR(K)=NE(K) : NEXT K
10420 GOSUB 8200 : REM get function of new vertex
10430 NE(M1)=SS
10440 :
10450 REM expand simplex if required
10460 IF NE(M1)<=S(LO(M1),M1) THEN FL=1 : GOSUB 9000 : REM expand
10470 :
10480 IF FL=1 THEN 10190
10490 IF NE(M1)<=S(HI(M1),M1) THEN FL=1 : GOSUB 8400
10500 IF FL=1 THEN 10190
10510 :
10520 REM contraction if still trying
10530 PRINT "CONTRACTION"
10540 FOR I=1 TO M : NE(I)=BE*S(HI(M1),M1) + (1-BE)*CE(I) : NEXT I
10550 :
10560 FOR K=1 TO M : PR(K)=NE(K) : NEXT K
10570 GOSUB 8200
10580 NE(M1)=SS
10590 :
10600 IF NE(M1)<=S(HI(M1),M1) THEN FL=1 : GOSUB 8400
10610 IF FL=1 THEN 10190
10620 :
10630 REM shrink simplex if all else fails
```

```
10640 PRINT "SHRINK"
10650 FOR I=1 TO M1
10660        FOR J=1 TO M : S(I,J)=( S(I,J) + S(LO(M1),J) )*BE : NEXT J
10670        IS=I
10680        FOR K=1 TO M : PR(K)=S(IS,K) : NEXT K
10690        GOSUB 8200
10700        S(IS,M1)=SS
10710 NEXT I
10720 GOTO 10190
10730 :
11000 REM  print out results
11010 PRINT "PROCEDURE HAS CONVERGED"
11020 FOR K=1 TO M : PRINT "PARAMETER #" :K: " = ":PR(K) : NEXT K
11030 END
```

```
   0 REM P6.3
   5 :
  10 : GOTO 10000
  15 :
  20 REM Powell's quadratic interpolation method
  30 REM    - finds minimum along a line
 500 REM function evaluation
 510 REM example from Walsh (1975) p 97
 520 F=0
 530 F=F+P(1)^4
 540 F=F-P(2)*(P(1)^3)
 550 F=F-P(1)*P(1)*P(2)*P(3)
 560 F=F+P(1)*P(2)*P(3)*P(4)
 570 RETURN
 580 :
8000 REM find a new point for the search set
8010 FOR I=1 TO M
8020        P(I)=PC(I) + H(X)*D(I) + HH*D(I) : REM D( ) is search direction
8030 NEXT I
8040 GOSUB 500 : F1=F
8050 :
```

```
8060 IF F1<F(X) THEN F(X)=F1 : H(X)=H(X) + HH
8070 :
8080 FOR I=1 TO M
8090        P(I)=PC(I) + H(X)*D(I) - HH*D(I)
8100 NEXT I
8110 GOSUB 500 : F2=F
8120 F(X)=F2 : H(X)=H(X) - HH
8130 RETURN
8140 :
8200 REM update the search set
8210 HI=1:FH=F(1)
8220 IF F(2) > FH THEN HI=2 : FH=F(2)
8230 IF F(3) > FH THEN HI=3 : FH=F(3)
8240 FOR I=1 TO M : P(I)=PC(I) + HM*D(I) : NEXT I
8250 GOSUB 500 : FM=F
8260 F(HI)=FM : H(HI)=HM
8270 RETURN
8280 :
8300 REM  main interpolation routine
8310 CY=0 : REM reset counter
8320 DH=-H
8330 FOR J=1 TO 4
8340        FOR I=1 TO M : P(I)=PC(I) + DH*D(I) : NEXT I
8350        GOSUB 500 : H(J)=DH : F(J)=F
8360        DH=DH+H
8370 NEXT J
8380 :
8390 IF F(2)>=F(3) THEN F(1)=F(4) : H(1)=H(4) : REM pick the three points
8395 :
8400 REM calculate turning point HM using Powell's formulae
8410 CY=CY+1 : IF CY>EP THEN PRINT "POWELL ITERATION LIMIT" : RETURN
8420 Y0=H(2)*H(3)*( H(3) - H(2) )*F(1)
8430 Y0=Y0 + H(3)*H(1)*( H(1) - H(3) )*F(2)
8440 Y0=Y0 + H(1)*H(2)*( H(2) - H(1) )*F(3)
8450 :
8460 Y1=( H(2)*H(2) - H(3)*H(3) )*F(1)
8470 Y1=Y1+( H(3)*H(3) - H(1)*H(1) )*F(2)
8480 Y1=Y1+( H(1)*H(1) - H(2)*H(2) )*F(3)
8490 :
```

```
8500 Y2=( H(3) - H(2) )*F(1) + ( H(1) - H(3) )*F(2) + ( H(2) - H(1) )*F(3)
8510 :
8520 Z=( H(1) - H(2) )*( H(2) - H(3) )*( H(3) - H(1) )
8530 :
8540 Y0=Y0/Z : Y1=Y1/Z : Y2=Y2/Z
8550 :
8560 FP=ABS(Y2)<1E-50
8570 :
8580 IF FP<>0 THEN PRINT "POWELL SINGULARITY FAILURE" : RETURN
8590 HM=-.5*Y1/Y2
8600 :
8610 IL=1 : MI=ABS(HM - H(1)) : MX=MI : IH=1
8620 :
8700 REM  find furthest and nearest points to turning point
8710 FOR J=2 TO 3
8720         Z=ABS( HM - H(J) )
8730         IF Z<MI THEN MI=Z : IL=J
8740         IF Z>MX THEN MX=Z : IH=J
8750 NEXT J
8760 :
8800 REM choose next step
8810 IF Y2<0 THEN X=IH : GOSUB 8100 : PRINT "MAX" : GOTO 8400 : REM pt. is max
8820 IF Y2>0 AND MI>HH THEN X=IL : GOSUB 8100 : PRINT "MIN. TOO FAR" : GOTO 8400
8830 IF Y2>0 AND MI<ER THEN PRINT "SUCCESS" : RETURN : REM successful search
8840 GOSUB 8200 : PRINT "REFINING POSITION OF MINIMUM" : GOTO 8400
8850 :
10000 REM  main segment
10010 INPUT "FRACTIONAL STEP = ";H
10020 INPUT "FRACTIONAL STEP LIMIT = ";HH
10030 INPUT "ERROR BOUND = ";ER
10040 INPUT "MAXIMUM NUMBER OF ITERATIONS = ";EP
10050 PRINT
10060 INPUT "NUMBER OF PARAMETERS = ";M : PRINT
10070 PRINT "GIVE STARTING POINT"; : PRINT
10080 :
10090 FOR I=1 TO M
10100         PRINT "P(";I;") = "; : INPUT "";PC(I)
10110 NEXT I
10120 PRINT
```

```
10130 PRINT "GIVE SEARCH DIRECTION" : PRINT
10140 :
10150 FOR I=1 TO M
10160         PRINT "D(";I;") = "; : INPUT "";D(I)
10170 NEXT I
10180 :
10190 D=0 : FOR I=1 TO M : D=D + D(I)*D(I) : NEXT I : D=SQR(D)
10200 :
10210 GOSUB 8300 : REM quadratic interpolation
10220 :
10230 REM successful search or premature termination
10240 :
10250 IF FP<>0 OR CY>EP THEN PRINT "LAST PARAMETER VALUES WERE" : GOTO 10310
10260 FOR I=1 TO M : P(I)=PC(I) + HM*D(I) : NEXT I
10270 GOSUB 500 : FM=F
10280 IF FM>F(IL) THEN FM=F(IL)
10290 PRINT : PRINT
10300 PRINT "MINIMUM VALUE = ";FM;"   AT " : PRINT
10310 FOR I=1 TO M
10320         PRINT "P(";I;") = ";P(I)
10330 NEXT I
10340 END
```

```
   0 REM  P6.4
  10 :
  20 REM Gauss-Newton gradient method for function optimization
  30 :
  40 TL=1E-6 : REM tolerance value for Gauss-Jordan pivoting
  50 :
  60 : GOTO 10000
 492 :
 494 REM example from Walsh (1975) p 140
 500 REM evaluate function
 510 F=2*P(1)^4 + P(2)^2 - 4*P(1)*P(2) + 5*P(2)
 520 RETURN
 530 :
 600 REM calculate gradient vector
 610 D1(1)=8*P(1)^3 - 4*P(2)
 620 D1(2)=2*P(2) - 4*P(1) + 5
 630 RETURN
 640 :
 700 REM calculate Hessian matrix
 710 D2(1,1)=24*P(1)^2
 720 D2(1,2)=-4
 730 D2(2,1)=-4
 740 D2(2,2)=2
 750 RETURN
 760 :
8000 REM solve for parameter corrections DP( ) by Gauss-Jordan elimination
8010 M1=M+1
8020 FOR I=1 TO M : D2(I,M1)=D1(I) : NEXT I
8030 REM "augment" D2 with extra column as RHS of equation set
8040 :
8050 FOR C=1 TO M
8060        MD=0
8070        FOR CC=C TO M : REM search down rest of column C
8080               D=D2(CC,C) : AD=ABS(D)
8090               IF AD>MD THEN MD=D : MC=CC : REM find largest element
8100        NEXT CC
8110        :
8120        IF AD<TL THEN PRINT "ZERO PIVOT" : END
8130        :
```

```
8140        IF MC=C THEN 8200 : REM largest element in current row - no swap
8150        FOR CC=C TO M1
8160               D2=D2(C,CC) : D2(C,CC)=D2(MC,CC) : D2(MC,CC)=D2
8170               REM D2 and D2() are different variables
8180        NEXT CC
8190        :
8200        FOR J=C+1 TO M1 : D2(C,J)=D2(C,J)/MD : NEXT J
8210        REM skip element in current and previous columns
8220        FOR I=1 TO M
8230               IF I=C THEN 8260 : REM current row already processed
8240               MU=D2(I,C)
8250               FOR J=C+1 TO M1 : D2(I,J)=D2(I,J) - MU*D2(C,J) : NEXT J
8260        NEXT I
8270 NEXT C
8280 RETURN
9990 :
10000 REM main segment
10010 INPUT "NUMBER OF PARAMETERS = ";M : PRINT
10020 FOR J=1 TO M
10030       PRINT "PARAMETER #" ;J;
10040       INPUT "STARTING VALUE = ";P(J)
10050       PRINT
10060 NEXT J
10070 PRINT : PRINT
10080 INPUT "CONVERGENCE FACTOR = ";EP
10090 :
10100 GOSUB 500 : REM evaluate function
10110 CY=0
10120 :
10130 REM iteration loop starts here
10140 CY=CY+1 : TF=F : REM count iterations and remember function value
10150 :
10160 GOSUB 600 : REM get derivative values
10170 FOR J=1 TO M : D1(J)=-D1(J) : NEXT J : REM  search "downhill"
10180 :
10190 GOSUB 700 : REM set up Hessian matrix
10200 :
10210 GOSUB 8000 : REM solve for DP()
10220 FOR J=1 TO M
```

```
10230        DP(J)=D2(J,M1)
10240        P(J)=P(J)+DP(J) : REM take a step - DP only here for clarity
10250 NEXT J
10260 :
10270 GOSUB 500 : REM get new function value
10280 :
10290 IF CY>1 AND ABS(TF-F)<EP THEN 11000 : REM stopping rule applied
10300 PRINT CY;"ITERATIONS","FUNCTION VALUE IS ";F
10310 :
10320 GOTO 10130 : REM try again
10330 :
11000 PRINT "PROCEDURE HAS CONVERGED"
11010 FOR I=1 TO M
11020        PRINT "PARAMETER # ";P(I)
11030 NEXT I
11040 PRINT "FUNCTION VALUE IS ";F
11050 END
```

```
   0 REM  P6.4A
  10 :
  20 REM Gauss-Newton gradient method for least squares fitting
  30 :
  40 TL=1E-6 : REM tolerance value for Gauss-Jordan pivoting
  50 :
  60 DIM DF(5,50),Y(50),F(50),X(50) : REM if M>9 then DIM for P(),D2(,),D1()
  70 GOTO 10000
  80 :
 100 INPUT "NUMBER OF DATA VALUES = ":ND
 110 FOR I=1 TO ND
 120         INPUT "X-VALUE, Y-VALUE ":X(I),Y(I)
 130 NEXT I
 140 RETURN
 150 :
5000 REM subroutine to calculate Hessian matrix for a least squares problem
5010 REM function to be fitted is P(1)*EXP( -P(2)*X(I) )  X(I) is independent
5020 REM variable and P(1),P(2) are parameters   F(I) is fitting function
5030 FOR I=1 TO ND : REM  ND is number of data values
5040         DF(1,I)=EXP( - P(2)*X(I) ) : REM can insert other functions here
5060         DF(2,I)= - P(1)*X(I)*EXP( - P(2)*X(I) ) : REM if functn. has hard
5070         F(I)=P(1)*EXP( - P(2)*X(I) ) : REM derivatives use differences
5080 NEXT I
5090 FOR J=1 TO M
5100         FOR K=J TO M
5110                 S=0
5120                 FOR I=1 TO ND
5130                         S=S + DF(J,I)*DF(K,I)
5140                 NEXT I
5150                 D2(J,K)=S : D2(K,J)=S : REM symmetrical Hessian matrix
5160         NEXT K
5170 NEXT J
5180 FOR J=1 TO M
5190         S=0 : F=0
5200         FOR I=1 TO ND
5210                 FY=Y(I) - F(I) : REM value of residual
5220                 F=F + FY*FY : REM sum of squares
5230                 S=S + DF(J,I)*FY
5240         NEXT I
```

```
 5250        D1(J)=S : REM  RHS values in set of linear equations for DP()
 5260 NEXT J
 5270 RETURN
 5280 :
 7990 REM lines 8000 to 8280 solve the system for DP() - the same as in P6.4
 9990 :
10000 REM main segment
10010 INPUT "NUMBER OF PARAMETERS = ":M : PRINT
10020 FOR J=1 TO M
10030        PRINT "PARAMETER #" ;J:
10040        INPUT "STARTING VALUE = ":P(J)
10050        PRINT
10060 NEXT J
10070 PRINT : PRINT
10080 INPUT "CONVERGENCE FACTOR = ":EP
10090 :
10100 GOSUB 100
10110 CY=0
10120 :
10130 REM iteration loop starts here
10140 CY=CY+1 : TF=F : REM count iterations and remember function value
10150 :
10160 GOSUB 5000 : REM get derivative values
10200 :
10210 GOSUB 8000 : REM solve for DP()
10220 FOR J=1 TO M
10230        DP(J)=D2(J,M1)
10240        P(J)=P(J)+DP(J) : REM take a step - DP only here for clarity
10250 NEXT J
10260 :
10280 :
10290 IF CY>1 AND ABS(TF-F)<EP THEN 11000 : REM stopping rule applied
10300 PRINT CY:"ITERATIONS","SUM OF SQUARES IS ":F : REM F is not F()
10310 :
10320 GOTO 10130 : REM try again
10330 :
11000 PRINT "PROCEDURE HAS CONVERGED"
11010 FOR I=1 TO M
11020        PRINT "PARAMETER # ":P(I)
```

```
11030 NEXT I
11040 PRINT "ESTIMATED STANDARD DEVIATION IS ":SQR( F/(ND - M) ) : REM ND>M
11050 PRINT "X-VALUES           Y-VALUES          FITTED VALUES"
11060         FOR I=1 TO ND : PRINT X(I);" ":Y(I);" ":F(I) : NEXT I
11070 END
11080 REM given good starting values, this method converges well but
11090 REM if not it is prone to jump "into the blue". It may be useful to
11100 REM put limits on the P() and/or F to warn of this and prevent overflow
```

```
  0 REM  P6.5
 10 :
 20 REM  Davidon-Fletcher-Powell's quasi-Newton method of optimization
 30 REM  Uses Broyden's complementary formula for greater stability
 40 :
 50 EI=30 : EP=30 : REM limits on iteration of main and Powell routine
 60 H=.1 : HH=1 : ER=.0001 : REM values needed for Powell's interpolation
 70 GOTO 10000
 80 :
500 REM function evaluation using example from Walsh (1975) p 117
501 REM function has a minimum value of 0 at (0,0)
510 F=0
520 F=F + P(1)*P(1)
530 F=F - P(2)*P(1)
540 F=F + 3*P(2)*P(2)
550 RETURN
560 :
600 REM evaluate derivatives
610 G(1)= 2*P(1) - P(2)
620 G(2)= - P(1) + 6*P(2)
630 RETURN
```

```
 640 :
 650 REM alternative estimation of derivatives by central differences
 651 REM this is Stewart's modification
 660 FOR I=1 TO M
 670         P(I)=P(I) + DP(I)
 680         GOSUB 500 : FP=F
 690         P(I)=P(I) - 2*DP(I)
 700         GOSUB 500 : G(I)=.5*(FP-F)/DP(I)
 710 NEXT I
 720 RETURN
 730 :
 800 REM convergence test
 810 FL=0
 820 FOR I=1 TO M : P(I)=PC(I) : NEXT I
 830 GOSUB 500 : F1=F
 840 FOR I=1 TO M : P(I)=PN(I) : NEXT I
 850 GOSUB 500 : F2=F
 860 FL=ABS(F1-F2)<TL : REM flag is non-zero if convergence test is true
 870 RETURN
 880 :
7990 REM lines 8000 to 8840 comprise Powell's quadratic interpolation to
7995 REM search along a given direction for a minimum
7997 :
8001 REM put in program statements for Powell's method here
9590 :
9600 REM  update the matrix using the Broyden complementary DFP formula
9610 FOR I=1 TO M
9620         Z=0 : FOR J=1 TO M : Z=Z + MC(I,J)*GA(J) : NEXT J
9630         Z(I)=Z
9640 NEXT I
9650 :
9660 BE=0 : Z1=0
9670 FOR I=1 TO M : BE=BE + GA(I)*Z(I) : Z1=Z1 + GA(I)*S(I) : NEXT I
9680 BE=1 + BE/Z1
9690 :
9700 FOR I=1 TO M
9710         FOR J=1 TO M
9720                 A(I,J)=S(I)*S(J)*BE : REM this is an "outer product"
9730         NEXT J
```

```
9740 NEXT I
9750 :
9760 FOR I=1 TO M
9770        Z=0 : FOR J=1 TO M : Z=Z + GA(J)*MC(J,I) : NEXT J
9780        Z(I)=Z
9790 NEXT I
9800 :
9810 FOR I=1 TO M : FOR J=1 TO M : B(I,J)=S(I)*Z(J) : NEXT J : NEXT I
9820 :
9830 FOR I=1 TO M
9840        FOR J=1 TO M
9850               ZZ=0
9860               FOR K=1 TO M
9870                      ZZ=ZZ + MC(I,K)*GA(K)*S(J)
9880               NEXT K
9890               ZZ(I,J)=ZZ
9900        NEXT J
9910 NEXT I
9920 :
9930 FOR I=1 TO M
9940        FOR J= 1 TO M
9950               MC(I,J)=MC(I,J) + ( A(I,J) - B(I,J) - ZZ(I,J) )/Z1
9960        NEXT J
9970 NEXT I
9980 :
9990 RETURN
9995 :
10000 REM  main segment
10001 :
10010 INPUT "NUMBER OF PARAMETERS = ";M : PRINT
10020 PRINT "ARE YOU SUPPLYING DERIVATIVES? Y/N?" : INPUT "";D$
10030 PRINT : PRINT "GIVE STARTING POINT" : PRINT
10040 :
10050 FOR I=1 TO M
10060        PRINT "P(";I;") = "; : INPUT "";PC(I)
10070        IF D$ ="Y" OR D$="y" THEN GOTO 10090
10080        PRINT "INCREMENT FOR DERIVATIVE ESTIMATE = "; : INPUT "";DP(I)
10090 NEXT I
10100 :
```

```
10110 PRINT : INPUT "TOLERANCE = ";TL
10120 :
10130 FOR I=1 TO M
10140         FOR J=1 TO M : MC(I,J)=0 : NEXT J
10150         MC(I,I)=1 : REM  initialize MC( , ) to unit matrix
10160 NEXT I
10170 :
10180 FOR I=1 TO M : P(I)=PC(I) : NEXT I
10190 IF D$="n" OR D$="N" THEN GOSUB 650 : GOTO 10210
10200 GOSUB 600
10210 FOR I=1 TO M : GC(I)=G(I) : NEXT I
10220 :
10300 REM iteration loop starts here
10310 :
10320 FOR I=1 TO M
10330         D=0
10340         FOR J=1 TO M
10350                 D=D - MC(I,J)*GC(J) : REM -ve sign gives "downhill"
10355                 REM search since MC( , ) is unit matrix in first
10357                 REM iteration  - first step is steepest descent
10360         NEXT J
10370         D(I)=D
10380 NEXT I
10390 :
10400 GOSUB 8300 : REM Powell search
10410 :
10420 IF FP<>0 OR CY>EP THEN 10700
10430 REM if FP is 0 search failed - restart by small random jump from PC( )
10440 :
10450 FOR I=1 TO M : S(I)=HM*D(I) : NEXT I
10460 FOR I=1 TO M : PN(I)=PC(I) + S(I) : NEXT I
10470 :
10480 GOSUB 800 : REM check for convergence
10490 KY=KY+1 : IF KY>EI THEN PRINT "ITERATION LIMIT EXCEEDED" : END
10500 IF FL<>0 THEN 10750
10510 REM some BASIC's use -1 instead of +1 for logical truth hence FL<>0
10520 :
10530 FOR I=1 TO M : P(I)=PN(I) : NEXT I
10540 IF D$="N" OR D$="n" THEN GOSUB 650 : GOTO 10560
```

```
10550 GOSUB 600
10560 FOR I=1 TO M : GN(I)=G(I) : NEXT I
10570 FOR I=1 TO M : GA(I)=GN(I) - GC(I) : NEXT I
10580 :
10590 GOSUB 9600 : REM  update MC( , )
10600 :
10610 FOR I=1 TO M : PC(I)=PN(I) : GC(I)=GN(I) : NEXT I
10620 :
10630 GOTO 10300
10640 :
10700 FOR I=1 TO M : PC(I)=PC(I) + H*(RND(1) - .5) : NEXT I : GOTO 10130
10710 :
10750 FOR I=1 TO M : PRINT "PARAMETER # ";I;" = ";P(I) : NEXT I : END
```

```
  0 REM P6.6
 10 :
 20 REM Least squares solution of overdetermined system of equations
 30 REM For system K C = A, Givens' Method is used to reduce the
 40 REM M by N matrix K to upper triangular form whilst modifying vector A
 50 REM Back substitution determines C
 60 REM It is still possible that back substitution will fail if columns
 70 REM of K are not independent - could then use Singular Value Decomposition.
 80 REM This is essentially the method of Golub.
 85 :
 90 D$=CHR$(4) : CR$=CHR$(13) : DR$=CR$ + D$
100 REM Apple II disc access control codes
110 :
120 DIM Y(5,250),E(250),Z(250),K(250,5),A(250)
130 REM if short of memory can rearrange to use Y(0, ) and K( ,0)
140 REM dimension other small arrays if N > 10
150 :
160 GOTO 10000
170 :
200 REM read a file
210 PRINT DR$"OPEN"F$,D2"
```

```
 220 PRINT DS"READ"FS
 230 :
 240 INPUT P1 : REM input the number of points from disc
 250 FOR KK=1 TO P1 : INPUT Z(KK) : NEXT KK
 260 PRINT DS"CLOSE"FS
 270 RETURN
 280 :
 500 REM input values for matrix K( , )
 510 INPUT "NUMBER OF COMPONENTS = ":N
 520 FOR I=1 TO N
 530        PRINT "PURE COMPONENT # " :I: " = ": : INPUT "":F$(I)
 540 NEXT I
 550 INPUT "EXPERIMENTAL = ":E$
 560 RETURN
 570 :
 600 REM set up K( , ) and A( )
 610 INPUT "NUMBER OF POINTS TO USE = ":M
 620 ST=INT(P1/M) : REM constant to determine interval of sampling
 630 FOR I=1 TO N
 640        FOR J=1 TO M
 650                K(J,I)=Y(I,J*ST)
 660        NEXT J
 670 NEXT I
 680 FOR J=1 TO M : A(J)=E(J*ST) : NEXT J
 690 RETURN
 695 :
 800 REM  read files
 810 FOR I=1 TO N
 820        F$=F$(I) : REM F$ and F$() not treated as same variable
 830        GOSUB 200
 840        FOR J=1 TO P1 : Y(I,J)=Z(J) : NEXT J
 850 NEXT I
 860 F$=E$ : GOSUB 200
 870 FOR J=1 TO P1 : E(J)=Z(J) : NEXT J
 880 RETURN
 890 :
1000 REM back substitution step
1010 IF ABS( K(N,N) )<TL THEN PRINT "SINGULAR UPPER TRIANGULAR MATRIX" : END
1020 C(N)=A(N) / K(N,N)
```

```
1030 FOR I=N-1 TO 1 STEP -1
1040        S=0 : FOR K=I+1 TO N : S=S+K(I,K)*C(K) : NEXT K
1050        C(I)=( A(I) - S ) / K(I,I)
1060 NEXT I
1070 RETURN
1080 :
2000 REM Gauss-Jordan elimination to invert matrix
2010 :
2020 FOR I=1 TO N
2030        FOR J=1 TO N : IM(I,J)=0 : NEXT J
2040        IM(I,I)=1 : REM fill diagonal of unit matrix with ones
2050 NEXT I
2060 :
2070 FOR C=1 TO N
2080        DD=M(C,C)
2090        IF ABS(DD)<TL THEN PRINT "ZERO PIVOT" : END
2100        FOR J=C+1 TO N : REM need not process current/previous columns
2110              M(C,J)=M(C,J)/DD
2120        NEXT J
2130        :
2140        FOR J=1 TO N : REM must treat IM( , ) as N RHS's
2150              IM(C,J)=IM(C,J)/DD
2160        NEXT J
2170        :
2180        FOR I=1 TO N
2190              IF I=C THEN 2270 : REM skip pivot row
2200              MU=M(I,C)
2210              FOR J=C+1 TO N
2220                    M(I,J)=M(I,J) - MU*M(C,J)
2230              NEXT J
2240              FOR J=1 TO N
2250                    IM(I,J)=IM(I,J) - MU*IM(C,J)
2260              NEXT J
2270        NEXT I
2280 NEXT C
2290 RETURN
2295 :
2500 REM form product of upper triangular matrix with its transpose
2510 FOR J=1 TO N
```

```
2520        FOR I=1 TO J : REM since upper triangle only go to J
2530                Z=0 : FOR K=1 TO I : Z=Z + K(K,J)*K(K,I) : NEXT K
2540                M(I,J)=Z : M(J,I)=Z : REM product is symmetric
2550        NEXT I
2560 NEXT J
2570 RETURN
2580 :
3000 REM Givens' Reduction
3010 FOR J=1 TO N
3020        FOR I=J+1 TO M
3030                KJ=K(J,J) : KI=K(I,J)
3040                D=SQR( KJ*KJ + KI*KI )
3050                IF D<TL THEN 3140 : REM skip if v. small - KI,KJ both 0
3060                C=KJ/D : S=KI/D : REM calculate SIN,COS rotation angle
3070                FOR L=I TO N : REM skip results to be 0
3080                        KJ= C*K(J,L) + S*K(I,L) : REM do rotation
3090                        KI=-S*K(J,L) + C*K(I,L)
3100                        K(I,L)=KI : K(J,L)=KJ : REM  avoid overwrite
3110                NEXT L
3120                AJ=C*A(J) + S*A(I) : AI=-S*A(J) + C*A(I)
3130                A(I)=AI : A(J)=AJ
3140        NEXT I
3150 NEXT J
3160 RETURN
3170 :
10000 REM main segment
10010 TL=1E-10 : REM  tolerance for Gauss-Jordan pivot and Givens' 0 check
10020 GOSUB 500 : REM input filenames
10030 GOSUB 800 : REM read files
10040 GOSUB 600 : REM set up K( , ),A( )
10050 GOSUB 3000 : REM reduce K( , ) to upper triangle - old values below
10060 GOSUB 1000 : REM solve for C( ) by back substitution
10070 :
10080 REM  Error analysis - see text
10090 :
10100 GOSUB 2500 : REM form M = Ut * U
10110 GOSUB 2000 : REM invert M
10120 :
10130 S=0 : FOR I=N+1 TO M : S=S + A(I)*A(I) : NEXT I
```

```
10140 REM remaining M-M elements of A( ) give us sum of squares
10150 S=S/(M-N) : REM variance
10160 FOR I=1 TO N
10170         FOR J=1 TO N : IM(I,J)=IM(I,J)*S : NEXT J
10180 NEXT I
10190 :
10200 PRINT
10210 FOR J=1 TO N
10220         SD(J)=SQR( IM(J,J) )
10230         PRINT C(J); " +/- " ;SD(J)
10240 NEXT J
10250 :
10260 REM  form variance-covariance matrix IM( , )
10270 PRINT
10280 FOR I=1 TO N
10290         FOR J=1 TO N : IM(I,J)=IM(I,J) / ( SD(I)/SD(J) ) : NEXT J
10300 PRINT
10310 NEXT I
10320 :
10330 END
```

References

Adby, P. R. and Dempster, M. A. H. (1974) *Introduction to Optimization Methods*. London, Chapman & Hall.

Atkins, P. W. (1982) *Physical Chemistry*, 2nd edn. Oxford, Oxford University Press.

Bartels, R. H., Beatty, J. C. and Barsky, B. A. (1987) *An Introduction to Splines for Use in Computer Graphics and Geometric Modelling*. Los Altos, Morgan Kaufmann.

Becsey, J. G., Berke, L. and Callan, J. R. (1968) Nonlinear least squares methods. *J. Chem. Education* **45**, 728–730.

Blackburn, J. A. (ed.) (1970) *Spectral Analysis*, New York, Dekker.

Bracewell, R. (1978) *The Fourier Transform and its Applications*. New York, McGraw-Hill.

Bradley, J. N., Hack, W., Hoyermann, K. and Wagner, H. G. (1973) Kinetics of the reaction of hydroxyl radicals with ethylene and with C_3 hydrocarbons. *J. Chem. Soc. Faraday Trans. I* **69**, 1889–1898.

Bromba, M. U. A. and Ziegler, H. (1979) Efficient computation of polynomial smoothing digital filters. *Analyt. Chem.* **51**, 1760–1762.

Brown, A. F. (1968) *Statistical Physics*. Edinburgh, Edinburgh University Press.

Buckingham, R. A. (1962) *Numerical Methods*. London, Pitman.

Burden, R. L., Faires, J. and Reynolds, A. C. (1981) *Numerical Analysis*. Prindle, Weber & Schmitt.

Caceci, M. S. and Cacheris, W. P. (1984) Fitting curves to data. *Byte* **9**, No. 5, 340–362.

Carley, A. F., and Morgan, P. H. (1989) *Advanced Computational Methods for the Chemical Sciences*, in preparation.

Carley, A. F., Rassias, S. and Roberts, M. W. (1983) The specificity of surface oxygen in the activation of adsorbed water at metal surfaces. *Surface Science* **135**, 35–51.

Carley, A. F., Roberts, M. W. and Tilley, R. J. D. (1989) in preparation.

Chandler, J. P., Hill, D. E. and Spivey, H. O. (1972) A program for efficient integration of rate equations and least squares fitting of chemical reaction data. *Computers and Biomedical Research* **5**, 515–534.

Chapra, S. C. and Canale, R. P. (1985) *Numerical Methods for Engineers*. New York, McGraw-Hill.

Churchhouse, R. F. (1978) *Numerical Analysis*. Cardiff, University College Cardiff Press.

Churchhouse, R. F. (ed.) (1981) *Numerical Methods* (Handbook of Applicable Mathematics, Volume III). Chichester, John Wiley.

Cohen, A. M., Cutts, J. F., Fielder, R., Jones, D. E., Ribbans, J. and Stuart, E. (1973) *Numerical Analysis*. London, McGraw-Hill.

Conway, G. R., Glass, N. R. and Wilcox, J. C. (1970) Fitting non-linear models to biological data by Marquardt's algorithm. *Ecology* **51**, 503–507.

Crank, J., McFarlane, N. R., Newby, J. C., Paterson, G. D. and Pedley, J. B. (1981) *Diffusion Processes in Environmental Systems*. London, Macmillan.

Daul, C., Schlaepfer, C. W., Mohos, B., Ammeter, J. and Gamp, E. (1981) Simulation of EPR spectra. *Computing Physics Commun.* **21**, 385–395.

Draper, N. R. and Smith, H. (1966) *Applied Regression Analysis*. New York, John Wiley.

Dumont, M. and Dufour, P. (1986) Monte Carlo simulation of surface reactions. *Computer Phys. Comm.* **41**, 1–19.

Duncan, G. T. (1978) An empirical study of jackknife-constructed confidence regions in nonlinear regression. *Technometrics* **20**, 123–129.

Fletcher, R. (ed.) (1969) *Optimization. Proceedings of the Symposium of the Institute of Mathematics and its Applications, University of Keele, England*, 1968. London, Academic Press.

Fletcher, R. (1970) A new approach to variable metric algorithms. *Computer Journal* **13**, 317–322.

Fletcher, R. and Powell, M. J. D. (1963) A rapidly converging descent method for minimization. *Computer Journal* **6**, 163–168.

Forsythe, G. E. (1957) Generation and use of orthogonal polynomials for data fitting with a digital computer. *J. Soc. Ind. Appl. Maths.* **5**, 74–88.

Francis, P. G. (1984) *Mathematics for Chemists*. London, Chapman & Hall.

Froeberg, C.-E. (1969) *Introduction to Numerical Analysis*. Reading, MA, Addison-Wesley.

Golden, P. J. and Deming, S. N. (1984) Sequential simplex optimization with laboratory microcomputers. *Lab. Microcomputer* **3**, 44–47.

Handbook of Chemistry and Physics (1974) Cleveland, CRC Press.

Handscomb, D. C. (ed.) (1966) *Methods of Numerical Approximation*. Oxford, Pergamon.

Hersh, R. and Griego, R. J. (1969) Brownian motion and potential theory. *Scientific American* **220**, 67–74.

Hosking, R. J., Joyce, D. C. and Turner, J. C. (1978) *First Steps in Numerical Analysis*. London, Hodder & Stoughton.

Johnson, K. J. (1980) *Numerical Methods in Chemistry*. New York, Marcel Dekker.

Jost, W. (1960) *Diffusion in Solids, Liquids and Gases*. New York, Academic Press.

Kantaris, N. and Howden, P. F. (1983) *The Universal Equation Solver*. Sigma Technical Press.

Knuth, D. E. (1981) *The Art of Computer Programming* Vol. 2: *Semi-numerical Algorithms*. Reading, MA, Addison-Wesley.

Krylov, V. I. (1962) *Approximate Calculation of Integrals*. Macmillan.

Lapidus, L. and Seinfeld, J. H. (1971) *Numerical Solution of Ordinary Differential Equations*. New York, Academic Press.

Lee, J. D. and Lee, T. D. (1982) *Statistics and Computer Methods in BASIC*, New York, Van Nostrand Reinhold.

Levenberg, K. (1944) A method for the solution of certain non-linear problems in least squares. *Quart. Appl. Math.* **2**, 164–168.

McCracken, D. D. (1955) The Monte Carlo method. *Scientific American* **192**, 90–96.

McCracken, D. D. and Dorn, W. S. (1964) *Numerical Methods and Fortran Programming*. New York, John Wiley.

Marquardt, D. W. (1963) An algorithm for least-squares estimation of nonlinear parameters. *J. Soc. Indust. Appl. Math.* **11**, 431–441.

Meiron, J. (1965) Damped least squares method for automatic lens design. *J. Opt. Soc. Am.* **55**, 1105–1109.

Metropolis, *et al.* (1980) *A History of Computing in the 20th Century*.

Meyer, H. A. (ed.) (1956) *Symposium on Monte Carlo Methods*, New York, John Wiley.

Milne, W. E. (1949) *Numerical Calculus*. Princeton, Princeton University Press.

Nash, J. C. (1979) *Compact Numerical Methods for Computers: Linear Algebra and Function Minimisation*. Bristol, Adam Hilger.

Nelder, J. A. and Mead, R. (1965) A Simplex method of function minimization. *Computer Journal* **7**, 308–313.

Pavelle, R., Rothstein, M. and Fitch, J. (1981) Computer algebra. *Scientific American* **245** No. 6, 102–113.

Perrin, C. L. (1970) *Mathematics for Chemists*. New York, Wiley–Interscience.

Phillips, E. G. (1960) *A Course of Analysis*. London, Cambridge University Press.

Pitha, J. and Jones, R. N. (1966) *Canadian J. Chem.* **44**, 3031–3050.

Powell, M. J. D. (1964) An efficient method of finding the minimum of a function of several variables without calculating derivatives. *Computer Journal* **7**, 155–162.

Pratt, W. K. (1978) *Digital Image Processing*. New York, John Wiley.

Ralston, R. and Wilf, H. S. (1960) *Mathematical Methods for Digital Computers*. New York, John Wiley.

Ramajaraman, V. (1971) *Computer Oriented Numerical Methods*. New Delhi, Prentice-Hall.

Ripley, B. (1983) Take your pick. *Personal Computer World* **6**, No. 9, 188–191.

Rosenbrock, H. H. (1960) An automatic method for finding the greatest or least value of a function. *Computer Journal* **3**, 175–184.

Savitzky, A. and Golay, J. E. (1964) Smoothing and differentiation of data by simplified least squares procedures. *Analyt. Chem.* **36**, 1627–1639.

Singer, J. (1964) *Elements of Numerical Analysis*. London, Academic Press.

Spendley, W. Hext, G. R. and Himsworth, F. R. (1962) Sequential applications of simplex designs in optimization and evolutionary operation. *Technometrics* **4**, 441–461.

Southworth, R. W. and Deleeuw, S. L. (1965) *Digital Computation and Numerical Methods*. New York, McGraw-Hill.

Steinier, J., Termonia, Y. and Deltour, J. (1972) Comments on smoothing and differentiation of data by simplified least squares procedures. *Analyst. Chem.* **44**, 1906–1909.

Stewart, G. W. (1967) A modification of Davidon's method to accept difference approximations of derivatives. *J. Assoc. Comput. Mach.* **14**, 72–83.

Swain, C. G. and Swain, M. S. (1980) A uniform random number generator that is reproducible, hardware-independent and fast. *J. Chem. Inf. Comput. Sci.* **20**, 56–58.

Vernin, G. and Chanon, M. (1986) *Computer Aids to Chemistry*. Chichester, Ellis Horwood.

Walsh, G. R. (1975) *Methods of Optimization*. London, John Wiley.

Wilkes, M. V. (1966) *A Short Introduction to Numerical Analysis*. Cambridge, Cambridge University Press.

Yakowitz, S. J. (1977) *Computational Probability and Simulation*. Reading, MA. Addison-Wesley.

List of programs

P1.1 Compute approximate value of a derviative 71
P1.2 Horner's scheme for efficient evaluation of a polynomial. 72
P1.3 Application of Wegstein's method to computing a square root 73
P1.4 Application of Regula Falsi to chemical equilibrium example. 74
P1.5 Application of Newton–Raphson method to chemical equilibrium example. 75
P1.6 Application of Newton–Raphson method to titration example 77
P1.7 Random walk simulation of 1-D diffusion 78
P1.8 Random walk solution of Laplace's equation 79
P1.9 Monte Carlo integration using the hit or miss method. 81
P1.10 Monte Carlo double integration by sample mean method — 2-D diffusion 82
P1.11 Monte Carlo simulation of 2-D diffusion example of P1.10 83

P2.1 Computation of a divided difference table 118
P2.2 Interpolation using Newton's forward formula 119
P2.3 Interpolation using Aitken's method. 120
P2.4 Spline interpolation 122

P3.1 Composite (repeated) Simpson's Rule for approximation to an integral 149
P3.2 Adaptive Simpson's Rule. 150
P3.3 Spline integration. 152
P3.4 2-D Simpson integration over rectangle. 154
P3.5 Adaptive Simpson's Rule applied to calculation of EPR spectrum. 156

P4.1 Tangent field plotter. 196
P4.2 Basic Euler method for solving an initial value problem 197
P4.3 Heun's method for first order initial value problems 199
P4.4 Runge–Kutta method with interval doubling and halving 201
P4.5 Adams' 4th order preditor-corrector method with Runge–Kutta starter 203
P4.6 Extrapolated trapezoidal rule for initial value problems 206
P4.7 P4.4 adapted for a system of first order differential equations 208
P4.8 Solution of diffusion equation by simple explicit method 212
P4.9 P4.8 adapted for solution of a diffusion-reaction equation. 213

P5.1 Gaussian pseudo-random number generator 246
P5.2 Best straight line by least squares 247
P5.3 Runs test for small number of runs in a series of data values 249
P5.4 P3.2 applied to runs test for large number of runs 251
P5.5 Least squares fit to polynomial 253
P5.6 Data fitting using Forsythe polynomials 255
P5.7 Computing power series coefficients from Forsythe polynomial coefficients 257
P5.8 Quadratic smoothing by method of Savitzky and Golay 259

P6.1 Hooke and Jeeves optimization method 299
P6.2 Simplex optimization 301
P6.3 Powell's quadratic interpolation method 305
P6.4 Gauss–Newton method for function minimization 309
P6.4a Gauss–Newton method applied to least squares fitting 312
P6.5 Davidon–Fletcher–Powell quasi-Newton method 314
P6.6 Least squares solution of overdetermined system of linear equations 318

Index

Page numbers in **bold** indicate main treatment of topics

Adam's fourth order method
  solving differential equations, 169
  program, 203
adaptive integration, 130
  and extrapolation methods, 175
  application to runs test, 251
  program, 150
  Simpson's Rule implementation, 132
Aitken's interpolation method, **90**
  program, 120
Applesoft
  BASIC commands, 8
  number handling, 21
approximation
  derivative, 14
  functions, 11
  polynomial, 17
  Taylor series, 18
  Weierstrass' Theorem, 18
arrays, **49**
  application to matrices, 51
  look-up time, 54
  multiple index, 50
augmented matrix, 56
autonomous differential equations, 177
axis system, 62

BASIC
  constructions, 8
  dialects, 9
  DIM statement, 49
  integer definition, 9
  interpreter, 9, 21, 54
  loops, 10, 54
  matrix arithmetic in, 60
  random number generator, 40
  REPEAT...UNTIL simulation, 201
Beer's Law, 67
best fit line, **222**
bilinear interpolation, 116
  evaluation of double integral, 137
bisection method, **30**
bivariate interpolation
  Langrangian formula, 114
  on a rectangular grid, 114
  planar, 115
Bolzano's Method, **30**
boundary conditions
  derivative, for partial differential equation, 187
  programming, 212
  spline, 112
  use of fictional cells, 186
boundary value problem, 41

calibration matrix, 68
Central Limit Theorem, **218**
  and Gaussian noise generator, 218
centroid of data points, 225
Chebyshev polynomials, 240
chemical equilibrium, program example, 74, 75
chemical kinetics, systems of differential equations, 177
column matrices, 61
combinations, program to calculate number of, 249
compiler, 10
components of a vector, 63
composite Simpson's Rule, 131
computer graphics, splines applied to, 108
constrained minimization, **291**
  barrier function methods, 292
  by sequential unconstrained gradient methods, 293
  penalty function methods, 293
contour plot, 262
convergence, 24
  apparent, in summation of harmonic series, 8
  of solution of partial differential equation, 188
  order of, 24
  solution of non-linear system in Heun's method, 168
  third order example, 35
correlation
  of parameters, 288, 291
  coefficient, Pearson's r-value, 226
Cotes formulae for numerical integration, 127

covariance matrix, 280
cubic interpolation method for linear search, 276
curve-fitting, *see also* least squares
  overlapping bands, 283
  polynomial, **234**

data transformation, 244, 247
Davidon-Fletcher-Powell method, **279**
  complementary updating formula, 280
  program using complementary updating formula, 314
  Stewart's 'direct' version, 280
definite integral, 15
  as an area, 16
degrees of freedom, 225, 289
derivatives, 13
  approximation of, 14, 143
  computation of, 71
  higher, 13
  partial 16, 146
  second, 13
  table of approximating formulae, 144
differences
  central, 144
  forward, 101
differential, 14
differential equations, **159**
  anlaytical integration of, 162
  autonomous, 177
  Euler's method, **165**
  example of system, 180
  extrapolation methods, 175
  Heun's method, 168
  instability of numerical solution, 165
  modified Euler method, 168
  multistep methods, 169
  parasitic numerical solutions, 184
  partial, **184**
  predictor-corrector methods, **168**
  radioactive decay, 162
  Runge-Kutta methods, **170**
  stiff systems of, **181**, 184
  systems of, **177**
  Taylor's series method, 162
  trapezoidal rule, 167
diffusion, **184**
  and reaction, 189
  cellular approximation, 185
  coefficient, 185, 188
  one-dimensional, simulation by Monte Carlo method, 39
  program to simulate diffusion with reaction, 213
  two dimensional, simulation by Monte Carlo method, 45
diffusion equation, 40, 186
  grid representation for solution, 186
  program for solution, 212
direct search minimization, **263**
  global grid method, 263
  Hooke and Jeeves method, **267**
  least squares solution of linear systems, **293**
  local grid method, 264
  one-at-a-time method, **266**
  simplex method, **269**
Dirichlet problem, 41
divergence of Newton-Raphson method, 34
divided differences, **86**
  applied to spline formulae, 111
  bivariate, 115
  computational example, 90
  higher order, 88
  limiting cases, 111
  program, 118
  recursive formation, 88
  table construction, 90
  table for specific heat data, 93
  zero order, 88
dot product, 62
double integration, as a two stage process, 139
double precision arithmetic, 237

Einstein-Smoluchowski relation, 40, 46, 83
  and diffusion equation, 188
EPR spectra, simulated by adaptive integration, 135, 156
equilibrium bond length, from numerical differentiation, 145
erf( ), 12
error, **20**
  absolute, 20, 23
  accumulated, 23
  and step size, 22
  arithmetic, 21
  computational, 21, 22
  due to simplification, 21
  empirical, 21
  estimates for differential equation solutions, 174
  estimation in Simpson's composite rule, 128
  experimental, 21
  Gaussian distribution of, 21
  growth of, 24
  in products or quotients, 23
  in sums or differences, 23
  interpolation and extrapolation, **102**
  matrix, 289
  propagation, 23
  relative, 20, 23
  round-off, 22, 23, 71
  Simpson's Rule, 127
  solution of partial differential equation, 188
  solving linear systems, 59
  truncation, 18, 71
error estimate for parameters
  in optimization methods, 280, **286**
  Monte Carlo method for, **289**
error function, erf( ), 12
Euler's method for ordinary differential equations, **165**
  effect of step size, 166
  program, 197
  systems of differential equations, **177**
extrapolation methods, for error cancelling, **175**

and adaptive quadrature, 175
based on trapesoidal rule, program, 206
Richardson method applied to differential equations, 169, 176

factorial, 18
false position, method of, **31**
Fick's law of diffusion, 184
finite difference equation, Kirchoff's first law, 193
first-order decay, 242
fitting data to models, **216**
goodness of fit, 217
flow graph
explanation of matrix operation, 68
in solution of partial differential equation, 187
Forsythe polynomials
in data fitting, **239**
program for data fitting, 255
function
definition of, 11
differentiable, 13
graph of, 12
minimization and maximization, 261
multivariable, 16
polynomial, 12

Gauss-Jordan method, **56**
applied to overlapping spectra, 70
Gauss-Newton method, **278**
application to least squares, 283
Levenberg's modification, **285**
Marquardt's modification, **285**
Meiron's modification, 284
program, 309
program for least squares fitting, 312
updating formula, 279
Gauss-Seidel iteration for linear systems, 194
Gaussian distribution, 40
of errors, 217
program for integral of by integrating series, 252
Gaussian elimination, **60**
for tridiagonal matrices, 122
solution of spline equations, 113
Gaussian pseudo-random numbers program, 246
Gaussian quadrature, 126
Givens' reduction method
for matrix reduction to upper-triangular form, 296
program to solve overdetermined liner system, 318
global error in solutions of ordinary differential equations, 174
global grid minimization, 263
goodness of fit, 217, **220**
gradient method for minimization, 263, **274**
constrained minimization, 293
Davidon-Fletcher-Powell method, **279**
Gauss-Newton, **278**
quasi-Newton methods, **279**
steepest descent, **276**
gradient vector (grad()), 277
in least squares, 283
GWBASIC commands, 9

harmonic series, 'summation' of, 8
Hessian matrix, 278
Heun's method for solution of ordinary differential equations, 168
as a Runge-Kutta formula, 173
program, 199
Hooke and Jeeves method, **267**
example, 280
program, 299
Horner's Rule
for evaluating polynomials, 19
in polynomial curve fitting, 237
program, 72

ill-condition, 24
in linear system arising from overlapped spectra, 70
of polynomial fitting, 237
implicit formulae in corrector stage of predictor-corrector method, 180
infinite integrals, 147, 148
initial value problem, 160
instability in solution of ordinary differential equations, 165
integration, **125**
comparison of methods, **131**
definition, 14
for spectrum simulation, **134**, **141**
hit or miss Monte Carlo, **81**
indefinite, 134
Monte Carlo, **43**
multiple, 17, **135**
non-rectangular region, 140
power series, 126
program for, 2-D Simpson's Rule, 154
program for spline method, 151
Simpson's Rule, **125**
spline, **130**
with infinite limits, 147
formulae
Boole's Rule, 129
closed, 127
composite, 127
mid-ordinate rule, 129
Milne's 129
Newton-Cotes, 127, 129
open, 129
partial range, 127, 129
Simpson's Rule, 127
Steffenson's, 129
trapezoidal rule, 129
interpolation, **85**
Aitken's method, **90**
bilinear, 116, 117, 137
bivariate, 114, 115
cubic, 276
multivariable, **114**
Newton's Forward Formula, 89, 119
non-polynomial, **105**

piecewise polynomial, 110
polynomial
Aitken's, 95
differentiation formulae from, 142
equispaced ordinate formulae, **100**
Gauss' formulae, 99
Lagrange's, 98
Newton's backward, 98
Newton's forward, 89
program for Aitken's method, 120
program for spline method, 122
quadratic, 143, **274**
rational, 105
spline, **108**
interpreter, BASIC, 9, 21, 54
inversion of a matrix Gauss-Jordan method, 59
iteration, 24
and accumulated error, 23
Gauss-Seidel, for partial differential equations, 194
Jacobi, for partial differential equations, 194

Jacobi iteration for linear systems, 194
jack-knife method for parameter error estimation, 289

Lagrange's interpolation polynomial, 98, 126
Laplace's equation, 41, 190, 193
least squares
best straight line, **222**, 233
fitting general models to data, **281**
fitting to data using orthogonal polynomials, **237**
fitting to data using polynomials, **234**
for systems of linear equations, **293**
Newton's method for overlapped spectra, 283
normal equations, 223, 235
principle, 222
program for best straight line, 247
program for fitting overlapped peaks, 312
program for fitting, using Forsythe polynomials, 255
program for polynomial fitting, 253
solution of linear system for overlapped peaks, 298
Levenberg's method, **285**
likelihood, principle of maximum, 286
Lin's method, **32**
local error, in solutions of ordinary differential equations, 174
local grid minimization, 264
loops
simulated REPEAT...UNTIL, 201
loops in BASIC, 10, 54
Lorenzian lineshape function, 131

Marquardt's method, **285**
matrices, **51**
non-square and overdetermined, 294
orthonormal, 294
partitioned, 295
rotation, 296
sparse, 194
upper-triangular, 294
matrix, **51**
augmented, 56
calibration, 68
column, 51, 61
covariance, 280
equations, 52
error, 289
flow-graph of a, 68
Hessian 278
identity, 53
inverse, 59
lower-triangular, 61
multiplication, 52
non-square, 70
operating on a vector, 65
upper-triangular form, 56
maxima, 14
maximization of a function, 261
Meiron's method, **285**
Milne's integration formula, 129
minima, 14
minimization
of potential energy of molecules, 143
of a function, **261**
topographical analogy, 263
Monte Carlo integration, **43**
hit or miss method, 44
integration on irregular region, 44
multidimensional integrals. 43
program for, 2-D integration, 82
program for hit or miss method, 81
sample means method, 45
Monte Carlo method, **38**
and simulation of diffusion, 188
as Nature's model, 39
for error estimates in parameters, 289
neutron diffusion, 39
one-dimensional diffusion, 39
partial differential equation, 43
program for Laplace's equation, 79
program for one-dimensional diffusion, 78
random walk, 39
two-dimensional diffusion, 45
usefulness of, 84
multiple integrals, 43, 135, 139
multistep methods for solving differential equations, 169
multivariable differentiation, **145**

narrow valley problem, 267, 276
natural spline, 122
Newton's forward formula, **86**
program, 119
Newton's method, *see* Gauss-Newton method
Newton-Cotes formulae
and predictor-corrector method, 169
for numerical integration, 127, 129
Newton-Raphson method, **32**
as limit of secant method, 33
program for chemical equilibrium, 75

program to solve titration equation, 77
noise, 240
and data fitting, 217
effect on integration of peak area, 134
non-linear algebraic equations sytems, of, 181
non-linear equations, **30**
bisection method, **30**
examples of solution, 36
Halley's method, **36**
Lin's method, **32**
Newton-Raphson method, **32**
Regula Falis method, **30**
secant method, **32**
Steffensen's method, **35**
successive approximations, **24**
Wegstein's method, **28**
non-polynomial models, **241**
normal distribution, 40
of errors, 217
normal equations
for least squares fitting, 223, 235
matrix notation in least squares, 236
numbers, binary and floating point, 21
numerical differentiation
least squares polynomial, 145
resolution enhancement by, 145

one-at-a-time minimization, 266
optimization methods, **261**;
see also gradient methods
see also direct search minimization
error estimates for parameters, 280
narrow valley problem, 267
orthogonal polynomials use in polynomial fitting, **237**
orthonormal matrices, 294

parabolic differential equation, program to solve numerically, 212
parameters
correlated, 288, 291
in data modelling, 216
parametric equations, relation to coupled differential equations, 178
parasitic solutions, of differential equations, 184
partial derivatives, 16
formulae for approximating, 146
in image analysis, 147
partial differential equations, **184**
extension to more space dimensions, 190
matrix-vector form, 187
oscillation and equilibrium, 190
parabolic, hyperbolic and elliptic, **190**
solution at a single point, 42
solution by Monte Carlo method, 42
stability and error propagation, 188
partial range integration formulae, 127, 129
partitioned matrices, 295
pattern move, in direct search, 267, 268
peak area
from composite trapezoidal rule, 129
program using composite Simpson's Rule, 132
Pearson's correlation coefficient, 226
periodic excitation and second order kinetics, 179
pivoting
full, by columns and rows, 61
partial, by rows, 59, 60
points of inflection, 14
Poisson's equation, 43
polar coordinates for special integration regions, 140
polynomials, 12; *see also* interpolation polynomial
approximation using, 70, 85
curve-fitting, **234**
smoothing (Savitzky-Golay), **240**
Chebyshev, 240
Forsythe's orthogonal, **239**
Horner's scheme, 19
program for evaluation, 72
Taylor expansion, 32
zeros of, 20
power series
integration of, 126
Taylor's, 18
power series coefficients, program for calculating from Forsythe polynomials, 257
predictor-corrector methods
for differential equation solution, **168**
program for fourth order method, 203
product
dot, 62
outer, 65
scalar, 62

quadratic interpolation
for linear search, **274**
for one-parameter minimization, 143
program for Powell's method, 305
quadrature, *see* integration
quasi-Newton methods, **279**
Davidon-Fletcher-Powell methods, **279**
general updating formula, 279

random number generator, **46**, 218
congruential method, 48
shift register method, 47
random numbers (pseudo-), 46
random walk
Laplace's equation, solution of, **41**
one-dimensional diffusion, **39**
program for Laplace's equation, 79
program for one-dimensional diffusion example, 78
two-dimensional diffusion, 83
randomness, spectral test of, 48
recursion, simulated in BASIC, 154
Regula Falsi, **30**
program, 74
relative machine precision, evaluation of estimate for, 76
residuals
in data fitting, 220
plots of, 226

randomness of, **228**
variation with fitted value, 227
vector of, 294
Richardson extrapolation, error cancelling in predictor-corrector methods, 169, 176
roots of equations, **30**
rotation matrices, 296
row vector, 65
Runge's phenomenon, 104
Runge-Kutta formulae
graphical illustration of, 172
starter for predictor-corrector methods, 169, 175, 203
Runge-Kutta method, **170**
program, 201
program for system of equations, 208
runs
Probability distribution for large run number, 232
test
for randomness of residuals, **229**
program for long data series, 251
program for short data series, 249

Savitzky and Golay method
least squares polynomial differentiation, 145
polynomial smoothing, **240**
program for quadratic smoothing, 259
Schroedinger's equation, 191
secant method, **32**
program, 74
second order kinetics, 162
signal averaging, 219
signal to noise ratio, 219
simplex method, **269**
flow chart, 271
program, 301
termination of search, 273
Simpson's Rule, **125**
adaptive formulation, 130
approximation to definite integral, 126
composite, 127
error estimation, 128
preogram for adaptive formulation, 150
program for composite rule, 149
unexpected accuracy of, 127
simultaneous linear equations, **51**
sin( ), Taylor series for, 18
singular point, 161
smoothing
by least square polynomials (method of Savitzky and Golay), **240**
over-, 241
program for quadratic smoothing, 259
spectra, overlapping, 67
splines
application to computer graphics, 108
boundary conditions, 112
continuity equations for, 112
end-points, 112
integration using, 130
interpolation using, **108**
natural, 113
program for spline interpolation, 122
thin bean least energy conformation, 113
matrix form of spline equations, 112
tridiagonal form of spline equations, 113
square root evaluation, 28
comparison of methods, 35
Haley's method, 36
program using Wegstein's method, 73
stability, of solution of partial differential equation, 188
standard deviation
of fit, 225
of parameters, 224
steepest descent method, **276**
narrow valley problem, 276
descent method, updating formula, 277
Steffensen's integration formula, 129
Steffensen's method, **35**
stiffness
in systems of ordinary differential equations, **181**
numerical example of, 183
straight line best fit through data, **222**
successive approximations, **24**
adaptive, 28
and predictor-corrector method, 168
divergence of, 25
example of, 27
extensions to, 28
for square roots, 28
self-correcting nature of, 25
with attenuating function, **29**
sum of squares function, 222, 282
superposition principle, and linearity, 188
symbol manipulation language, automated analytical derivative by, 162
synthetic division, 20
systems of linear equations, **54**
Gaussian elimination, **60**
Gauss-Jordan method, **55**
Gauss-Seidel method, 66
iterative methods, 66
Jacobi method, 66
solution by least squares, **293**
upper-triangular form, 56

tangent field diagrams, **159**
program, 196
Taylor expansion, 32
Taylor polynomial for approximate data fitting, 244
Taylor series, 18
solving differential equations, 162
stepsize in ordinary differential equation solution, 162
truncation of, 18, 22
total differential, 17
and errors, 23
and Rung-Kutta formulae, 170
transformation of data, 244
transpose of a matrix, 65

trapezoidal rule
  integration formula, 129
  related to spline integration, 131
  two variables, 138
tridiagonal matrix spline equations, 113
triple integration, 139
truncation, chopping and rounding, 22
truncation, of series, 18, 22
truncation, error in predictor-corrector methods, 169
TurboBASIC commands, 9
turning points, 14
  conditions for, 17

Universal Equation Solver, **29**
upper-triangular matrix, 294
  reduction to by Given's method, 296

variable metric methods, *see* quasi-Newton methods
vectors, **51**, **61**
  angle between, 62
  components of, 63
  dot (scalar) product of, 62
  independence of axis system, 62
  lengths of, 62
  moduli of, 62
  rotation and scaling of by matrix, 65
  row, 65
  unit, 62
volume, relation to double integral, 137
von Neumann, 38

Wegstein's method, **28**
  program, 73
Weierstrass Approximation Theorem, 18
weights in least squares, 222, 282

zeros of polynomials, 20

Program index

Adams' 4th order method, 203
adaptive formulation of Simpson's Rule, 150
Aitken's method of interpolation, 120

best straight line by least squares, 247

chemical equilibrium calculation by Newton-Raphson method, 74, 75
combinations, number of, 249
complementary updating formula, Davidon-Fletcher-Powell method using, 314
composite Simpson's Rule, 149
  for peak area calculation, 132

data fitting using Forsythe polynomials, 255
Davidon-Fletcher-Powell method using complementary updating formula, 314
derivative, approximation to, 71
differential equations
  Adams' 4th order method, 203
  Euler's method, 197
  extrapolated trapezoidal rule, 206
  Heun's method, 199
  Runge-Kutta method
    for systems of ordinary differential equations, 208
    with interval doubling and halving, 201
diffusion, calculation for diffusion with reaction, 213
diffusion equation solution, 212
divided difference table, computation of, 118
double integration
  by Monte Carlo method, 82
  by Simpson's Rule, 154

error function, evaluation of
  by integrating series, 252
  by adaptive Simpson integration, 251
Euler's method for initial value problem, 197
evaluation of polynomials by Horner's scheme, 72
extrapolated trapezoidal rule for initial value problems, 206

fitting overlapped peaks using least squares, 312
fitting by least squares using Forsythe polynomials, 255
Forsythe polynomials
  calculation of power series coefficients from, 257
  for data fitting, 255

Gauss-Newton method
  for function minimization, 309
  for least squares fitting, 312
Gaussian distribution, integrating series to evaluate error function, 252
Gaussian pseudo-random number generator, 246
Givens' reduction method to solve overdetermined linear system, 318

Heun's method for initial value problem, 199
hit or miss method for Monte Carlo integration, 81
Hooke and Jeeves optimization method, 299
Horner's scheme for evaluation of a polynomial, 72

integral of Gaussian function, 252
integration
  adaptive, 150
  composite Simpson's Rule method, 149
  double integration by Simpson's Rule, 154
  Monte Carlo, 81, 82
  spline method, 152
interpolation
  Aitken's method, 120
  Powell's quadratic method, 305
  spline method, 122
  using Newton's forward formula, 119

Laplace's equation, solution of
  Monte Carlo method, 79
  random walk, 79
least squares
  best straight line, 247
  data fitting using Forsythe polynomials, 255
  fitting overlapped peaks, 312
  Gauss-Newton method for, 312

polynomial fitting, 253
solution of overdetermined system of linear equations, 318

minimization
Davidon-Fletcher-Powell quasi-Newton method, 314
Gauss-Newton method
for function minimization, 309
for least squares fitting, 312
Hooke and Jeeves method, 299
simplex method, 301
Monte Carlo integration
double integral, 82
hit or miss method, 81
Monte Carlo method
Laplace's equation, 79
one-dimensional diffusion, 78
two-dimensional diffusion, 83

Newton-Raphson method
chemical equilibrium example, 75
to solve titration equation, 77

one-dimensional diffusion
Monte Carlo method for simulating, 78
random walk simulation, 78
overdetermined linear system, solution by Givens' reduction method, 318

parabolic differential equation, numerical solution of, 212
peak area calculation using composite Simpson's Rule, 133
polynomial fitting by least squares, 253
polynomials, evaluation by Horner's scheme, 72
Powell's method of quadratic interpolation, 305
power series coefficients, calculation from Forsythe polynomials, 257
predictor-corrector methods, fourth order method, 203

quadratic interpolation using Powell's method, 305
quadratic smoothing by the method of Savitzky and Golay, 259

random (pseudo-) number generator, 246
random walk
solution of Laplace's equation. 79
one-dimensional diffusion example, 78
two-dimensional diffusion example, 83
Regula Falsi method for non-linear equations, 74
Rung-Kutta method
for systems of ordinary differential equations, 208
with interval doubling and halving, 201
runs test
long data series, 251
short data series, 249
Savitzky and Golay method for quadratic smoothing, 259
simplex optimization, 301
Simpson's Rule
adaptive formulation, 150
adaptive method applied to calculation of EPR spectra, 156
composite rule, 149
double integration over a rectangle, 154
smoothing, quadratic, 259
spline interpolation, 122
spline method
integration, 152
interpolation, 122
square root evaluation by Wegstein's method, 73
systems of ordinary differential equations solved by Runge-Kutta method, 208

tangent field plotter, 196
titration equation solution by Newton-Raphson method, 77
trapezoidal rule (extrapolated), for initial value problems, 208

Wegstein's method, applied to square root evaluation, 73

TDR